ALSO BY GEORGE DYSON

*Baidarka*

*Project Orion*

*Darwin Among the Machines*

# TURING'S CATHEDRAL

# TURING'S CATHEDRAL

## THE ORIGINS OF THE DIGITAL UNIVERSE

George Dyson

Pantheon Books · New York

Pantheon Books and colophon are registered trademarks of Random House, Inc.

Library of Congress Cataloging-in-Publication Data
Dyson, George, [date]
Turing's cathedral : the origins of the digital universe / George Dyson.
p.   cm.
Includes index.
Summary: "In a revealing account of John von Neumann's realization of Alan Turing's
Universal Machine, George Dyson vividly illuminates the nature of digital computers, the
lives of those who brought them into existence, and how code took over the world."
ISBN 978-0-375-42277-5 (hardback)
1. Computers—History. 2. Turing machines. 3. Computable functions.
4. Random access memory. 5. Von Neumann, John, 1903–1957.
6. Turing, Alan Mathison, 1912–1954. I. Title.
QA76.17.D97 2012        004'.09—dc23        2011030265

www.pantheonbooks.com

*Case photograph: Alan Mathison Turing, 1951, by Elliot & Fry,*
copyright © National Portrait Gallery, London
*Jacket design by Peter Mendelsund*

Printed in the United States of America
First Edition

2   4   6   8   9   7   5   3   1

*It was not made for those who sell oil or sardines . . .*

—G. W. Leibniz

# CONTENTS

# POINT SOURCE SOLUTION

*I am thinking about something much more
important than bombs. I am thinking about
computers.*

—John von Neumann, 1946

There are two kinds of creation myths: those where life arises out of
the mud, and those where life falls from the sky. In this creation myth,
computers arose from the mud, and code fell from the sky.

In late 1945, at the Institute for Advanced Study in Princeton, New
Jersey, Hungarian American mathematician John von Neumann gathered a small group of engineers to begin designing, building, and
programming an electronic digital computer, with five kilobytes of
storage, whose attention could be switched in 24 microseconds from
one memory location to the next. The entire digital universe can be
traced directly to this 32-by-32-by-40-bit nucleus: less memory than is
allocated to displaying a single icon on a computer screen today.

Von Neumann's project was the physical realization of Alan Turing's Universal Machine, a theoretical construct invented in 1936. It
was not the first computer. It was not even the second or third computer. It was, however, among the first computers to make full use of
a high-speed random-access storage matrix, and became the machine
whose coding was most widely replicated and whose logical architecture was most widely reproduced. The stored-program computer, as
conceived by Alan Turing and delivered by John von Neumann, broke
the distinction between numbers that *mean* things and numbers that
*do* things. Our universe would never be the same.

Working outside the bounds of industry, breaking the rules of
academia, and relying largely on the U.S. government for support, a
dozen engineers in their twenties and thirties designed and built von
Neumann's computer for less than $1 million in under five years. "He
was in the right place at the right time with the right connections with
the right idea," remembers Willis Ware, fourth to be hired to join the

engineering team, "setting aside the hassle that will probably never be resolved as to whose ideas they really were."[1]

As World War II drew to a close, the scientists who had built the atomic bomb at Los Alamos wondered, "What's next?" Some, including Richard Feynman, vowed never to have anything to do with nuclear weapons or military secrecy again. Others, including Edward Teller and John von Neumann, were eager to develop more advanced nuclear weapons, especially the "Super," or hydrogen bomb. Just before dawn on the morning of July 16, 1945, the New Mexico desert was illuminated by an explosion "brighter than a thousand suns." Eight and a half years later, an explosion one thousand times more powerful illuminated the skies over Bikini Atoll. The race to build the hydrogen bomb was accelerated by von Neumann's desire to build a computer, and the push to build von Neumann's computer was accelerated by the race to build a hydrogen bomb.

Computers were essential to the initiation of nuclear explosions, and to understanding what happens next. In "Point Source Solution," a 1947 Los Alamos report on the shock waves produced by nuclear explosions, von Neumann explained that "for very violent explosions . . . it may be justified to treat the original, central, high pressure area as a point."[2] This approximated the physical reality of a nuclear explosion closely enough to enable some of the first useful predictions of weapons effects.

Numerical simulation of chain reactions within computers initiated a chain reaction among computers, with machines and codes proliferating as explosively as the phenomena they were designed to help us understand. It is no coincidence that the most destructive and the most constructive of human inventions appeared at exactly the same time. Only the collective intelligence of computers could save us from the destructive powers of the weapons they had allowed us to invent.

Turing's model of universal computation was one-dimensional: a string of symbols encoded on a tape. Von Neumann's implementation of Turing's model was two-dimensional: the address matrix underlying all computers in use today. The landscape is now three-dimensional, yet the entire Internet can still be viewed as a common tape shared by a multitude of Turing's Universal Machines.

Where does time fit in? Time in the digital universe and time in our universe are governed by entirely different clocks. In our universe, time is a continuum. In a digital universe, time (T) is a countable number of discrete, sequential steps. A digital universe is bounded at the beginning, when T = 0, and at the end, if T comes to a stop. Even in a perfectly deterministic universe, there is no consistent method to predict the ending in advance. To an observer in our universe, the digital universe appears to be speeding up. To an observer in the digital universe, our universe appears to be slowing down.

Universal codes and universal machines, introduced by Alan Turing in his "On Computable Numbers, with an Application to the Entscheidungsproblem" of 1936, have prospered to such an extent that Turing's underlying interest in the "decision problem" is easily overlooked. In answering the *Entscheidungsproblem*, Turing proved that there is no systematic way to tell, by looking at a code, what that code will do. That's what makes the digital universe so interesting, and that's what brings us here.

It is impossible to predict where the digital universe is going, but it is possible to understand how it began. The origin of the first fully electronic random-access storage matrix, and the propagation of the codes that it engendered, is as close to a point source as any approximation can get.

ACKNOWLEDGMENTS

# IN THE BEGINNING WAS
# THE COMMAND LINE

*Intuition of truth may not Relish soe much
as Truth that is hunted downe.*

—Sir Robert Southwell to William Petty, 1687

In 1956, at the age of three, I was walking home with my father, physicist Freeman Dyson, from his office at the Institute for Advanced Study in Princeton, New Jersey, when I found a broken fan belt lying in the road. I asked my father what it was. "It's a piece of the sun," he said.

My father was a field theorist, and protégé of Hans Bethe, former wartime leader of the Theoretical Division at Los Alamos, who, when accepting his Nobel Prize for discovering the carbon cycle that fuels the stars, explained that "stars have a life cycle much like animals. They get born, they grow, they go through a definite internal development, and finally they die, to give back the material of which they are made so that new stars may live."[1] To an engineer, fan belts exist between the crankshaft and the water pump. To a physicist, fan belts exist, briefly, in the intervals between stars.

At the Institute for Advanced Study, more people worked on quantum mechanics than on their own cars. There was one notable exception: Julian Bigelow, who arrived at the Institute, in 1946, as John von Neumann's chief engineer. Bigelow, who was fluent in physics, mathematics, and electronics, was also a mechanic who could explain, even to a three-year-old, how a fan belt works, why it broke, and whether it came from a Ford or a Chevrolet.

A child of the Depression, Bigelow never threw anything away. The Institute for Advanced Study, occupying the site of the former Olden Farm, owned a large, empty barn, where surplus parts and equipment from the construction of von Neumann's computer were stored amid bales of hay, spring-tooth harrows, and other remnants of the farm's working life. I was one of a small band of eight- to ten-year-olds who spent our free time exploring the Institute Woods, and occasionally visited the barn. A few beams of sunlight perforated the roof through dust raised by pigeons that fluttered away from us overhead.

Julian's cache of war surplus electronics had already been scavenged for needed parts. We had no idea what most of it was—but that did not stop us from dismantling anything that would come apart. We knew that Julian Bigelow had built a computer, housed in a building off-limits to children, just as we knew that Robert Oppenheimer, who lived in the manor house belonging to the barn, had built an atomic bomb. On our expeditions into the woods, we ignored birds and mammals, hunting for frogs and turtles that we could capture with our bare hands. It was still the age of reptiles to us. The dinosaurs of computing, in contrast, were warm-blooded, but the relays and vacuum tubes we extracted from their remains had already given up their vital warmth.

I was left with an inextinguishable curiosity over the relics that had been abandoned in the barn. "Institutes like nations are perhaps happiest if they have no history," declared Abraham Flexner, the founding director of the Institute for Advanced Study, in 1936. It was thanks to this policy of Dr. Flexner's, maintained by his successors, including Oppenheimer, regarding the history of the Institute in general, and the history of the Electronic Computer Project in particular, that much of the documentary material behind this book lay secreted for so many years. "I am reasonably confident that there is nothing here that would interest him," Carl Kaysen, Oppenheimer's successor, noted in response to an inquiry, in 1968, about records concerning the von Neumann computer project from a professor of electrical engineering at MIT.[2]

Thanks to former director Phillip Griffiths, with support from trustees Charles Simonyi and Marina von Neumann Whitman, I was invited to spend the 2002–2003 academic year as Director's Visitor at the Institute for Advanced Study, and granted access to files that in some cases had not seen the light of day since 1946. Historical Studies–Social Science Librarian Marcia Tucker and Archivist Lisa Coats began working to preserve and organize the surviving records of the Electronic Computer Project, and Kimberly Jacobsen transcribed thousands of pages of documents that are only sparsely sampled here. Through the efforts of current director Peter Goddard, and a gift from Shelby White and the Leon Levy Foundation, a permanent Archives

Center has now been established at the IAS. Archivists Christine Di Bella, Erica Mosner, and all the staff at the Institute, especially Linda Cooper, helped in every capacity, and the current trustees, especially Jeffrey Bezos, have lent continuing encouragement and support.

Many of the surviving eyewitnesses—including Alice Bigelow, Julian Bigelow, Andrew and Kathleen Booth, Raoul Bott, Martin and Virginia Davis, Akrevoe Kondopria Emmanouilides, Gerald and Thelma Estrin, Benoît Mandelbrot, Harris Mayer, Jack Rosenberg, Atle Selberg, Joseph and Margaret Smagorinsky, Françoise Ulam, Nicholas Vonneumann, Willis Ware, and Marina von Neumann Whitman— took time to speak with me. "You're within about five years of not having a testifiable witness," Joseph Smagorinsky warned me in 2004.

In 2003 the Bigelow family allowed me to go through the boxes of papers that Julian had saved. In one box, amid Office of Naval Research technical reports, World War II vacuum tube specification sheets, Bureau of Standards newsletters, and even a maintenance manual for the ENIAC, stamped RESTRICTED, was a sheet of lined paper that had evidently been crumpled up and thrown away, then uncrumpled and saved. It had been turned sideways, and had one line of handwriting across the top of the page, as follows:

Orders: Let a word (40bd) be 2 orders, each order = C(A) = Command (1–10, 21–30) • Address (11–20, 31–40)

The use of *bd* for *binary digit* dates this piece of paper from the beginning of the von Neumann project, before the abbreviation of *binary digit* to *bit*.

"In the beginning," according to Neal Stephenson, "was the command line." Thanks to Neal, and many other supporters, especially those individuals and institutions who allowed me into their basements, I spent an inordinate amount of time, over the past eight years, immersed in the layers of documents that were deposited when the digital universe was taking form. From Alex Magoun at RCA to Willis Ware at RAND, and many other keepers of institutional memory in between—including the *Annals of the History of Computing* and the Charles Babbage Institute's oral history collection—I am indebted to those who saved records that otherwise might not have been preserved. To a long list of historians and biographers— including William Aspray, Armand Borel, Alice Burks, Flo Conway, Jack Copeland, James Cortada, Martin Davis, Peter Galison, David

Alan Grier, Rolf Herken, Andrew Hodges, Norman Macrae, Brian Randell, and Jim Siegelman—I owe more than is acknowledged here. All books owe their existence to previous books, but among the antecedents of this one should be singled out (in chronological order) Beatrice Stern's *History of the Institute for Advanced Study, 1930–1950* (1964), Herman Goldstine's *The Computer from Pascal to von Neumann* (1972), Nicholas Metropolis's *History of Computing in the Twentieth Century* (1980), Andrew Hodges's *Alan Turing: The Enigma* (1983), Rolf Herken's *The Universal Turing Machine: A Half-Century Survey* (1988), and William Aspray's *John von Neumann and the Origins of Modern Computing* (1990).

Julian Bigelow and his colleagues designed and built the new computer in less time than it took me to write this book. I thank Martin Asher, John Brockman, Stefan McGrath, and Katinka Matson for their patience in allowing this. The Bigelow family, the Institute for Advanced Study, Françoise Ulam, and, especially, Marina von Neumann Whitman provided access to the documents that brought this story to life. Gabriella Bollobás translated a large body of correspondence, interpreting not only the nuances of the Hungarian language, but also the emotional and intellectual context of Budapest at that time. Belá Bollobás, Marion Brodhagen, Freeman Dyson, Joseph Felsenstein, Holly Given, David Alan Grier, Danny Hillis, Verena Huber-Dyson, Jennifer Jacquet, Harris Mayer, and Alvy Ray Smith offered comments on early drafts. Akrevoe Kondopria Emmanouilides, who typed and proofread the Institute for Advanced Study Electronic Computer Project's progress reports as a teenager in 1946, found errors that might otherwise have been missed.

Finally, thanks to those who sponsored the work that forms the subject of this book. "While old men in congresses and parliaments would debate the allocation of a few thousand dollars, farsighted generals and admirals would not hesitate to divert substantial sums to help oddballs in Princeton, Cambridge, and Los Alamos," observed Nicholas Metropolis, reviewing the development of computers after World War II.[3]

Early computers were built in many places, leaving fossils that remain well preserved. But what, exactly, once everything else was in place, sparked the chain reaction between address matrix and order codes, spawning the digital universe in which we are all now immersed?

*C(A)* is all it took.

# PRINCIPAL CHARACTERS

**Katalin (Lili) Alcsuti (1910–1990):** John von Neumann's younger cousin and granddaughter of von Neumann's maternal grandfather, Jacob Kann (1854–1914).

**Hannes Alfvén (1908–1995):** Swedish American magnetohydrodynamicist and author (under pseudonym Olof Johannesson) of *The Tale of the Big Computer.*

**Frank Aydelotte (1880–1956):** Second director of the Institute for Advanced Study (IAS), 1939–1947.

**Louis Bamberger (1855–1944):** Newark, New Jersey, department store magnate and founder, with sister Carrie Fuld, of the IAS.

**Nils Aall Barricelli (1912–1993):** Norwegian Italian mathematical biologist and viral geneticist; at the IAS in 1953, 1954, and 1956.

**Julian Himely Bigelow (1913–2003):** American electronic engineer and collaborator, with Norbert Wiener, on antiaircraft fire control during World War II; chief engineer of the IAS Electronic Computer Project (ECP), 1946–1951.

**Andrew Donald Booth (1918–2009):** British physicist, X-ray crystallographer, inventor, and early computer architect; visited the IAS ECP in 1946 and 1947.

**Kathleen (née Britten) Booth (1922– ):** Computational physicist and member of J. D. Bernal's Biomolecular Structure Group; visitor at the IAS ECP in 1947; author of *Programming for an Automatic Digital Calculator* (1958).

**Arthur W. Burks (1915–2008):** American ENIAC (Electronic Numerical Integrator and Computer) project engineer, philosopher, logician, and "scribe" for the IAS preliminary design team in 1946.

**Vannevar Bush (1890–1974):** Analog computer pioneer, director of the U.S. Office of Scientific Research and Development during World War II, and lead administrator of the Manhattan Project.

**Jule Gregory Charney** (1917–1981): American meteorologist and leader of the IAS Meteorology Project between 1948 and 1956.

**Richard F. Clippinger** (1913–1997): American mathematician and computer scientist; supervised the retrofit of the ENIAC to stored-program mode in 1947.

**Hewitt Crane** (1927–2008): American electrical engineer and IAS ECP member, 1951–1954; subsequently lead scientist at Stanford Research Institute.

**Freeman J. Dyson** (1923–): British American mathematical physicist; arrived at the IAS as a Commonwealth Fellow in September 1948.

**Carl Henry Eckart** (1902–1973): American physicist, first director of Scripps Institution of Oceanography, and fourth husband of Klára (Klári) von Neumann.

**John Presper Eckert** (1919–1995): American electronic engineer, ENIAC developer, and cofounder, with John Mauchly, of the Electronic Control Company, manufacturers of BINAC and UNIVAC.

**Akrevoe (née Kondopria) Emmanouilides** (1929–): Administrative secretary to Herman Goldstine on the ENIAC project at the Moore School and on the IAS ECP during 1946–1949.

**Gerald Estrin** (1921–): IAS ECP member 1950–1956, with a leave of absence to direct the construction of the WEIZAC, a first-generation sibling of the MANIAC, at the Weizmann Institute in Rehovot, Israel, 1953–1955.

**Thelma Estrin** (1924–): Electronic engineer, member of the IAS ECP 1950–1956, and wife of Gerald Estrin.

**Foster** (1915–1999) **and Cerda** (1916–1988) **Evans:** Los Alamos physicists and husband-and-wife thermonuclear programming team; at the IAS in 1953 and 1954.

**Richard P. Feynman** (1918–1988): American physicist and member of the wartime Los Alamos computing group.

**Abraham Flexner** (1866–1959): American schoolteacher, educational reformer, and founding director of the IAS, 1930–1939.

**Simon Flexner** (1863–1946): American philanthropist, Rockefeller Foundation officer, and older brother of Abraham Flexner.

**Stanley P. Frankel** (1919–1978): American physicist, student of Robert Oppenheimer, and Los Alamos colleague of Richard Feynman; member of the original ENIAC and IAS thermonuclear calculation team; minicomputer design pioneer.

**Kurt Gödel** (1906–1978): Moravian-born Austrian logician; arrived at the IAS in 1933.

**Herman Heine Goldstine (1913–2004):** American mathematician, U.S. Army officer, ENIAC administrator, and associate director of the IAS ECP during 1946–1956.

**Irving John (Jack) Good (born Isadore Jacob Gudak; 1916–2009):** British American Bayesian statistician, artificial intelligence pioneer, cryptologist, and assistant to Alan Turing during the British code-breaking effort in World War II.

**Leslie Richard Groves (1896–1970):** U.S. Army general, commander of Los Alamos during World War II, and, later, research director at Remington Rand.

**Verena Huber-Dyson (1923–):** Swiss American logician and group theorist; arrived at the IAS as a postdoctoral fellow in 1948.

**James Brown Horner (Desmond) Kuper (1909–1992):** American physicist and second husband of Mariette (Kővesi) von Neumann.

**Herbert H. Maass (1878–1957):** Attorney and founding trustee of the IAS.

**Benoît Mandelbrot (1924–2010):** Polish-born French American mathematician; invited by von Neumann to the IAS to study word frequency distributions in 1953.

**John W. Mauchly (1907–1980):** American physicist, electrical engineer, and cofounder of the ENIAC project.

**Harris Mayer (1921–):** American Manhattan Project physicist and collaborator with Edward Teller and John von Neumann.

**Richard W. Melville (1914–1994):** Lead mechanical engineer for the IAS ECP, 1948–1953.

**Nicholas Constantine Metropolis (1915–1999):** Greek American mathematician and computer scientist, early proponent of the Monte Carlo method, and leader of Los Alamos computing group.

**Bernetta Miller (1884–1972):** Pioneer aviatrix; administrative assistant at the IAS, 1941–1948.

**Oskar Morgenstern (1902–1977):** Austrian American economist, coauthor of *Theory of Games and Economic Behavior* (1944).

**Harold Calvin (Marston) Morse (1892–1977):** American mathematician; sixth professor to be hired at the IAS.

**Maxwell Herman Alexander Newman (1897–1984):** British topologist, computer pioneer, and mentor to Alan Turing.

**J. Robert Oppenheimer (1904–1967):** Physicist; Los Alamos National Laboratory director during World War II and director of the IAS, 1947–1966.

**William Penn (1644–1718):** Quaker agitator and son of Admiral Sir

William Penn (1621–1670); founder of Pennsylvania and early proprietor of the land on which the IAS was later built.

James Pomerene (1920–2008): American electronic engineer; member of the IAS ECP, 1946–1955; replaced Julian Bigelow in 1951 as chief engineer.

Irving Nathaniel Rabinowitz (1929–2005): Astrophysicist and computer scientist; member of the IAS ECP, 1954–1957.

Jan Rajchman (1911–1989): Polish American electronic engineer; inventor of resistor-matrix storage and RCA's Selectron memory tube.

Lewis Fry Richardson (1881–1953): British pacifist, mathematician, electrical engineer, and early proponent of numerical weather prediction.

Robert Richtmyer (1910–2003): American mathematical physicist and nuclear weapons design pioneer.

Jack Rosenberg (1921–): American electronic engineer and IAS ECP member 1947–1951.

Morris Rubinoff (1917–2003): Canadian American physicist and electronic engineer; IAS ECP member, 1948–1949.

Martin Schwarzschild (1912–1997): German American astrophysicist and developer of early stellar evolution codes.

Atle Selberg (1917–2007): Norwegian American number theorist; arrived at the IAS in 1947.

Hedvig (Hedi; née Liebermann) Selberg (1919–1995): Transylvanian-born mathematics and physics teacher; wife of Atle Selberg, collaborator with Martin Schwarzschild, and lead coder for the IAS ECP.

Claude Elwood Shannon (1916–2001): American mathematician, electrical engineer, and pioneering information theorist; visiting member of the IAS (1940–1941).

Ralph Slutz (1917–2005): American physicist and member of the IAS ECP, 1946–1948; supervised construction of the SEAC (Standards Eastern Automatic Computer), the first of the IAS-class designs to become operationally complete.

Joseph Smagorinsky (1924–2005): American meteorologist; at the IAS, 1950–1953.

Lewis L. Strauss (1896–1974): American naval officer, businessman, IAS trustee, and head of U.S. Atomic Energy Commission.

Leó Szilárd (1898–1964): Hungarian American physicist, reluctant nuclear weapon pioneer, and author of *The Voice of the Dolphins*.

Edward Teller (1908–2003): Hungarian American physicist and lead-ing advocate of the hydrogen (or "super") bomb.

Philip Duncan Thompson (1922–1994): U.S. Air Force meteorological liaison officer, assigned to the IAS ECP, 1948–1949.

Bryant Tuckerman (1915–2002): American topologist and computer scientist; member of the IAS ECP, 1952–1957.

John W. Tukey (1915–2000): American statistician at Princeton Uni-versity and Bell Labs; coined the term *"bit."*

Alan Mathison Turing (1912–1954): British logician and cryptologist; author of "On Computable Numbers" (1936).

Françoise (née Aron) Ulam (1918–2011): French American editor and journalist; wife of Stanislaw Ulam.

Stanislaw Marcin Ulam (1909–1984): Polish American mathematician and protégé of John von Neumann.

Oswald Veblen (1880–1960): American mathematician, nephew of Thorstein Veblen, and first professor appointed to the IAS in 1932.

Theodore von Kármán (1881–1963): Hungarian American aerody-namicist, founder of Jet Propulsion Laboratory (JPL).

John von Neumann (born Neumann János; 1903–1957): Hungarian American mathematician; fourth professor appointed to the IAS, in 1933; founder of the IAS ECP.

Klára (née Dán) von Neumann (1911–1963): Second wife of John von Neumann; married in 1938.

Margit (née Kann) von Neumann (1880–1956): Mother of John von Neumann.

Mariette (née Kővesi) von Neumann (1909–1992): First wife of John von Neumann; married in 1929.

Max von Neumann (born Neumann Miksa; 1873–1928): Investment banker, lawyer, and father of John von Neumann.

Michael von Neumann (born Neumann Mihály; 1907–1989): Physi-cist, and younger brother of John von Neumann.

Nicholas Vonneumann (born Neumann Miklos; 1911–2011): Patent attorney, and youngest brother of John von Neumann.

Willis H. Ware (1920–): American electrical engineer and member of the IAS ECP, 1946–1951, subsequently at RAND.

Warren Weaver (1894–1978): American mathematician, self-described "Chief Philanthropoid" at the Rockefeller Foundation, and direc-tor of the Applied Mathematics Panel of the U.S. Office of Scien-tific Research and Development during World War II.

Marina (née von Neumann) Whitman (1935–): Economist, U.S. presidential adviser, and daughter of John von Neumann and Mariette Kővesi von Neumann.

**Norbert Wiener** (1894–1964): American mathematician and founder, with Julian Bigelow and John von Neumann, of what would become known as the Cybernetics Group.

**Eugene P. Wigner** (born Wigner Jenő; 1902–1995): Hungarian American mathematical physicist.

**Frederic C. Williams** (1911–1977): British electronic engineer; World War II radar pioneer and developer, at the University of Manchester, of the "Williams" cathode-ray storage tube, and of the Manchester "Mark 1," the first operational stored-program computer to utilize it.

**Vladimir Kosma Zworykin** (1889–1982): Russian-born American television pioneer and director of the Princeton laboratories of RCA.

# TURING'S CATHEDRAL

# ONE

# 1953

*If it's that easy to create living organisms,*
*why don't you create a few yourself?*

—Nils Aall Barricelli, 1953

AT 10:38 P.M. on March 3, 1953, in a one-story brick building at the end of Olden Lane in Princeton, New Jersey, Italian Norwegian mathematical biologist Nils Aall Barricelli inoculated a 5-kilobyte digital universe with random numbers generated by drawing playing cards from a shuffled deck. "A series of numerical experiments are being made with the aim of verifying the possibility of an evolution similar to that of living organisms taking place in an artificially created universe," he announced.[1]

A digital universe—whether 5 kilobytes or the entire Internet—consists of two species of bits: differences in space, and differences in time. Digital computers translate between these two forms of information—structure and sequence—according to definite rules. Bits that are embodied as structure (varying in space, invariant across time) we perceive as memory, and bits that are embodied as sequence (varying in time, invariant across space) we perceive as code. Gates are the intersections where bits span both worlds at the moments of transition from one instant to the next.

The term *bit* (the contraction, by 40 bits, of *"binary digit"*) was coined by statistician John W. Tukey shortly after he joined von Neumann's project in November of 1945. The existence of a fundamental unit of communicable information, representing a single distinction between two alternatives, was defined rigorously by information theorist Claude Shannon in his then-secret *Mathematical Theory of Cryptography* of 1945, expanded into his *Mathematical Theory of Communication* of 1948. "Any difference that makes a difference" is how cybernetician Gregory Bateson translated Shannon's definition into informal terms.[2] To a digital computer, the only difference that makes a difference is the difference between a zero and a one.

3

That two symbols were sufficient for encoding all communication had been established by Francis Bacon in 1623. "The transposition of two Letters by five placeings will be sufficient for 32 Differences [and] by this Art a way is opened, whereby a man may expresse and signifie the intentions of his minde, at any distance of place, by objects . . . capable of a twofold difference onely," he wrote, before giving examples of how such binary coding could be conveyed at the speed of paper, the speed of sound, or the speed of light.[3]

That zero and one were sufficient for logic as well as arithmetic was established by Gottfried Wilhelm Leibniz in 1679, following the lead given by Thomas Hobbes in his *Computation, or Logique* of 1656. "By Ratiocination, I mean *computation*," Hobbes had announced. "Now to compute, is either to collect the sum of many things that are added together, or to know what remains when one thing is taken out of another. *Ratiocination*, therefore is the same with *Addition* or *Substraction*; and if any man adde *Multiplication* and *Division*, I will not be against it, seeing . . . that all Ratiocination is comprehended in these two operations of the minde."[4] The new computer, for all its powers, was nothing more than a very fast adding machine, with a memory of 40,960 bits.

In March of 1953 there were 53 kilobytes of high-speed random-access memory on planet Earth.[5] Five kilobytes were at the end of Olden Lane, 32 kilobytes were divided among the eight completed clones of the Institute for Advanced Study's computer, and 16 kilobytes were unevenly distributed across a half dozen other machines. Data, and the few rudimentary programs that existed, were exchanged at the speed of punched cards and paper tape. Each island in the new archipelago constituted a universe unto itself.

In 1936, logician Alan Turing had formalized the powers (and limitations) of digital computers by giving a precise description of a class of devices (including an obedient human being) that could read, write, remember, and erase marks on an unbounded supply of tape. These "Turing machines" were able to translate, in both directions, between bits embodied as structure (in space) and bits encoded as sequences (in time). Turing then demonstrated the existence of a Universal Computing Machine that, given sufficient time, sufficient tape, and a precise description, could emulate the behavior of any other computing machine. The results are independent of whether the instructions are executed by tennis balls or electrons, and whether the memory is stored in semiconductors or on paper tape. "Being digital should be of more interest than being electronic," Turing pointed out.[6]

Von Neumann set out to build a Universal Turing Machine that would operate at electronic speeds. At its core was a 32-by-32-by-40-bit matrix of high-speed random-access memory—the nucleus of all things digital ever since. "Random access" meant that all individual memory locations—collectively constituting the machine's internal "state of mind"—were equally accessible at any time. "High speed" meant that the memory was accessible at the speed of light, not the speed of sound. It was the removal of this constraint that unleashed the powers of Turing's otherwise impractical Universal Machine.

Electronic components were widely available in 1945, but digital behavior was the exception to the rule. Images were televised by scanning them into lines, not breaking them into bits. Radar delivered an analog display of echoes returned by the continuous sweep of a microwave beam. Hi-fi systems filled postwar living rooms with the warmth of analog recordings pressed into vinyl without any losses to digital approximation being introduced. Digital technologies—Teletype, Morse code, punched card accounting machines—were perceived as antiquated, low-fidelity, and slow. Analog ruled the world.

The IAS group achieved a fully electronic random-access memory by adapting analog cathode-ray oscilloscope tubes—evacuated glass envelopes about the size and shape of a champagne bottle, but with walls as thin as a champagne flute's. The wide end of each tube formed a circular screen with a fluorescent internal coating, and at the narrow end was a high-voltage gun emitting a stream of electrons whose aim could be deflected by a two-axis electromagnetic field. The cathode-ray tube (CRT) was a form of analog computer: varying the voltages to the deflection coils varied the path traced by the electron beam. The CRT, especially in its incarnation as an oscilloscope, could be used to add, subtract, multiply, and divide signals—the results being displayed directly as a function of the amplitude of the deflection and its frequency in time. From these analog beginnings, the digital universe took form.

Applying what they had learned in the radar, cryptographic, and antiaircraft fire-control business during the war, von Neumann's engineers took pulse-coded control of the deflection circuits and partitioned the face of the tube into a 32-by-32 array of numerically addressable locations that could be individually targeted by the electron beam. Because the resulting electric charge lingered on the coated glass surface for a fraction of a second and could be periodically refreshed, each 5-inch-diameter tube could store 1,024 bits of information, with the state of any specified location accessible at any time. The transition from analog to digital had begun.

The IAS computer incorporated forty cathode-ray memory tubes, with memory addresses assigned as if a desk clerk were handing out similar room numbers to forty guests at a time in a forty-floor hotel. Codes proliferated within this universe by taking advantage of the architectural principle that a pair of 5-bit coordinates ($2^5 = 32$) uniquely identified one of 1,024 memory locations containing a string (or "word") of 40 bits. In 24 microseconds, any specified 40-bit string of code could be retrieved. These 40 bits could include not only data (numbers that mean things) but also executable instructions (numbers that do things)—including instructions to modify the existing instructions, or transfer control to another location and follow new instructions from there.

Since a 10-bit order code, combined with 10 bits specifying a memory address, returned a string of 40 bits, the result was a chain reaction analogous to the two-for-one fission of neutrons within the core of an atomic bomb. All hell broke loose as a result. Random-access memory gave the world of machines access to the powers of numbers—and gave the world of numbers access to the powers of machines.

The computer building's plain concrete-block core had been paid for jointly by the U.S. Army's Ordnance Department and the U.S. Atomic Energy Commission (AEC). To reconcile the terms of the government contract, specifying a temporary structure, with the sentiments of the neighboring community, the Institute for Advanced Study had paid the additional $9,000 (equivalent to about $100,000 today) to finish the building with a brick veneer.

There were close ties between the IAS and the AEC. J. Robert Oppenheimer was director of the IAS and chairman of the General Advisory Committee of the AEC. Lewis Strauss was chairman of the AEC and president of the IAS Board of Trustees. The freewheeling mix of science and weaponeering that had thrived at Los Alamos during the war had been transplanted to Princeton under the sponsorship of the AEC. "The Army contract provides for general supervision by the Ballistic Research Laboratory of the Army," it was noted on

November 1, 1949, "whereas the AEC provides for supervision by von Neumann."[7] As long as the computer was available for weapons calculations, von Neumann could spend the remaining machine time as he pleased.

In 1953, Robert Oppenheimer and Lewis Strauss—who had engineered Oppenheimer's appointment as director of the Institute in 1947, and would turn against him in 1954—were still on friendly terms. "There is a case of Chateau Lascombes waiting for you with my compliments at Sherry Wine & Spirits Co., 679 Madison Avenue (near 61st Street)," Strauss informed Oppenheimer on April 10, 1953. "I hope you and Kitty will like it."[8]

"We picked up the wine two days ago, and opened a bottle that night," Oppenheimer replied on April 22. "It was very good; and now Kitty and I can thank you, not merely for your kindness, but for the great pleasure that you have made us."[9] Robert and Kitty had drunk from the poisoned chalice. One year later, the man who had done so much to deliver the powers of atomic energy into the hands of the U.S. government, but had then turned against his masters to oppose the development of the hydrogen bomb, would be stripped of his security clearances after a dramatic hearing before the Atomic Energy Commission's Personnel Security Board.

While the computer was still under construction, a small team from Los Alamos, led by Nicholas Metropolis and Stanley Frankel, quietly took up residence at the Institute. There were two separate classes of membership at the IAS: permanent members, who were appointed for life by a decision of the faculty as a whole, and visiting members, who were invited by the individual schools, usually for one year or less. Metropolis and Frankel did not belong to either group and mysteriously just showed up. "All I was told was that what Metropolis came out for was to calculate the feasibility of a fusion bomb," remembers Jack Rosenberg, an engineer who had designed, built, and installed a hi-fi audio system in Albert Einstein's house for his seventieth birthday in 1949, using some of the computer project's spare vacuum tubes and other parts. "That's all I knew. And then I felt dirty. And Einstein said 'that's exactly what I thought they were going to use it for.' He was way ahead."[10]

The new machine was christened MANIAC (Mathematical and Numerical Integrator and Computer) and put to its first test, during the summer of 1951, with a thermonuclear calculation that ran for sixty days nonstop. The results were confirmed by two huge explosions in the South Pacific: Ivy Mike, yielding the equivalent of 10.4 million tons of TNT at Enewetak on November 1, 1952, and Castle Bravo, yielding 15 megatons at Bikini on February 28, 1954.

The year 1953 was one of frenzied preparations in between. Of the eleven nuclear tests, yielding a total of 252 kilotons, conducted at the Nevada Test Site in 1953, most were concerned not with trying to make large, spectacular explosions, but with understanding how the effects of more modest nuclear explosions could be tailored to trigger a thermonuclear reaction resulting in a deliverable hydrogen bomb.

Ivy Mike, fueled by 82 tons of liquid deuterium, cooled to minus 250 degrees in a tank the size of a railroad car, demonstrated a proof of principle, whereas Castle Bravo, fueled by solid lithium deuteride, represented a deployable weapon that could be delivered, in hours, by a B-52. It was von Neumann, in early 1953, who pointed out to the air force that rockets were getting larger, while hydrogen bombs were getting smaller. Delivery in minutes would be next.

The Americans had smaller bombs, but the Russians had larger rockets. Plotting the increasing size of rockets against the decreasing size of warheads, von Neumann showed that the intersection resulting in an intercontinental ballistic missile—a possibility he referred to as "nuclear weapons in their expected most vicious form"—might occur in the Soviet Union first.[11] The air force, pushed by Trevor Gardner and Bernard Schriever, formed a Strategic Missiles Evaluation Committee chaired by von Neumann, and the Atlas ICBM program, which had been limping along since 1946, was off the ground. The year 1953 was the first one in which more than $1 million was spent on guided missile development by the United States. "Guided" did not imply the precision we take for granted now. "Once it was launched, all that we would know is what city it was going to hit," von Neumann answered the vice president in 1955.[12]

Numerical simulations were essential to the design of weapons that were, as Oppenheimer put it, "singularly proof against any form of experimental approach." When Nils Barricelli arrived in Princeton in 1953, one large thermonuclear calculation had just been completed, and another was in the works. The computer was usually turned over to the Los Alamos group, led by Foster and Cerda Evans, overnight. It was agreed, on March 20, that "during the running of the Evans problem there would be no objection to using some time on Saturday and Sunday instead of operating from midnight to 8:00 a.m."[13] Barricelli had to squeeze his numerical universe into existence between bomb calculations, taking whatever late-night and early-morning hours were left.

During the night of March 3, 1953, as Barricelli's numerical organisms were released into the computational wilderness for the first time, Joseph Stalin was sinking into a coma in Moscow, following a stroke. He died two days later—five months short of witnessing the first Soviet hydrogen bomb test at Semipalatinsk. No one knew who or what would follow Stalin, but Lavrentiy Beria, director of the NKVD secret police and supervisor of the Soviet nuclear weapons program, was the heir apparent, and the U.S. Atomic Energy Commission made it their business to fear the worst. After Barricelli's "Symbiosis Problem" ran without misadventure overnight, the machine log notes "over to blast wave" on the morning of March 4. Later in the day the log simply reads "over to" followed by a pencil sketch of a mushroom cloud.

Three technological revolutions dawned in 1953: thermonuclear weapons, stored-program computers, and the elucidation of how life stores its own instructions as strings of DNA. On April 2, James Watson and Francis Crick submitted "A Structure for Deoxyribose Nucleic Acid" to *Nature*, noting that the double helical structure "suggests a possible copying mechanism for the genetic material." They hinted at the two-bits-per-base-pair coding whereby living cells read, write, store, and replicate genetic information as sequences of nucleotides we identify as A, T, G, and C. "If an adenine forms one member of a pair, on either chain, then on these assumptions the other member must be thymine; similarly for guanine and cytosine," they explained. "If only specific pairs of bases can be formed, it follows that if the sequence of bases on one chain is given, then the sequence on the other chain is automatically determined."[14]

The mechanism of translation between sequence and structure in biology and the mechanism of translation between sequence and structure in technology were set on a collision course. Biological organisms had learned to survive in a noisy, analog environment by repeating themselves, once a generation, through a digital, error-correcting phase, the same way repeater stations are used to convey intelligible messages over submarine cables where noise is being introduced. The transition from digital once a generation to digital all the time began in 1953.

The race was on to begin decoding living processes from the top down. And with the seeding of an empty digital universe with self-modifying instructions, we took the first steps toward the encoding of living processes from the bottom up. "Just because the special conditions prevailing on this earth seem to favor the forms of life which are based on organo-chemical compounds, this is no proof that it is not possible to build up other forms of life on an entirely different basis," Barricelli explained.[15] The new computer was assigned two problems: how to destroy life as we know it, and how to create life of unknown forms.

What began as an isolated 5-kilobyte matrix is now expanding by over two trillion transistors per second (a measure of the growth in processing and memory) and five trillion bits of storage capacity per second (a measure of the growth in code).[16] Yet we still face the same questions that were asked in 1953. Turing's question was what it would take for machines to begin to think. Von Neumann's question was what it would take for machines to begin to reproduce.

When the Institute for Advanced Study agreed, against all objections, to allow von Neumann and his group to build a computer, the concern was that the refuge of the mathematicians would be disturbed by the presence of engineers. No one imagined the extent to which, on the contrary, the symbolic logic that had been the preserve of the mathematicians would unleash the powers of coded sequences upon the world. "In those days we were all so busy doing what we were doing we didn't think very much about this enormous explosion that might happen," says Willis Ware.

Was the explosion an accident or deliberately set? "The military wanted computers," explains Harris Mayer, the Los Alamos physicist who was working with both John von Neumann and Edward Teller at the time. "The military had the need and they had the money but they didn't have the genius. And Johnny von Neumann was the genius. As soon as he recognized that we needed a computer to do the calculations for the H-bomb, I think Johnny had all of this in his mind."[17]

TWO

# Olden Farm

*It was the Lenni Lenape! It was the tribes
of the Lenni Lenape! The sun rose from
water that was salt, and set in water that
was sweet, and never hid himself from
their eyes. . . . It was but yesterday that the
children of the Lenape were masters of the
world.*

—James Fenimore Cooper, 1826

PRINCETON, NEW JERSEY, in summer has been described as "like the inside of a dog's mouth." The area's original inhabitants, the Lenni Lenape ("original people" or "Men of Men") abandoned the interior of New Jersey in summer and headed either to the Jersey Shore or to encampments on the estuaries of Delaware Bay. "From thence [June] to this present Month [August]," reported William Penn during his first summer on the Delaware River, in 1683, "we have had *extraordinary* heats."[1] Penn had landed in Delaware Bay on October 27, 1682, after a passage (from Deal, England) of fifty-nine days aboard the *Welcome,* during which smallpox broke out and thirty-one of his ninety-nine fellow colonists died. Penn, who had survived smallpox at the age of three, ministered to the sick during the voyage and arrived in excellent health.

The Lenni Lenape, a subset of the Algonquin people, were called Delawares by the Dutch, Swedish, and English settlers who arrived in the seventeenth century, following the visit of the Italian navigator Giovanni da Verrazzano in 1524. The Lenape met the newcomers with diplomacy, but the colonists had technology and immunity on their side. "What is the Matter with us Indians," asked Chief Tenoughan of the Schuylkill River, according to Penn, "that we are thus sick in our own Air, and these Strangers well?"[2]

The Lenni Lenape around Princeton belonged to the Unami nation, identified with the Turtle clan. European observers were never quite sure whether the Unami belonged to the Turtle, or the Turtle to the Unami. New Jersey was home to eleven species of turtle, adapted to all conditions, from hibernating at the bottom of frozen ponds to basking in the midsummer sun. To an American snapping turtle, a species unchanged for sixty million years, only the blink of a double-lidded eye separates us from William Penn.

In 1609, Henry Hudson, representing the Dutch East India Company, explored the Newark estuary before sailing up the river that bears his name. In 1614, Cornelius Jacobsen Mey, also Dutch, explored Delaware Bay and entered the Delaware River, navigable as far as the site of present-day Trenton, or Delaware Falls. Upon his restoration in 1660, Charles II challenged the Dutch claims to North America, granting the entire territory between Virginia and New France to his brother, the Duke of York (later James II), in 1664. The territory was named New York, and a portion of it, between the Delaware and Hudson rivers, was subdivided further, with the west half (bordering the Delaware River and Delaware Bay) assigned to Lord Berkeley and the east half (bordering the Hudson River and the Atlantic Ocean) assigned to Sir George Carteret. The province was named New Jersey, and soon fell into the hands of Quakers, or the Society of Friends.

In 1675, Lord Berkeley sold his interest in West New Jersey for £1,000 to John Fenwick and Edward Byllynge, two Quakers whose subsequent dispute over the property was referred for arbitration to William Penn. Fenwick sailed for America with his family and a group of fellow Quakers aboard the *Griffin* (or *Griffith*), founding a colony at "a pleasant rich spot" (Salem) on the Delaware, while Byllynge fell into debt and eventually transferred his interest to a group of creditors that now included Penn.[3]

Penn, who had studied law in London, took the lead in drafting a constitution for the new colony, issued in 1676 as the "Concessions and Agreements of the Proprietors, Freeholders, and Inhabitants of the Province of West New Jersey." One hundred years before the Declaration of Independence, this document established a representative democracy with freedom of religion and assembly, trial by jury, economic liberty, and other principles that would later be incorporated into the constitutions of Pennsylvania and eventually the United States. Penn also joined a partnership that purchased East New Jersey at auction in 1682 for £3,400 from Carteret's estate.

William Penn was the rebellious son of Admiral Sir William Penn, who led the English fleet in two wars against the Dutch and captured Jamaica (for Cromwell) in 1655. During the English Civil War, he sided with Parliament against the king, but secretly offered to switch sides and subsequently became a favorite of the king's brother James. The younger Penn was sent to Oxford at age fifteen and soon expelled, for conducting religious services in his room and refusing to attend chapel or wear a gown. After a two-year tour of Europe he was entrusted with the management of his father's estates in Ireland, where he took up with the Quakers, a rapidly growing nonconformist sect. He was promptly arrested and imprisoned, the first of some seven times. "Mr. William Pen[n], who is lately come over from Ireland, is a Quaker again, or some very melancholy thing," noted Samuel Pepys on December 29, 1667.

Upon his return to London, Penn began pamphleteering, with "The Sandy Foundation Shaken" (questioning the Trinity) landing him in the Tower of London for eight months (during which he wrote the book *No Cross, No Crown* and several inflammatory tracts). In August 1670 he was again arrested in London, this time with William Mead, for preaching in the street after the Quaker meetinghouse in Grace-church Street was padlocked by the authorities, who charged that Penn and Mead "unlawfully and tumultuously did Assemble and Con-gregate themselves together, to the Disturbance of the Peace of the said Lord the King."[4]

After two weeks in Newgate Prison, Penn and Mead pled not guilty. "We did not make the Tumult, but they that interrupted us," argued Penn. " 'Tis very well known that we are a peaceable People, and can-not offer Violence to any Man."[5] The jury's verdict was not guilty, upon which they, too, were imprisoned, for contempt of court. "You shall not be dismist till we have a Verdict, that the Court will accept; and you shall be lock'd up, without Meat, Drink, Fire, and Tobacco; you shall not think thus to abuse the Court; we will have a Verdict, by the help of God, or you shall starve for it," they were instructed by the Crown.[6] This miscarriage of justice provoked an outcry that led to the release of Penn and Mead, along with their jurors, and precipitated a change in English law. Penn would soon be back in prison, sentenced to six months for refusing to take an oath of allegiance to the king, in 1671.

Admiral Sir William Penn died in 1670, leaving an unresolved £16,000, officially for £11,000 in "victualling expenses," plus interest, owed to him by the Crown. Rumors lingered that the admiral had covered the king for a gambling debt. His son William petitioned the king, in 1680, for a settlement, proposing that the Crown grant him "a tract of land in America, lying north of Maryland, on the east bounded with Delaware River, on the west, limited as Maryland is, and northward to extend as far as plantable, which is altogether Indian."[7] Charles and James said yes—resolving the debt, and exporting Penn. The colony of Pennsylvania, with its capital, Philadelphia, was the result.

Penn arrived in 1682, assumed the governorship, and traveled extensively into the surrounding wilderness, learning the Lenni Lenape language well enough to converse without an interpreter and contrasting the justice and equality he found among the Indians to the injustices and inequalities he had left behind. "I find them . . . of a deep natural sagacity," he wrote to his friend Robert Boyle of the Royal Society in 1683. "The low dispensation of the poor Indian out shines the lives of those Christians, that pretend an higher."[8]

To the west of Pennsylvania lay open wilderness, while to the east the wilds of New Jersey were now squeezed between two growing populations centered upon Philadelphia and New York. The most direct connection between the two settlements was overland across the "waist" of New Jersey, between the head of navigation on the Delaware River (near present-day Trenton, upstream from Philadelphia) and the head of navigation on the Raritan River (near present-day New Brunswick, upstream from New York). This well-worn Lenni Lenape footpath became, in succession, a trail passable by horses, a wagon road, the "King's Highway" for stagecoaches, and finally State Routes 27 and 206.

In 1683 a settler named Henry Greenland opened a tavern near the midpoint of the wagon road, and around this nucleus a village began to form. The proprietors of East and West New Jersey met at Greenland's Tavern in 1683 to decide their common boundary, and this put Prince-Town on the map, while the nearby wilderness attracted a small group of Quakers seeking to distance themselves as far as possible from the secular influences of Philadelphia and New York. Halfway between the Raritan and the Delaware, and just to the south of the overland trail, was a small stream named Wapowog by the Lenape, flowing through land that was taken up in 1693, as his share of the original grant to the proprietors of West New Jersey, by

William Penn. Six close-knit Quaker families, with Penn as an absentee partner, founded a colony here in 1696, and named it Stony Brook. Instead of a tavern, they built a Quaker meetinghouse.

The patriarchs of these families were Benjamin Clarke, William Olden, Joseph Worth, John Hornor, Richard Stockton, and Benjamin FitzRandolph, with Stockton becoming the largest landowner thanks to 5,500 acres purchased for £900 in 1701 from William Penn (who reserved 1,050 acres "as to said William Penn shall seem meet and convenient" for himself).[9] Benjamin Clarke purchased 1,200 acres between Stony Brook, the Province Line, the present Stockton Street, and the present Springdale Road, in 1696, conveying 400 acres (including the future site of the Institute for Advanced Study) to his brother-in-law William Olden and deeding 9.6 acres in trust for the Friends meetinghouse and cemetery in 1709. The meetinghouse was finished in 1726, and the settlers also opened a school and built water-driven mills. By 1737 a stage wagon was running between Trenton and New Brunswick twice a week, and Prince-Town had grown to accommodate those who stopped to change horses or spend the night.

With no institution of higher learning between Yale University, in New Haven, Connecticut, and the College of William and Mary, in Williamsburg, Virginia, a College of New Jersey was established (by Presbyterians) in 1746, holding classes first in Elizabeth, and then in Newark, before being deeded 10 acres of cleared land and 200 acres of woodland by the Quaker community in Princeton in January 1753. The first students arrived in 1756, and, in January 1774, with revolution brewing, they demonstrated their support for the cause of independence by burning the steward's winter store of tea.

When war arrived in Princeton at the end of November 1776, the American side was in disarray, with George Washington's exhausted forces in retreat back to Pennsylvania after a series of defeats that included Brooklyn Heights (on Long Island), White Plains, and Fort Washington (on Manhattan Island in New York).

General Washington and some three thousand men arrived in Princeton on the night of December 1, pursued by Lord Cornwallis's forces and their retinue of Hessian conscripts, who were looting and pillaging along the way. Washington regrouped in Princeton for a week before retreating to Trenton and finally to safety across the Delaware, while the British gathered at Trenton in pursuit.

On Christmas Eve, Washington slipped back across the Delaware (with 2,400 men) and launched a surprise attack in a snowstorm at eight o'clock on the morning of the twenty-sixth. He then returned to Pennsylvania, replenishing his forces as best he could until New Year's Day, when, with some 5,000 men (more than half of them irregulars), he reassembled at Trenton, preparing to meet Cornwallis's forces, who were on the advance from Princeton. This led to an inconclusive standoff along the banks of the Assunpink Creek. The night of the second was bitterly cold, freezing the muddy roads and allowing Washington's army, with their artillery, to escape under cover of darkness along the back roads toward Stony Brook.

Sunrise of January 3 found the main column of the American Army near the Friends meetinghouse, marching toward the village of Princeton over a route that skirted the shallow ditch between Olden Farm and the banks of Stony Brook, passing across what is now the open field behind the Institute for Advanced Study and then veering right toward the village of Princeton through what is now the Springdale Golf Club and was then Stockton's farm. The route passed over a low patch of ground that, 170 years later, would become the site of the Electronic Computer Project building at the end of Olden Lane, and this is where the main column was when the Battle of Princeton began.

Washington had dispatched his close friend General Hugh Mercer and about 350 men to backtrack up Stony Brook along Quaker Road and destroy the bridge at Worth's Mills (where Route 206 crosses the brook today). Mercer's party, however, was discovered by British forces heading from Princeton to Trenton to join the expected battle there, and a brief, intense engagement ensued that left about 50 Americans dead and 150 wounded (against 24 dead, 58 wounded, and 194 prisoners on the British side). General Mercer, surrounded, refused to surrender and, mistaken for General Washington, was bayoneted and left for dead. Washington, with reinforcements, rallied the survivors, drove the British from the field, and stormed their headquarters in the college's Nassau Hall, while those British not taken prisoner retreated toward New Brunswick through the hills. Mercer regained consciousness and survived for nine days in the Clarke farmhouse, converted to a field hospital, close to where he fell. A physician by profession, he realized that although his head injuries were survivable, his abdominal wounds were not. But the tide of war had turned. Enlistments swelled Washington's forces, the insurgency gained popular support, and the British left New Jersey for New York.

Olden Farm, after a brief role in the first American revolution, would remain undisturbed until the mathematicians arrived and began working on the next.

# THREE

# Veblen's Circle

*What could be wiser than to give people who
can think the leisure in which to do it?*

—Walter W. Stewart to Abraham Flexner, 1939

ON MAY 2, 1847, newlyweds Thomas and Kari Veblen, who spoke barely a word of English, left their home in the landlocked Valdres district of Norway to emigrate to America, leaving behind a severe economic depression and the body of their infant son. The voyage took nineteen weeks, with a shipboard fever costing Thomas his health, and all passengers under the age of six their lives. With Wisconsin on the verge of statehood, the Veblens arrived in Milwaukee on September 16, and after Kari nursed her husband back to health, Thomas, a cabinetmaker, built a house in the village of Port Ulao, in Ozaukee County, on Lake Michigan's western shore. In September 1848, Andrew Anders Veblen, the first of their eleven American children, was born. While the nine children who survived their pioneer existence were growing up, the Veblens moved three more times: first to Sheboygan County in 1849, then to Manitowoc County in 1854, and finally to Rice County, Minnesota, in 1864. At every move, Thomas Veblen built his own house, including outbuildings and barns, and cleared his own land.

The Veblen children worked long hours on the farm. Kari's father, Thorstein Bunde, had lost the family property in Norway to the machinations of unscrupulous lawyers and had died under the burdens that followed, when Kari was five. Determined to secure a better future in America, the Veblens sent all their children to college, including their four daughters and two subsequently distinguished sons. Andrew Veblen became a professor of mathematics and physics at the University of Iowa, while Thorstein Veblen, born in 1857, became an influential social theorist, best known for coining the phrase "conspicuous consumption" in his 1899 masterpiece *The Theory of the Leisure Class.*

Thorstein Veblen had a Darwinian eye, sharpened by growing up on the edge of the wilderness, for the coevolution of corporations, financial instruments, and machines. Although respected as an economist, he struggled financially for much of his life, and his only significant personal investments, in the California raisin business, failed. In 1888 and 1889 he retreated to his wife's farm in Stacyville, Iowa, and translated the eleventh-century Norse epic *Laxdæla Saga* ("an ethnological document of a high order") into English, but failed to find a publisher until 1925.

In a string of books—including *The Theory of Business Enterprise* (1904), *The Instinct of Workmanship and the State of the Industrial Arts* (1914), *An Inquiry into the Nature of Peace and the Terms of Its Perpetuation* (1917), *The Higher Learning in America: A Memorandum on the Conduct of Universities by Business Men* (1918), *The Vested Interests and the Common Man* (1919), and *Absentee Ownership and Business Enterprise in Recent Times: The Case of America* (1923)—Thorstein applied evolutionary economics, a field he pioneered, to the problems of society looming large at the time. He helped found the New School of Social Research, the *Journal of Political Economy*, and the Technocracy movement. His books were widely read, but his warnings widely disregarded, and he died, discouraged, in Menlo Park, California, on the eve of the Great Depression, in 1929. "He heard members of his family, long since dead, speak to him in Norwegian," a neighbor noted near the end.[1]

Oswald Veblen, Thorstein's nephew and the first of Andrew Veblen's eight children, attended public schools in Iowa City, followed by the University of Iowa, where he was awarded one prize in sharpshooting and another prize in math. He took time off from his studies to travel down the Iowa and Mississippi rivers in the style of Huckleberry Finn, and remained an avid outdoorsman until the day he died. He was tall, lithe, and always looked like he had just come in from the woods. "I don't ever remember seeing him in anything that looked new," says Herman Goldstine. Albert Tucker adds that "he always had a fourth button on his coat because he was so tall and slim." An attachment to the soil ran deep in his Norwegian blood. "He is a most excellent person," warned Abraham Flexner, the Institute for Advanced Study's first director, "but the word 'building' or 'farm' has an intoxicating effect upon him."[2]

After obtaining a degree in mathematics at age eighteen in 1898, Oswald Veblen stayed on as a teaching assistant in physics for one year and then left for Harvard, earning a second BA in 1900, before heading to the University of Chicago (where Thorstein was an assistant professor of political science) for his PhD. His thesis, on the foundations of geometry, led to his recruitment by Princeton University in 1905, during a period of expansion after Woodrow Wilson, the future president of the United States, was appointed president of the university in 1902.

The College of New Jersey had been renamed Princeton University in 1896, with ambitions to expand into graduate education and scientific research. Wilson, the first president of the college who was not a clergyman, began by hiring "preceptors"—junior faculty expected to work closely with undergraduates—and encouraged faculty research. While Princeton remained off-limits to black students (until the U.S. Navy's V-12 program broke the barrier in 1942) and women (until 1969), Wilson did appoint the first Catholic and the first Jew to the faculty. The number of Jewish students reached twenty-three in the class of 1925.

Wilson left the university to become governor of New Jersey in 1911 and president of the United States in 1913. War was declared against Germany in April of 1917, at the beginning of his second term. Some 138 of Princeton's faculty joined the armed services before the war came to an end, with Veblen among the first round of volunteers. Commissioned as a captain in the Army Reserve, and later promoted to major, he was assigned to the Army Ordnance Department's Office of Ballistic Research in Sandy Hook, New Jersey, just in advance of its transfer to the Aberdeen Proving Ground in Maryland, a thirty-five-thousand-acre military reservation on the shores of Chesapeake Bay.

The Aberdeen of 1918, its makeshift roads a sea of mud, was a precursor to the Los Alamos of 1943. Its mission was to enlist American science and industry against the German war machine, but by the time the Proving Ground was operational, the war in Europe was drawing to a close. According to Thorstein Veblen, the United States had entered the war, belatedly, only to ensure that the transnational interests of the industrialists would be protected against any social upheavals that peace in Europe might unleash. Oswald Veblen, a sharpshooter at heart, had no doubts about how to help, and improving the accuracy of guns raised none of the moral questions that would later be raised by the development of atomic bombs. After completing basic training, according to Herman Goldstine, Veblen kept busy at Sandy Hook, in advance of the move to Aberdeen, by "leaning out of an airplane dropping bombs from his hands trying to see how this whole bombing thing would go."[3]

A stalemate in the trenches, the Great War became a battle for bigger and better guns. Artillery fire caused some three-quarters of all casualties, with aircraft and bombs remaining on the sidelines until World War II. The United States entered the war with horse-drawn artillery and fired the last shot from a 155-mm howitzer nicknamed "Calamity Jane." New, long-range artillery and shells were being rushed into production and delivered to Aberdeen to be tested before being shipped overseas to the American Expeditionary Force.

The first test round was fired, in the midst of the worst winter on record, on January 2, 1918. Veblen arrived on January 4. With the same ease with which Oppenheimer would later assume command at Los Alamos, he rose to the occasion, assuming command of the entire ballistics group at Aberdeen. As the eldest of eight children, he found that leadership came naturally, while his willingness to shoulder his share of physical hardship on the firing ranges won the loyalty of his men.

"Veblen's tremendous influence was almost imperceptible at the time it was happening," explained his Princeton colleague Albert Tucker. "He had a rather hesitant way of speaking, very tentative and diffident," adds fellow topologist Deane Montgomery, "but he really was an extremely forceful man." To Klára (Klári) von Neumann he was "a tall, gaunt man, sporting a shyness which made him stutter in his speech, yet a formidable opponent when anybody crossed his path." Herman Goldstine, who credits Veblen for von Neumann's being allowed to build his computer at the Institute for Advanced Study in Princeton, remembers him as "the kind of guy who would keep dripping water on the stone until finally it eroded."[4]

Since the time of Archimedes and his siege engines, military commanders had brought in the mathematicians when they needed help. The problem facing Veblen was as old as gunnery itself: If you aim a gun in a given direction, and load it with a given shell, where will the shell land? Or, if you want to hit a given target with a given shell, where should the gun be aimed? According to Newton and Galileo, the path of a projectile was calculable, but in practice it was difficult to predict the behavior of a shell in flight.

With the introduction of breech-loading, rifled artillery, accuracy improved to where it became possible to test-fire a gun a fixed number of times, distributing the shots across a range of distances, and then use a mathematical model to fill in a complete firing table (or range table) from there. Range, speed, and altitude had increased to where the flight of the shell was affected by factors ranging from the changing density of the atmosphere to the rotation of the earth. Preparing the range tables required enormous numbers of calculations, largely performed by hand. The gap between what the models predicted and where the shells landed was narrowed, as far as possible, by the ballistic coefficient, an empirically derived constant that was rarely as constant as it should have been, and "was made to carry a very heavy burden," in the words of Veblen's colleague Forest Ray Moulton.[5]

Veblen organized the teams of human computers who were placed under his command, introducing mimeographed computing sheets that formalized the execution of step-by-step algorithms for processing the results of the firing range tests. It took the entire month of February to fire the first forty shots, yet by May his group was firing forty shots each day, and the growing force of human computers was keeping up. Veblen recruited widely, with a knack for discovering future mathematicians and making the best use of their talents during the war.

One of his recruits was Norbert Wiener, a twenty-four-year-old mathematical prodigy well trained after two years of postdoctoral study in Europe, but socially awkward and discouraged by the failures of his first teaching job. Even the army had rejected him, for poor eyesight and an inability to fire a rifle or maintain control of a horse. When Veblen located Wiener, he was living in Albany, New York, and writing articles for the *Encyclopedia Americana* to scrape by. "I received an urgent telegram from Professor Oswald Veblen at the new Proving Ground at Aberdeen, Maryland," Wiener later recalled. "This was my chance to do real war work . . . I took the next train to New York, where I changed for Aberdeen."

Wiener was transformed by the Proving Ground. "We lived in a queer sort of environment, where office rank, army rank, and academic rank all played a role, and a lieutenant might address a private under him as 'Doctor,' or take orders from a sergeant," he wrote. "When we were not working on the noisy hand-computing machines which we knew as 'crashers,' we were playing bridge together after hours using the same computing machines to record our scores. We went swimming together in the tepid, brackish waters of Chesapeake Bay, or took walks in the woods."

"Whatever we did, we always talked mathematics," Wiener explained. "Much of our talk led to no immediate research." Wiener found that the Proving Ground "furnished a certain equivalent to that cloistered but enthusiastic intellectual life which I had previously experienced at the English Cambridge, but at no American university." Veblen had gathered a community that would redefine American mathematics in the years between World War I and World War II. "For many years after the First World War," wrote Wiener, "the overwhelming majority of significant American mathematicians was to be found among those who had gone through the discipline of the Proving Ground. Thus the public became aware for the first time that we mathematicians had a function to perform in the world."[6]

After the Armistice was signed in November of 1918, Veblen took a four-month tour of Europe to debrief his counterparts about their experiences during the war. Many had already returned to academia, giving him the opportunity to observe the state of European mathematics firsthand. Göttingen, Berlin, Paris, and Cambridge were the centers of the mathematical world, while Harvard, Chicago, and Princeton were still far from catching up. Veblen returned to Princeton determined both to replicate the success of the European institutions and to recapture some of the informal mathematical camaraderie of the Proving Ground.

He set three immediate goals: to sponsor postdoctoral fellowships for promising young mathematicians, to free existing professors from crushing teaching loads, and to promote cross-fertilization between mathematics and other fields. "It has frequently happened that an attempt to solve a physical problem has resulted in the creation of a new branch of mathematics," he wrote to Simon Flexner, director of the Rockefeller Institute for Medical Research, urging the Rockefeller Institute to extend its existing National Research Council fellowships, focused on physics and chemistry, to include mathematical research.[7]

Veblen's proposal was adopted, and four months later he returned to Simon Flexner with a more ambitious request. "The way to make another step forward," he suggested, "is to found and endow a Mathematical Institute. The physical equipment of such an institute would be very simple: a library, a few offices, and lecture rooms, and a small amount of apparatus such as computing machines." Veblen sketched out a mathematical Utopia, somewhere between the High Tables of Cambridge or Oxford and the computing shacks of the Proving Ground. "The main funds of such an institute," he emphasized, "should be used for the salaries of men and women whose business is mathematical research."[8]

Simon Flexner answered that "I wish that sometime you might speak with my brother, Mr. Abraham Flexner, of the General Education Board."[9] The Rockefeller Foundation's General Education Board, chartered by Congress in 1903 for "the promotion of education within the United States of America, without distinction of race, sex, or creed," focused on high school education in the American South, but was free to support higher education of any kind.

Simon and Abraham Flexner, the fifth and seventh of nine children, were born in Louisville, Kentucky, and ended up at the Rockefeller Foundation by very different paths. Their father, Moritz Flexner, born in Bohemia in 1820, was a Jewish immigrant peddler who settled in Louisville in 1854, carrying his wares on his back until he saved four dollars to buy a horse. Simon Flexner, born in 1863, dropped out of school after the seventh grade, drifting in no particular direction until a job in a drugstore and a near-fatal brush with typhoid sparked an interest in microbiology that led to a career in medicine and, eventually, to his becoming an authority on infectious diseases and director of the Rockefeller Institute for Medical Research. Established on a 425-acre farm in Princeton, and later expanded to 800 acres, the Rockefeller Institute was the leading microbiological research institute in the United States.

Abraham Flexner, born in 1866 and the only Flexner to be sent to college, was sponsored by his eldest brother, Jacob, after the death of Moritz, in attending Johns Hopkins University, graduating in 1886. He returned to Louisville to teach Latin and Greek at Louisville High School, and then opened his own school, after establishing his reputation by flunking an entire class. He credited his parents with being "shrewd enough to realize that their hold upon their children was strengthened by the fact that they held them with a loose rein," a principle that guided his educational philosophy, even though "to be sure, we shall thus free some harmless cranks."[10]

"A small hawk-like wiry man with a wonderful twinkle in his eye and a front of obviously false modesty that immediately made you suspect the strength and power, the cunning and cleverness that were hidden behind a delightful sense of humor, Flexner was not a scholar himself," explained Klári von Neumann, "but had a very practical mind which conceived the idea that there should be a place where men whose only tools of work were their brains could spend time entirely on their own, with no obligation of teaching or looking after students; a place in a milieu of relaxed thinking, a place where they could talk to each other if they felt like it but if not, each was respectfully left alone."[11]

In 1898, Abraham Flexner married a former student, Anne Crawford, who became a successful Broadway playwright (*The Marriage Game, All Soul's Eve, Mrs. Wiggs of the Cabbage Patch*), allowing them to leave Louisville behind. He sold his school in 1905, going first to Harvard, where he earned an MA in philosophy in 1906, and then to Germany, where he wrote a scathing critique of American higher education, published in 1908. The Carnegie Foundation then commissioned him to compile a report on medical education, where standards were even worse. He visited some 155 medical schools, and his exposure of their deficiencies resulted in the closure of two-thirds of the medical schools in the United States. In 1911 he was commissioned by John D. Rockefeller Jr. to conduct a "thorough and comprehensive" study of prostitution in Europe, an assignment that took him to twenty-eight cities in twelve countries, from London to Budapest. His report, published in 1914, was acclaimed in the United States and he was made a Chevalier of the Legion of Honor in France. In 1913 he joined the Rockefeller Foundation's General Education Board, his influence rising until he was ousted in 1928.

Veblen's proposal to the Flexners met with no immediate response, but Princeton University was eventually awarded $1 million by the General Education Board, contingent on the university's securing an additional $2 million in matching funds. Veblen deferred the credit to Woodrow Wilson and Henry Fine. "I should not think that my little dream of a Math institute should be stressed too much," Veblen later admitted, looking back. "The building of a mathematical department in Princeton University which was completely out of proportion with the standing of the University in other fields of scholarship had been going on under the leadership of H. B. Fine since 1885."[12]

Henry Burchard Fine, the son of a Presbyterian minister from rural Pennsylvania, had entered the College of New Jersey in 1876, and became close friends with Woodrow Wilson while serving as editor of the *Princetonian* in his senior year. After obtaining a PhD in Leipzig in 1885, he returned to Princeton and was appointed dean of faculty by Woodrow Wilson in 1903. Fine chose Veblen for one of the new preceptorial positions, and built up the core of the Princeton mathematical group. He hired promising young mathematicians, supported their research, and made no objections when they were called somewhere else.

Fine's brother John founded the Princeton Preparatory School (on the east side of town) for boys, while his sister, May, founded Miss Fine's School (on the west side of town) for girls. After Wilson's election to the U.S. presidency, Henry Fine declined an appointment as ambassador to Germany because he believed that teaching undergraduates should come first. Fine and Wilson were both friends with fellow alumnus Thomas Davies Jones, who enjoyed a lucrative Chicago law practice, a controlling interest in the Mineral Point Zinc Company, and self-described "superfluous wealth." The Jones family put up the entire $2 million to match the contribution from the General Education Board, more than enough to realize Veblen's ambitions for mathematics at Princeton—but Fine began distributing the money to other departments first. Things changed suddenly at the end of 1928.

In 1913 the former Lenni Lenape footpath through Princeton had become part of the first transcontinental motorway across the United States. The Lincoln Highway, beginning at Times Square in New York City and terminating at an overlook above Point Lobos in San Francisco, followed the route of the old King's Highway between Princeton and Kingston, and was fully paved between New York and Philadelphia by 1922. In the late afternoon of December 21, with darkness falling, a driver heading toward Kingston failed to see a seventy-year-old bicyclist making a left turn. The cyclist was Henry Fine, entering the driveway of his brother's school. The driver, Mrs. Cedric A. Bodine, whose husband owned a funeral home in Kingston, was held under a charge of manslaughter, while a series of memorials, with Nassau Hall's bells tolling, mourned Princeton's loss through Christmas Day.

Thomas Jones and his niece Gwethalyn pledged an additional $500,000 to build (and maintain) a new mathematics building in memory of Fine. At the time of Veblen's arrival in Princeton, the mathematicians shared a few small offices in Palmer Hall. "The principle upon which Fine Hall was designed," according to Veblen, "was to make a place so attractive that people would prefer to work in the rooms provided in this building rather than in their own homes."[13] Jones, believing that "nothing is too good for Harry Fine," instructed Veblen to construct a building that "any mathematician would be loath to leave."

Half a million dollars (equivalent to over $6 million today) went a long way in 1929. Fine Hall opened in October of 1931, with no detail overlooked: from the showers and locker room in the basement ("members of the department who wish to avail themselves of the nearby tennis courts or the gymnasium will not find it necessary to return to their homes to dress") to the top-floor library with natural lighting, a central atrium, and a passageway to encourage mingling with the physicists in adjacent Palmer Hall. "There are nine offices with fireplaces and fifteen without," reported Veblen. "Overstuffed chairs and davenports take the place of chairs and desks and the classrooms are fitted out after the manner of private studies," reported *Science* magazine. The rooms were paneled in American oak, with concealed chalkboards and built-in filing cabinets. Equations for gravitation, relativity, quantum theory, five perfect solids, and three conic sections were set into leaded glass windows, and the central mantelpiece featured a carving of a fly traversing the one-sided surface of a Möbius strip. "Every little door knob, every little gargoyle, every little piece of stained glass that has a word on it, was something that Veblen personally supervised," noted Herman Goldstine in 1985.[14]

In April 1930, Veblen wrote to Albert Einstein requesting permission to inscribe a remark Einstein had made in Princeton in 1921—"Raffiniert ist der Herr Gott aber Boshaft ist Er nicht" (translated at the time as "God is clever, but not dishonest")—above the fireplace in the Professors' Lounge. "It was your reply when someone asked you if you thought that [Dayton C.] Miller's results would be verified," Veblen explained. "I hope you will not object to our using this 'child of your wit.'"[15] Einstein replied that "Lord" or "God" might be misconstrued, suggesting that what he really meant was "Nature conceals her secrets in the sublimity of her law, not through cunning."[16]

With the opening of Fine Hall, and Veblen's appointment as Henry Burchard Fine Professor of Mathematics (also funded by the Jones family), Veblen's position, along with that of mathematics at Princeton, appeared secure. Three misfortunes—the Great Depression, the rise of Nazism in Europe, and the approach of the Second World War—then combined with yet another windfall to allow Veblen's dream of an autonomous mathematical institute to be fulfilled.

During the three centuries since Henry Hudson explored the Newark estuary in 1609, the nonnative population of Newark had grown from 61 dissident Puritan settlers in 1666 to just under 350,000, almost doubling between 1890 and 1910. Among the new arrivals was Louis Bamberger, born in 1855, above his father's dry goods store, into a family of Jewish merchants who had immigrated to Baltimore from

Bavaria in 1823. After starting work in his maternal uncle's store at age fourteen, Louis served as a buyer in New York City, scraping together the money to buy the stock of a bankrupt dry goods firm in 1892. Selling it out of a rented storefront in a blighted neighborhood of Newark, he realized enough of a profit to open his own business, taking on as partners his sister Carrie; her husband, Louis Frank; and their close friend Felix Fuld.

By 1928, Bamberger's department store occupied 1 million square feet, with 3,500 employees and over $32 million in annual sales. The Amazon.com of its time, Bamberger's featured price tags on all merchandise, no-questions-asked money-back guarantees, toll-free telephone numbers, job security, and an on-site public library for employees. The eight-floor flagship store on Market Street in Newark included a 500-watt radio station, WOR, and introduced what is now the Macy's Thanksgiving Day Parade.

The four business partners, with no children, lived together on the outskirts of Newark, in South Orange, on a thirty-acre estate. After Louis Frank died in 1910, Carrie married Felix Fuld, who died in January of 1929. The two surviving Bambergers decided it was time to retire, and in June of 1929 negotiated a sale to R. H. Macy and Co. that closed in September, just six weeks before the stock market crash. They received 146,385 shares in Macy's, whose share price reached a high of $225 on November 3 before sinking to a low of $17 in 1932. The Bambergers, who took $11 million of the proceeds in cash, distributed $1 million among 225 employees who had served fifteen years or more, and enlisted their chief accountant, Samuel D. Leidesdorf, and legal adviser, Herbert H. Maass, to help decide what they should do with the rest.

"Because they had prospered to such an extent in the City of Newark, they were determined that whatever they did should benefit either the City of Newark or the State of New Jersey," Maass recalled in 1955. Their intention was to establish a medical college, giving preference to Jewish faculty and students, on their South Orange estate. Maass and Leidesdorf were referred to Abraham Flexner, whom they called on in his office at the Rockefeller Medical Foundation in December of 1929. "His advice to us was that there were ample medical school facilities in the United States," remembers Maass. Flexner, who had dismissed the American University as "an educational department store" in an *Atlantic Monthly* article in 1925, seized his chance. "Toward the end of our first conversation," according to Maass, "he asked us, 'Have you ever dreamed a dream?' "[17]

Flexner remembers having been "working quietly one day when the telephone rang and I was asked to see two gentlemen who wished to discuss with me the possible uses to which a considerable sum of money might be placed."[18] He had been preaching about the deficiencies of higher education for many years, and when Maass and Leidesdorf visited, he had the proofs of his forthcoming book, *Universities: American, English, German*, sitting on his desk. His guests took a copy with them when they left.

The book, expanding upon the Rhodes lectures Flexner had delivered at Oxford in 1928, gave a depressing account of higher education in America, concluding with a call for "the outright creation of a school or institute of higher learning" where "mature persons, animated by intellectual purposes, must be left to pursue their own ends in their own way . . . be they college graduates or not." Flexner argued that this "free society of scholars" should be governed by scholars and scientists, not administrators, and even "the term 'organization' should be banned."[19]

Maass was "fascinated," and Leidesdorf was "impressed." A series of luncheon meetings with the Bambergers was arranged, and before Louis and Carrie departed for their customary winter retreat to the Biltmore in Phoenix, Flexner drafted a codicil to their last wills and testaments. "Having made an extensive survey of the field, guided by expert advice," he wrote on their behalf, "we are presently of the opinion that the best service we can render mankind is to establish and endow a graduate college which shall be . . . free from all the impedimenta which now surround graduate schools because of the undergraduate activities connected therewith."[20]

The Bambergers were still cautiously approaching a decision, but there was no indecision on Flexner's part. He renewed contact with Oswald Veblen, who was immersed in the construction of Fine Hall. Veblen sensed something was up. He reported to Flexner on the progress being made in Princeton, adding that "I think my mathematical institute which has not yet found favor may turn out to be one of the next steps." Flexner took the bait. "What would American scholars and scientists do if some fellow or some foundation set up a 'sure enough' institution of learning?" he asked. "Is it necessary to carry the mill-stone of the college about the neck of the graduate school?"[21]

On May 20, 1930, with Flexner appointed as the first director at a salary of $20,000 per year (equivalent to over $250,000 today), a certificate of incorporation was signed for "the establishment, at or in the vicinity of Newark, New Jersey, of an institute for advanced study, and for the promotion of knowledge in all fields." The Bambergers committed $5 million to start things off. "So far as we are aware," they announced in their initial letter of instruction to the trustees, "there is no institution in the United States where scientists and scholars devote themselves at the same time to serious research and to the training of competent post-graduate students entirely independently of and separated from both the charms and the diversions inseparable from an institution the major interest of which is the teaching of undergraduates."[22]

For the first two years, the Institute existed only on paper, envisioned by Abraham Flexner as "a paradise for scholars who, like poets and musicians, have won the right to do as they please."[23] It was one thing to criticize higher education, as Flexner had been doing for twenty-two years, and another to replace it with something else. Creating a paradise, even with $5 million during the Great Depression, was easier said than done.

Flexner spent six months consulting with leading intellectuals and educational administrators across Europe and the United States. The notion of paradise varied from one person to the next. Classicists advised Flexner to start with classics, physicists with physics, historians with history, and mathematicians with mathematics. British biologist Julian Huxley advised mathematical biology, arguing that "there is in biology a lamentable lack of general appreciation of a great deal of systematic and descriptive work." The Bambergers wanted to start with economics and politics, which they hoped would "contribute not only to a knowledge of these subjects but ultimately to the cause of social justice which we have deeply at heart."[24]

Some thought the Institute should be closely associated with an existing university; others thought it should be far removed. "It is the multiplicity of its purposes that makes an American University such an unhappy place for a scholar," advised Veblen. "If you can resist all temptations to do the other good things that might be attempted, your adventure will be a success."[25]

Was a paradise for scholars even possible? Historian Charles Beard predicted "death—intellectual death—the end of many a well-appointed monastery in the Middle Ages." Future Supreme Court justice Felix Frankfurter, who scrawled, "NEWS FROM PARADISE. Not my style," across a letter from Flexner, pointed out that "for one thing, the natural history of paradise is none too encouraging as a precedent. Apparently it was an excellent place for one person, but it was fatal even for two."[26]

Flexner, who had given away some $600 million over the course of his association with the Rockefeller Foundation, believed that most educational funding had too many strings attached. Now was his chance to try something else. "I should think of a circle, called the Institute for Advanced Study," he envisioned in 1931. "Within this, I should, one by one, as men and funds are available—and only then—create a series of schools or groups—a school of mathematics, a school of economics, a school of history, a school of philosophy, etc. The 'schools' may change from time to time; in any event, the designations are so broad that they may readily cover one group of activities today, quite another group, as time goes on."[27]

"The Institute is, from the standpoint of organization, the simplest and least formal thing imaginable," he explained. "Each school is made up of a permanent group of professors and an annually changing group of members. Each school manages its own affairs as it pleases; within each group each individual disposes of his time and energy as he pleases . . . The results to the individual and to society are left to take care of themselves."[28] Flexner believed that knowledge, not profit, must be the goal of research. "As a matter of history, the scientific discoveries that have ultimately inured to the benefit of society either financially or socially have been made by men like Faraday and Clerk Maxwell who never gave a thought to the possible financial profit of their work," he wrote to the editors of *Science* in 1933, protesting against universities that were beginning to file for patents on their research. This did not mean that benefits should not be expected from pure research. In a *Harper's Magazine* essay titled "The Usefulness of Useless Knowledge," Flexner described the thinking behind the Institute for Advanced Study and argued that "the pursuit of these useless satisfactions proves unexpectedly the source from which undreamed-of utility is derived."[29]

After considering a school of economics, both to accommodate the Bambergers and because "the plague is upon us, and one cannot well study plagues after they have run their course," Flexner decided to start with mathematics. "Mathematics is singularly well suited to our beginning," he explained to the trustees. "Mathematicians deal with intellectual concepts which they follow out for their own sake, but they stimulate scientists, philosophers, economists, poets, musicians, though without being at all conscious of any need or responsibility to do so." There were practical advantages to the field as well: "It requires little—a few men, a few students, a few rooms, books, blackboard, chalk, paper, and pencils."[30]

There were two other reasons to start with math. Flexner needed to make a strong first impression, and the ranking of talent across mathematics was less subjective than in other fields. He deferred to Veblen as to candidates, explaining to the trustees that "mathematicians, like cows in the dark, all look alike to me." Second, Flexner knew that to satisfy the Bambergers, and secure the balance of their estate, he had to deliver immediate results. There was an existing mathematical utopia, Fine Hall, available off the shelf. Could Flexner sell this to the Bambergers? "Everybody who wants a teacher of mathematics comes shopping to Princeton," he wrote to them in Arizona, "just as the people who know what they are after go to L. Bamberger & Company in Newark."[31]

The Bambergers remained, as expressed in their original letter, "mindful of our obligations to the community of Newark" and still intended the Institute to be located "in the vicinity of such City." South Orange was "in the vicinity of Newark," but was Princeton? "Mr. Bamberger and Mrs. Fuld so clearly intended Newark and its immediate environment that I would hesitate to adopt any other view unless they first modified their letter," stated Maass. "Enclosed is a current road map for the State of New Jersey," added Edgar Bamberger, a nephew of Louis who had joined the board of trustees. "You will note that circles have been drawn at ten mile radii, with South Orange Village as a center. Princeton, you will notice, is roughly 35–40 miles by road from South Orange. Kindest personal regards." Flexner, reluctant to confront the Bambergers directly, began to substitute "in the State of New Jersey" for "in the vicinity of Newark" in the documents being circulated in preparation for the Institute's launch. The Bambergers, persuaded by his insistence that "it might be difficult to get able lecturers to come to Newark," eventually acquiesced.[32]

On June 5, 1932, Oswald Veblen was appointed to the first professorship (effective October 1, 1932) followed by Albert Einstein (effective October 1, 1933). They were joined by John von Neumann, Hermann Weyl, and James Alexander in 1933, and Marston Morse in 1934. "I had to bear in mind the importance of getting together a group, all the members of which would not grow old at the same time," Flexner explained to Veblen at the end of 1932. "You and Einstein are in the early 50's, Weyl in the middle 40's, Alexander in the early 40's—so that we have protected ourselves against any such fate as befell the deacon's one horse shay, which, as you remember, fell to pieces all at once without showing any signs of decay during one hundred years."[33] Von Neumann, when hired in January of 1933, had just turned twenty-nine. Princeton University agreed to provide a temporary home for the new institute in Fine Hall.

The Nazis launched their purge of German universities in April 1933, and the exodus of mathematicians from Europe—with Einstein leading the way to America—began just as the Institute for Advanced Study opened its doors. "The German developments are going bad and worse, the papers today wrote of the expulsion of 36 university professors, ½ of the Göttingen mathematics and physics faculty," von Neumann reported to Flexner on April 26. "Where will this lead, if not to the ruin of science in Germany?"[34]

Flexner, the Bambergers, and Veblen had envisioned a refuge from the mind numbing departmental bureaucracy of American universities, not the humanitarian disaster from which their sanctuary now offered an escape. "The Institute was a beacon in the descending darkness," wrote Director Harry Woolf in 1980, reflecting on the first fifty years, "a gateway to a new life, and for a very few a final place within which to continue to work and transmit to others the style and the techniques of great learning from the other shore."[35] Veblen assumed the chairmanship of the Rockefeller Foundation's Emergency Committee for Displaced German Scholars, using Rockefeller money and the promise of temporary appointments at the Institute to counter the twin misfortunes of anti-Semitism in Europe and a depression in the United States.

The problem was how to squeeze displaced scholars into a shrinking job market without provoking the very anti-Semitism those scholars were trying to escape. The United States offered non-quota visas to teachers and professors, but with insufficient openings for American candidates, finding positions for the refugees, especially in Princeton, was a difficult sell. An invitation to the Institute for Advanced Study allowed Princeton University, historically resistant to Jewish students and faculty, to reap the benefits of the refugee scientists without incurring any of the associated costs. The arrival of Einstein helped open the door. Princeton, despite its role in the American Revolution, had become one of the more conservative enclaves in the United States, "a quaint and ceremonious little village of puny demigods on stilts," as Einstein described it to the Queen of Belgium in 1933.[36]

Veblen pushed not only for academic positions but for land: enough to establish a refuge for wildlife as well as a refuge for ideas. "There is no educational institution in the United States which has not in the beginning made the mistake of acquiring too little rather than too much land," he wrote to Flexner, urging the acquisition of "a sufficiently large plot of land, which would thus be kept free from objectionable intruders." Flexner, who favored investing in scholarship over real estate, was gradually persuaded. "I have it in mind now to go down to Princeton quietly for a week or so for the purpose of familiarizing myself with the general situation, for that may help us in our final choice," he reported in October 1932. "I should like to be away from undergraduate activities and close to graduate activities."[37]

There was no turning back, once word leaked out that the Institute was looking for a home. "The fact that we propose to locate in the vicinity of Princeton is now a matter of public knowledge to such an extent that, I believe, we are being made the victim of a distinct firming up in prices, and that we had better attempt to come to an early decision," Maass advised Flexner in November 1932. "If we are going to have inflation, would it not be well to speed up the land question?" Veblen argued. "At least two of the proposed sites seem good to me." Despite the Bambergers complaining about "a policy of acquiring so much land for an institution that proclaimed not size but highest standards," Veblen persisted, and by 1936 some 256 acres had been purchased for a total of $290,000, including the 200-acre Olden Farm.[38] The property included Olden Manor (the former William Olden house, now the director's residence), a cluster of farm workers' houses at the end of Olden Lane, and a large working barn.

"I think it would be prudent for the present to keep the matter quiet," Flexner wrote to Bamberger in October 1935. "Though I do not wish to criticize either Mr. Maass or Professor Veblen, I think there is some danger that they will both be too enthusiastic about the acquisition of additional land." The Bambergers responded, as Maass reported to Flexner in December, by "[playing] Santa Claus by paying for the land."[39] Over the next few years, Veblen drove a series of tough bargains with Depression-strapped landowners to extend the Institute's holdings to a total of 610 acres, including the land bordering

Stony Brook that now constitutes the Institute Woods. "I have walked over the new property of the Institute several times since there has been a hard crust on top of the snow," he reported in early 1936. "This enables one to explore the woods down near the brook much better than one will be able to after the ground gets soft again."[40] Veblen arranged to have forty thousand evergreen seedlings from the state nursery in Washington Crossing planted on Institute property in April 1938.

In 1937, after the failure of negotiations that would have seen the Springdale Golf Club build a new clubhouse on Institute property and allow the Institute to occupy the old clubhouse (formerly Stockton's farm) on College Road, it was decided to begin construction of an Institute headquarters in the middle of Olden Farm, on level ground about halfway between Olden Manor and Stony Brook. The Bambergers, long opposed to spending money on buildings, changed their minds. Being snubbed by the old guard at the golf club may have helped.

Fuld Hall was an expanded version of Fine Hall, transplanted away from the university and given room to grow. The mathematicians played chess in the common room while the trustees played cards in the board rooms upstairs. "They were all friends—original friends of Mr. Bamberger," recalls Herman Goldstine. "Maass was his lawyer; Leidesdorf was his pinochle-playing crony and accountant, and that's how they were on the board. These people like Lewis Strauss were all of the kind of Jewish merchant princes, if you will."[41]

The Institute academic year, divided into two terms, extended from October through April, with a generous winter break in between and no duties or responsibilities except to be in residence at the Institute during term—amounting to about half the year. "The other half of the year the staff will be technically on vacation," it was reported in 1933, "but Dr. Flexner has found that those engaged in research often do their best work while 'on vacation.'" Flexner believed in generous remuneration for the permanent faculty, noting that although wealth might invite distractions from academic work, "it does not follow that, because riches may harm him, comparative poverty aids him." This generosity was not extended to visitors, "for on high stipends members will be reluctant to leave." Despite the Depression, and the war, faculty salaries continued to go up. "Professor Earle, who is extremely grateful for the increase in his salary, has been driven by his conscience to raise the uneasy question as to whether such increases in salary as we have recently made and are making are legal," Frank Aydelotte, Flexner's successor, noted in 1945.[42]

Princeton University professors referred to "the Institute for Advanced Salaries," while Princeton University graduate students referred to "the Institute for Advanced Lunch." The Institute was the unacknowledged realization of Thorstein Veblen's original call (in 1918) for "a freely endowed central establishment where teachers and students of all nationalities, including Americans with the rest, may pursue their chosen work as guests of the American academic community at large."[43] Despite Flexner's claims to paradise, Veblen's mathematicians never quite regained the informal camaraderie that had permeated the computing shacks at Aberdeen or the early days of Fine Hall.

Flexner's own tenure was short-lived. He started out determined to avoid "dull and increasingly frequent meetings of committees, groups, or the faculty itself. Once started, this tendency toward organization and formal consultation could never be stopped."[44] However, after a number of unwelcome decisions on his part, including two permanent appointments to the ill-fated School of Economics and Politics, the faculty held a series of meetings that bordered on revolt. Flexner resigned on October 9, 1939, and was replaced (bypassing any ambitions on the part of Veblen) by his understudy, Frank Aydelotte, a Louisville schoolteaching colleague who had become president of Swarthmore College (and an influential Quaker sympathizer), and who, by not belonging to either camp, was able to keep the scientists and the humanists in line.

Despite his lifelong support of scholarship, Flexner never claimed to belong to the club himself. "I, alas, have never been a scholar, for two years at the Johns Hopkins between 1884–1886 do not produce scholarship—though they do produce and did produce a reverence for it which I am now leaving in safe keeping with you," he wrote when turning over the directorship to Aydelotte in 1939. Flexner was deeply hurt by the faculty rebellion that pushed him out. Von Neumann remained neutral, and never forgot that it was Flexner's lifeline that had allowed him to stay in the United States. The feelings were reciprocal. "Flexner's attitude towards Johnny was real fun to watch," says Klári. "It was a mixture of an uncle towards a favorite nephew and a circus trainer showing off the magnificent tricks that his trained lion can do."[45]

Aydelotte, who had been a trustee of the Institute from the beginning, presided with great diplomacy through World War II, granting leave to those, such as von Neumann, Veblen, and Morse, who were engaged in the war effort, while preserving a refuge for those who were not. "In these grim days when the lights are going out all over Europe and when such illumination as we have in the blackout is likely to come, figuratively as well as literally, from burning cities set on fire by incendiary bombs, some men might question the justification for the expenditure of funds on Humanistic Studies—on epigraphy and archaeology, on paleography and the history of art," he reported in May 1941. He could not announce that the Institute was already engaged in behind-the-scenes support for work on atomic bombs, but he did announce the Institute's unwavering support for "the critical study of that organized tradition which we call civilization and which it is the purpose of this war to preserve. We cannot, and in the long run will not, fight for what we do not understand."[46]

Aydelotte was succeeded by J. Robert Oppenheimer in 1947, who presided until 1966. Whereas Flexner and Aydelotte had both been skilled teachers and educational administrators, but not scientists, Oppenheimer was both a first-rate scientist and a skilled administrator, as well as a connoisseur of history and art. The poet T. S. Eliot, who had been invited by Aydelotte and listed *The Cocktail Party* (1950) as his only "publication related to IAS residence," arrived at Oppenheimer's "intellectual hotel" as the first Director's Visitor for the fall term of 1948.[47]

The School of Mathematics opened in 1933, followed by the School of Humanistic Studies in 1934 and the School of Economics and Politics in 1935. The School of Historical Studies (amalgamating the humanists and economists) was formed in 1949, and the School of Natural Sciences was formed as an offshoot from Mathematics in 1966. The School of Social Science was established in 1973. Every decade or so there had been an attempt to smuggle a biologist into the Institute, starting with J. B. S. Haldane in 1936. "Haldane was interested in the applications of mathematics to biological phenomena," Veblen explained to Flexner, arguing that bringing in a biologist would not require starting another school. "The exact field to which he proposed to make the applications is genetics." Haldane declined the invitation, saying that he "was going to Spain to help with the defense of Madrid

against a gas attack threatened by German and Italian invaders."[48]
Sixty years later, a temporary group in theoretical biology was established in 1999, and a permanent Center for Systems Biology in 2005.

Two different Institutes have managed to coexist. "One, which was
adopted more by the historical school," according to Deane Montgomery, "is that it's a group of great scholars who occasionally communicate with the public and who have great thoughts. They tended
more to think of it as a lifetime fellowship for themselves." Veblen,
adds Montgomery, "said he and Einstein and Weyl didn't feel up
to that."[49] The other Institute was the annually changing group of
mostly young visitors at the beginning of their careers, interspersed
with occasional established scholars taking a year off. Benoît Mandelbrot, who arrived at von Neumann's invitation in the fall of 1953 to
begin a study of word frequency distributions (sampling the occurrence of *probably, sex,* and *Africa*) that would lead to the field known
as fractals, notes that the Institute "had a clear purpose and a rather
strange structure in which to assemble people: heavenly bodies in
residence, and then nobody, nobody, nobody, and then mostly young
people. Now it has a much more balanced distribution in terms of age
and fame." Mandelbrot got along wonderfully with von Neumann,
admiring how he "had accumulated a number of people who were
not part of the Princeton pigeon holes," while observing that among
the visiting scholars, "everybody else had the dreadful feeling that this
may be the best year of their life, so why wasn't it more enjoyable?"[50]
The freedom from day-to-day responsibilities came at the expense
of a pervasive and sometimes crippling expectation to do something
remarkable with one's year off.

Although it was Veblen who "conceived the whole project," in
the assessment of physicist P. A. M. Dirac, he remained in the background, as he had at the Proving Ground. In 1959, Oppenheimer
wrote to Veblen asking permission to change the name of one of the
Institute's private roads, a small cul-de-sac that overlooked the Battlefield, from Portico to Veblen Lane. Oppenheimer's notes record
Veblen's response: "Said no. Would rather wait until dead."[51] The road
is named Veblen Circle today.

At the time of the founding of the IAS, mathematics was divided
into two kingdoms: pure mathematics, and applied. With the arrival
of von Neumann, the distinctions began to fall. "The School of Mathematics has a permanent establishment which is divided into three
groups, one consisting of pure mathematics, one consisting of the-

oretical physicists, and one consisting of Professor von Neumann," Freeman Dyson explained to a review committee in 1954.[52]

A third kingdom of mathematics was taking form. The first kingdom was the realm of mathematical abstractions alone. The second kingdom was the domain of numbers applied, under the guidance of mathematicians, to the real world. In the third kingdom, the digital universe, numbers would assume a life of their own.

# FOUR

# Neumann János

*We are Martians who have come to Earth
to change everything—and we are afraid we
will not be so well received. So we try to keep
it a secret, try to appear as Americans . . . but
that we could not do, because of our accent.
So we settled in a country nobody ever has
heard about and now we are claiming to be
Hungarians.*

—Edward Teller, 1999

JOHN-LOUIS VON NEUMANN (Margittai Neumann János Lajos to his fellow Hungarians), the first child of Max Neumann and Margit (Gitta) Kann, was born in Budapest on December 28, 1903, the year that Oswald Veblen received his PhD. The nation of Hungary, the city of Budapest, and the von Neumann family were all on the ascent.

The establishment of the dual Austro-Hungarian monarchy in 1867 had brought an interlude of peace and prosperity, and a lifting of restrictions against Jews, to a country best known, according to Klári von Neumann, "for the gallantry of its men, the beauty of its women, and last, but not least, for its hopelessly unhappy and unlucky history."[1] When the towns of Buda and Pest, on opposite sides of the Danube, were amalgamated in 1873, the new Hungarian capital, now rivaling Vienna as the cultural and economic center of the Austro-Hungarian Empire, became the fastest-growing city in Europe. There were more than six hundred coffeehouses, three of the world's most rigorous high schools, and the first subway system on the European continent in young Neumann János's Budapest.

Max Neumann, born Neumann Miksa in 1873, grew up in Pécs, south of Budapest, and became a lawyer and investment banker specializing in the mix of technical knowledge and financial resources that drove Hungary's modernization in the years before World War I. He married into the Jacob Kann family, whose Kann-Heller agricultural machinery supply business (originally a supplier of millstones and later a pioneer, similar to the American Sears-Roebuck, in direct sales) occupied the ground floor of a prominent building at 62 Váczi Ut in Budapest. Max and Margit moved into an eighteen-room apartment on the top floor, surrounded by Margit's three sisters and their families, who occupied the remainder of the two upper floors, with a branch of the Heller family on the second floor below. Max commissioned a stained-glass window to commemorate his children, with John (born in 1903) symbolized by a rooster, Michael (born in 1907) symbolized by a cat, and Nicholas (born in 1911) symbolized by a hare. "When we visited Budapest for the first time in about 1983, it was still a Communist regime," says Nicholas, "but the janitors happily and courteously received us, and the window was still there."[2]

In 1913, Max was awarded a hereditary title by Emperor Franz Joseph, for "meritorious service in the financial field." The family name was changed to Margittai Neumann, or von Neumann in Germanized form. After the death of Max in 1928, all three of his sons converted to Catholicism ("for sake of convenience, not conviction," says Nicholas) and immigrated to the United States. Michael reverted to Neumann, while Nicholas adopted Vonneumann. John preserved von Neumann, while remaining simply "Jancsi" to his Hungarian, and "Johnny" to his American, friends.[3]

"There is no question that nobility, in 1913, was not the same kind of nobility which existed in the feudal ages," explains Nicholas, who enlisted in the U.S. Army in 1942, followed by several years in the Office of Strategic Services (OSS) before settling down to a career in patent law. "Whether [my father] purchased it or not was also somewhat beside the point. It was a reward for achievements in the economic life of Hungary. It wasn't the feudal age." What was important, emphasizes Nicholas, is that "Father believed in the life of the mind."[4]

Max installed a private library where John read voraciously while growing up, consuming Wilhelm Oncken's forty-four-volume *Allgemeine Geschichte in Einzeldarstellungen* (*General History*) in its entirety, and citing it in detail, from memory, when asked. He studied the thousand-year span of the Byzantine Empire, a subject that remained with him even as his mathematical abilities evaporated at the end of his life. "Its power and organization fascinated him," Stan Ulam recalls. According to Herman Goldstine, he "was able on once reading a book or article to quote it back verbatim," even after a period of years. "On one occasion I tested his ability by asking him to tell me how the *Tale of Two Cities* started. Whereupon, without any pause, he immediately began to recite the first chapter and continued until asked to stop."[5]

Von Neumann's happy childhood stands in sharp contrast to the global conflicts that would dominate his adult life. Children wandered freely between the adjacent households, while the clouds of war gathered outside. "One of the games played by the children, this one under John's leadership," remembers Nicholas, "consisted of 'battles' drawn symbolically or abstractly on graph paper, with castles, highways, fortifications, etc., represented by filling in or connecting the squares of the graph paper. The aim was to demonstrate and practice ancient strategies. There was no emotional content in assigning the roles of the participants in the confrontation, or of the victors and the vanquished."[6] In both World War I and World War II, Hungary ended up on the losing side.

Preparation for the Hungarian gymnasium (or high school) began at home. The von Neumann children (and cousins) were attended by both French and German governesses, with private tutors for subjects including Italian, fencing, and chess. John became fluent in Latin, Greek, German, English, and French. During World War I the children learned English from two British nationals, Mr. Thompson and Mr. Blythe, who had been held as enemy aliens in Vienna, but who, with Max's assistance, "had no difficulties in having their place of 'internment' officially moved to Budapest."[7]

After the war, Hungary was governed for 133 days by the Communist regime of Béla Kun. "I am violently anti-communist," von Neumann declared on his nomination to membership in the U.S. Atomic Energy Commission in 1955, "in particular since I had about a three-months taste of it in Hungary in 1919."[8] Thanks to Max's influence, the family retained occupancy of their house, after escaping to the safety of a summer home on the Adriatic near Venice during the worst of the upheavals in Budapest. "Under the guiding principle of equal facilities to all, the big apartments were broken up," remembers Nicholas, who was seven years old when a committee including a Communist Party official, a soldier of the regular army, and the janitor of the house arrived to administer the reallocation. "And on the piano under a weight, my father put a bundle of British pound notes, I don't know how much," says Nicholas. "The Communist official with the red armband promptly went there, took it, and the committee left and we remained in the apartment."[9]

At mealtimes children were treated as adults. "It was still customary at that time for the entire family to gather for a relatively full and lengthy lunch, after which we returned to our respective job, work, or studies until dinner time," Nicholas explains. Max entertained frequently at home, and according to Nicholas, "we got introduced to the 'secrets' of making business contacts and of management with executive powers in father's banking house."[10] Max was shrewd but kind. Nicholas remembers when their chauffeur—who had been using their expensive French Renault on his own for unauthorized purposes—wrecked the car and expected to lose his job. Max said nothing, and arranged with the Renault dealer for a replacement and repairs.

Max believed in demonstrating practical examples of the industrial applications of finance. "If these activities involved financing of a newspaper enterprise, the discussion was about the printing press and he brought home and demonstrated samples of type," says Nicholas. "Or if it was a textile enterprise, e.g., the 'Hungaria Jacquard Textile Weaving Factory,' the discussion centered around the Jacquard automatic loom. It probably does not take much imagination to trace this experience to John's later interest in punched cards!"[11]

For a wealthy middle-class Jewish youth from Pécs to join the Hungarian nobility was unusual, but not unprecedented, in fin de siècle Budapest. A window of liberalization had opened following the 1867 Compromise with Austria, and closed with the rise of Béla Kun and the late-1919 counterrevolution, under Admiral Horthy, that brought in the "Numerus Clausus," requiring that university enrollment reflect the composition of the general population, effectively returning to a quota against Jews entering academic and professional life. By that time, families such as the Kanns and Neumanns had been assimilated into the Hungarian upper class.

Unscathed by both the Red Terror of Béla Kun and the White Terror that followed, Max regained his position as a banker, joining the investment bank of Adolf Kohner and Sons. He opened doors to high and otherwise inaccessible places with the same ease and charm with which his son would later open the doors to power in the United States. "The essence of his philosophical, scientific, and humanitarian heritage was to do the impossible, that which was never done before," says Nicholas of what John learned from their father Max. "His approach was doing not just what was never done before but what was considered as impossible to be done."[12]

Hungarians had been facing the impossible for eleven hundred years, with few resources except a strategic location that had been occupied by the Roman, Ottoman, Russian, Holy Roman, Habsburg, Napoleonic French, Nazi German, and Soviet empires in turn. Von Neumann, according to Stan Ulam, credited Hungarian intellectual achievements to "a subconscious feeling of extreme insecurity in individuals, and the necessity of producing the unusual or facing extinction."[13] The Hungarian language, a branch of the Finno-Ugrian family incomprehensible to outsiders and closely related only to Finnish and Estonian, fortified Hungary against encroachment by its neighbors, while prompting Hungarian intellectuals to adopt German as a medium of exchange. To survive in a non-Hungarian-speaking world, Hungarians turned to the universal languages of music, mathematics, and the visual arts. Budapest, the city of bridges, produced a string of geniuses who bridged artistic and scientific gaps. In both mathematics and cinema it was said, "You don't have to be Hungarian, but it helps."

Von Neumann's talents stood out, even in Budapest. "Johnny's most characteristic trait was his boundless curiosity about everything and anything, his compulsive ambition to know, to understand any problem, no matter on what level," Klári recalls. "Anything that would tickle his curiosity with a question mark, he could not leave alone; he would sulk, pout and be generally impossible until, at least to his own satisfaction, he had found the right answer." He was able to disassemble any problem and then reassemble it in a way that rendered the answer obvious as a result. He had an ability, "perhaps somewhat rare among mathematicians," explains Stan Ulam, "to commune with the physicists, understand their language, and to transform it almost instantly into a mathematician's schemes and expressions. Then, after following the problems as such, he could translate them back into expressions in common use among physicists."[14]

Any subject was fair game. "I refuse to accept however, the stupidity of the Stock Exchange boys, as an explanation of the trend of stocks," he remarked to Ulam in 1939. "Those boys are stupid alright, but there must be an explanation of what happens, which makes no use of this fact." This question led to his *Theory of Games and Economic Behavior,* written with Oskar Morgenstern during the war years, with von Neumann giving the project his diminishing spare time and Morgenstern contributing "the period of the most intensive work I've ever known."[15]

"Johnny would get home in the evening after having zig-zagged through a number of meetings up and down the Coast," Klári recalls.

"As soon as he got in, he called Oskar and then they would spend the better half of the night writing the book. . . . This went on for nearly two years, with continuous interruptions of one kind or another. Sometimes they could not get together for a couple of weeks, but the moment Johnny got back, he was ready to pick up right where they stopped, as if nothing had happened since the last session."[16]

After threats of cancellation by Princeton University Press over the manuscript's escalating length, *Theory of Games and Economic Behavior* was finally published in 1944. Taking 673 pages to make their case, von Neumann and Morgenstern detailed how a reliable economy can be constructed out of unreliable parts, placing the foundations of economics, evolution, and intelligence on common mathematical ground. "Unifications of fields which were formerly divided and far apart," they counseled in their introduction, "are rare and happen only after each field has been thoroughly explored." Game theory was adopted first by military strategists, and the economists followed. Von Neumann "darted briefly into our domain," commented mathematical economist Paul Samuelson, looking back after fifty years, "and it has never been the same since."[17]

Klári remembers John being "as inept with his hands as he was adroit with his mind," and as a chemistry student he was considered a danger to glassware in the lab. He was drawn to "impossible" questions—predicting the weather, understanding the brain, explaining the economy, constructing reliable computers from unreliable parts. "It was a matter of pride with him to consider the weightiest questions in the spirit of a simple puzzle," says Klári, "as if he was challenging the world to give him any puzzle, any question, and then, with the stop-watch counting time, see how fast, how quickly and easily he could solve it."[18]

Edward Teller believed that "if a mentally superhuman race ever develops, its members will resemble Johnny von Neumann," crediting an inexplicable "neural superconductivity," and adding that "if you enjoy thinking, your brain develops. And that is what von Neumann did. He enjoyed the functioning of his brain."[19] If there wasn't anything to puzzle over, his attention wandered off. According to Herman Goldstine, "nothing was ever so complete as the indifference with which Johnny could listen to a topic or paper that he did not want to hear."[20]

As a child, von Neumann was at the head of his class in mathematics, history, languages, and science—everything except music and sports. Even in his youth, Klári remembers, "he already gave the impression of someone roly-poly, not middle-aged flabby, but babyishly plump and round like a child's drawing of the man on the moon." He was nonathletic, but enjoyed walking. "We had to hike a long way in order to see the 'Bear's Bathtub,' the 'Bridal Veil' in Yosemite, the 'Devil's Cauldron' in Yellowstone, the 'Devil's Tower' somewhere in the Dakotas," reported Klári. "These and others with similar names were the spots that triggered his curiosity to the point where, on several occasions, we made detours of hundreds of miles by car and sometimes even walked several miles just because his curiosity was aroused by these fancy names." He was attracted to stairs, and although "awkward at it . . . he liked to run upstairs two at a time," remembers Cuthbert Hurd, who met von Neumann while at Oak Ridge during the Manhattan Project and later became director of computing at IBM.[21]

Klári tried to interest John in skiing, but after two or three attempts, "He very simply and without any rancor offered divorce. . . . If being married to a woman, no matter who she was, would mean that he had to slide around on two pieces of wood on some slick mountainside, he would definitely prefer to live alone and take his daily exercise, as he put it, 'by getting in and out of a pleasantly warm bathtub.' "[22]

He rarely appeared without a suit and tie, a habit he attributed to being mistaken for a student when he arrived to teach at Princeton at the age of twenty-six, but otherwise relished the informalities of American life. "In addition to being a hard worker, Johnny seemed to be a bon vivant par excellence, always ready, for instance, for a wild drive to neighboring Spanish cafés in search of the peppery enchiladas he loved," Françoise Ulam recalls, adding that "Stan thought they probably reminded him of Hungarian goulash!" According to Klári, he was intensely superstitious. "A drawer could not be opened unless it was pushed in and out seven times, the same with a light-switch, which also had to be flipped seven times before you could let it stay."[23]

He evidenced "a chameleon-like ability to adapt to the people he was with," remembers Herman Goldstine, and never claimed he could not explain something to someone who did not understand the math. "It was just like being out on glass, it was so smooth. He somehow knew exactly how to get you through the forest. Whenever he gave a lecture, it was so lucid, it was like magic, it all seemed so simple you didn't need to take notes." Nicholas remembers his brother returning to Budapest to deliver a lecture on quantum mechanics and giving a nontechnical summary to the extended family before his talk. "The light theories of Dirac are not easy to explain," Nicholas points out.[24]

"The first thing that struck me about him were his eyes—brown, large, vivacious, and full of expression," remembers Stan Ulam, who first met von Neumann in Warsaw in 1935. "His head was impressively large. He had a sort of waddling walk." Ulam found him congenial, lighthearted, and "far from remote or forbidding," but noted that "he was ill at ease with people who were self-made or came from modest backgrounds. He felt most comfortable with third- or fourth generation wealthy Jews." Despite his sense of humor, "there seemed to be some sort of thin screen, or veil, a kind of restraint between him and others," noted Colonel Vincent Ford, von Neumann's attaché on the Air Force Strategic Missiles Evaluation Committee, and assistant while he was confined to Walter Reed Army Medical Center during the last year of his life. "He seemed to be a part of this world in one sense . . . and not a part of it in another."[25]

The frontline engineers on the computing project, who were treated as warmly by von Neumann as they were treated coolly by the other professors at the IAS, were nonetheless intimidated when he visited them in the machine room or at the bench. "The likelihood of getting actual numerical results was very much larger if he was not in the computer room, because everybody got so nervous when he was there," says Martin Schwarzschild. "But when you were in real thinking trouble, you would go to von Neumann and nobody else."[26]

"We can all think clearly, more or less, some of the time," says fellow Hungarian American mathematician Paul Halmos, "but von Neumann's clarity of thought was orders of magnitude greater than that of most of us, all the time." His was a calculating, logical intelligence, and "he admired, perhaps envied, people who had the complementary qualities, the flashes of irrational intuition that sometimes change the direction of scientific progress," Halmos adds. "Perhaps the consciousness of animals is more shadowy than ours and perhaps their perceptions are always dreamlike," physicist Eugene Wigner recalled in 1964. "On the opposite side, whenever I talked with von Neumann, I always had the impression that only he was fully awake."[27]

Von Neumann compensated for these superhuman abilities with an earthy sense of humor and tireless social life, and tried, with mixed success, to blend in on a normal human scale. "You would tell him something garbled, and he'd say, 'Oh, you mean the following,' and it would come back beautifully stated," says his former protégé Raoul

Bott. "He couldn't tell really very good people from less good people," Bott adds. "I guess they all seemed so much slower."[28]

In 1914, at age ten, von Neumann entered the Lutheran Gymnasium, one of Budapest's three elite high schools that offered competing eight-year curricula and that supported a small group of serious mathematicians who combined teaching with original research. He drew the attention of the legendary mathematics teacher László Rátz, who, according to classmate (and future economist) William Fellner, "expressed to Johnny's father the opinion that it would be nonsensical to teach Johnny school mathematics in the conventional way." Rátz had a gift for identifying mathematical talent and encouraging it to grow. "How can you know that this precocious 10-year-old will someday become a great mathematician?," asks Eugene Wigner. "You really cannot. Yet somehow Rátz did know this. And he knew it very quickly."[29]

Under the guidance of Joseph Kürschák at the University of Budapest, and the private tutoring of Gabriel Szegö, Michael Fekete, and Leopold Fejér, as well as Rátz, John began serious training in mathematics at the age of thirteen. His first published paper was written (with Fekete as coauthor) at the age of seventeen, and by the time of his high school graduation in 1921, he was recognized as a mathematician of professional rank. Yet his father doubted that mathematics alone offered a viable career path.

Theodore von Kármán—the Hungarian aerodynamicist who established the Jet Propulsion Laboratory in Pasadena, built the first supersonic wind tunnel, assumed the first chairmanship of the Air Force Scientific Advisory Board, and "invented consulting," according to von Neumann—remembers how "a well-known Budapest banker came to see me with his seventeen-year-old son. . . . He had an unusual request. He wanted me to dissuade young Johnny from becoming a mathematician. 'Mathematics,' he said, 'does not make money.'"

"I talked with the boy," von Kármán continues. "He was spectacular. At seventeen he was already involved in studying on his own the different concepts of infinity, which is one of the deepest problems of abstract mathematics. . . . I thought it would be a shame to influence him away from his natural bent."[30] A compromise was reached where von Neumann enrolled in the chemical engineering program at the Eidgenössische Technische Hochschule (ETH) in Zurich "to prepare him for what was then a reasonable profession," while simultaneously enrolling as a student of mathematics at both the University of Berlin and the University of Budapest. For the next four years he divided his time between Zurich and Berlin, attending classes in chemistry while working independently in mathematics and returning to Budapest at the end of each term for examinations that he passed without attending class. He received his degree in chemical engineering from the ETH in Zurich in 1925, followed by his doctorate in mathematics from Budapest.

His thesis, on the axiomatization of set theory, was the result of work begun in his freshman year. Abraham Fraenkel, editor of the *Journal für Mathematik* in 1922–1923, remembers receiving "a long manuscript of an author unknown to me, Johann von Neumann, with the title 'Die Axiomatisierung der Mengenlehre' ('The Axiomatization of Set Theory'). I don't maintain that I understood everything, but enough to see that this was an outstanding work and to recognize 'ex ungue leonem' (the claw of the lion)."[31] The paper was published in 1925 under the title "Eine Axiomatisierung der Mengenlehre" ("An Axiomatization of Set Theory") and expanded in 1928 with the *An* changed back to *The*.

Axiomatization is the reduction of a subject to a minimal set of initial assumptions, sufficient to develop the subject fully without new assumptions having to be introduced along the way. The axiomatization of set theory formed the foundations, mathematically, of everything else. An ambitious previous attempt, by Bertrand Russell and Alfred North Whitehead, despite 1,984 pages extending across three volumes, still left fundamental questions unresolved. Von Neumann started fresh. "The conciseness of the system of axioms is surprising," comments Stan Ulam. "The axioms take only a little more than one page of print. This is sufficient to build up practically all of the naive set theory and therewith all of modern mathematics . . . and the formal character of the reasoning employed seems to realize Hilbert's goal of treating mathematics as a finite game."[32]

The mathematical landscape of the early twentieth century was dominated by Göttingen's David Hilbert, who believed that from a strictly limited set of axioms, all mathematical truths could be reached by a sequence of well-defined logical steps. Hilbert's challenge, taken up by von Neumann, led directly both to Kurt Gödel's results on the incompleteness of formal systems of 1931 and Alan Turing's results on the existence of noncomputable functions (and universal computation) of 1936. Von Neumann set the stage for these two revolutions, but missed taking the decisive steps himself.

Gödel proved that within any formal system sufficiently powerful to include ordinary arithmetic, there will always be undecidable statements that cannot be proved true, yet cannot be proved false. Turing proved that within any formal (or mechanical) system, not only are there functions that can be given a finite description yet cannot be computed by any finite machine in a finite amount of time, but there is no definite method to distinguish computable from noncomputable functions in advance. That's the bad news. The good news is that, as Leibniz suggested, we appear to live in the best of all possible worlds, where the computable functions make life predictable enough to be survivable, while the noncomputable functions make life (and mathematical truth) unpredictable enough to remain interesting, no matter how far computers continue to advance.

In his axiomatization of set theory, "one can divine the germ of von Neumann's future interest in computing machines," says Ulam, speaking with hindsight from 1958. "The economy of the treatment seems to indicate a more fundamental interest in brevity than in virtuosity for its own sake. It thereby helped prepare the grounds for an investigation of the limits of finite formalism by means of the concept of 'machine.'"[33]

Von Neumann's style was now set. He would approach a subject, identify the axioms that made it tick, and then, using those axioms, extend the subject beyond where it was when he showed up. "What made it possible for him to make so many contributions in so many different parts of mathematics?" asks Paul Halmos. "It was his genius at synthesizing and analyzing things. He could take large units, rings of operators, measures, continuous geometry, direct integrals, and express the unit in terms of infinitesimal little bits. And he could take infinitesimal little bits and put together large units with arbitrarily prescribed properties. That's what Johnny could do, and what no one else could do as well."[34]

Along with his doctorate in 1926—after an oral examination at which David Hilbert was reported to have asked a single question: "In all my years I have never seen such beautiful evening clothes: pray, who is the candidate's tailor?"—von Neumann received a Rockefeller Fellowship to work with Hilbert at Göttingen, a lifeline from America at a time when positions in Europe were scarce.[35] He published twenty-five papers in the next three years, including a 1928 paper on the theory of games (with its minimax theorem proving the existence of good strategies, for a wide class of competitions, at the saddle point between convex sets) as well as the book *Mathematical Foundations of Quantum Mechanics,* described by Klári as his "permanent passport to the world of science," and, eighty years later, still in print. In 1927 he was appointed a Privatdozent (or associate professor) at the University of Berlin, and transferred to the University of Hamburg in 1929.

By this time Nazism was on the rise in Europe and depression was descending over the United States. Oswald Veblen was recruiting for the Princeton University mathematics department in preparation for the move to new quarters in Fine Hall, and, as Klári puts it, "in his search for talent he found Johnny . . . and used every means of persuading first the University, then the Institute, into appointing this young, relatively unknown Hungarian."[36] The initial invitation, from the university, was for a visiting lectureship to be shared with Eugene (Jenő) Wigner, with the two Hungarians dividing their time between Europe and the United States. To the gatekeepers at Princeton, half of two Hungarians was more acceptable than hiring one Hungarian full-time.

"One day I received a cable offering a visiting professorship at about eight times the salary which I had at the Institute of Technology in Berlin," Wigner recalls. "I thought this was an error in transmission. John von Neumann received the same cable, so we decided that maybe it was true, and we accepted." The pay was $3,000 for the semester, with $1,000 for travel—a small fortune at the time.

Newly married to Mariette Kővesi, the daughter of a prominent physician who was the director of the Jewish Hospital of Budapest, von Neumann arrived in Princeton in February of 1930 and, according to Wigner, "felt at home in America from the first day." Upon arrival in New York City, Wigner (then Wigner Jenő) and von Neumann "agreed that we should try to become somewhat American: that he would call himself 'Johnny' von Neumann, while I would be 'Eugene' Wigner."[37]

Von Neumann received a permanent appointment to Princeton University in 1931. "Johnny was among the very early birds to leave," says Klári, "voluntarily resigning his excellent academic position long before the Nazis had the power to force him out." His decision was made on economic as well as political grounds. "The economical crisis in Germany is very acute now," he wrote to Veblen in January of

1931, "and as people do not like to be alone in their miseries, there is much talk about the bad state of things in America. Is anything of this kind true?"[38]

In January of 1933 (using funds that had been earmarked for an indecisive Hermann Weyl), Abraham Flexner offered von Neumann a professorship at the Institute, where he joined Oswald Veblen, Albert Einstein, and James Alexander, who were already settled in Fine Hall, the Institute's interim home. The starting salary was $10,000 (higher than at the university), with benefits that included help with purchasing a house (or building one on Institute land that was being subdivided along Battle Road at the edge of Olden Farm). The Institute term, beginning in October and ending by early May, left time for a return to Europe for the summer months. The Flexners had a summer retreat at Magnetawan, two hundred miles north of Toronto, in the Canadian woods, and the Veblens had theirs in Maine. Einstein spent summers sailing on Long Island Sound, and Alexander, a passionate mountaineer who had soloed Alexander's Chimney on the East Face of Long's Peak in Colorado in 1922, spent his summers in the American West.

In the spring of 1933 the Nazis began dismissing Jewish professors from German universities, and von Neumann resigned his position in Berlin, followed by his membership in the German Mathematical Society in January of 1935. "He, more than anyone I know," adds Klári, "took it as a most personal affront that any nation, group of people, or individuals, could possibly prefer the base and unsophisticated philosophy of Nazism or any other 'ism' to such minds as Einstein, Hermann Weyl, Wolfgang Pauli, Schrödinger and many, many others including, last but not least, himself."[39]

According to Klári, "during the Thirties Johnny had criss-crossed the Atlantic at least twenty times," until the doors closed in 1939. "There is not much happening here except that people begin to be extremely proud in Hungary, about the ability of this country to run its revolutions and counter-revolutions in a much smoother and more civilized way than Germany," he reported to Veblen in April of 1933 from Budapest. "I did not hear anything about changes or expulsions in Berlin, but it seems that the 'purification' of universities has only reached till now Frankfurt, Göttingen, Marburg, Jena, Halle, Kiel, Königsberg—and the other 20 will certainly follow."[40] A line began to form at the exit toward the United States. The emergency Immigration Restriction Act of 1921 and the National Origins scheme of the

Reed-Johnson Act of 1924 limited immigrants from Hungary to an annual total of only 869. Exemptions were available to teachers or professors with full-time appointments, but full-time appointments were in short supply even for those already in the United States.

Veblen devoted all available resources—from the Bambergers, the Rockefeller Foundation, Princeton University, and a network of mathematics departments—to rescuing as many mathematicians as he could. Although von Neumann left for America ahead of the flood of refugees, and could easily have obtained a position elsewhere, he credited Veblen for his chance at a new life. "There was a very real affection between those two," says Klári. "Johnny, who lost his father when he was quite young, somehow transferred his filial affections to Veblen. He was convinced that had it not been for Veblen, he would have perished in the European mess."[41]

"His hatred, his loathing for the Nazis was essentially boundless," Klári adds. "They came and destroyed the world of this perfect intellectual setting. In quick order they dispersed the concentration of minds and substituted concentration camps where many of those who were not quick enough . . . perished in the most horrible ways." Von Neumann bore the traces of this for life. "There was a surface there of a very convivial sort," his daughter, Marina, explains, "which overlay what was fundamentally a rather cynical and pessimistic view of the world."[42]

"I feel the opposite of a nostalgia for Europe, because every corner I knew reminds me of the world, of the society, of the excitingly nebulous expectations of my childhood—and by childhood I mean childhood, which ended when I was 19 or 22 or something like that—of a world which is gone, and the ruins of which are no solace," von Neumann reported after his first postwar visit to Europe, in 1949. "My second reason for disliking Europe is the memory of my total disillusionment in human decency between 1933 and September 1938."[43]

Princeton was four and a half thousand miles away. The Institute for Advanced Study, an enclave within the enclave of Fine Hall, had been founded just in time. There were no quotas at the Institute, and Flexner, Veblen, and the Bambergers extended invitations as far as their budget could be stretched. Kurt Gödel was brought to Princeton (from Vienna) on a stipend of $2,400 for 1939 (although prevented from leaving until 1940); Stan Ulam (from Warsaw), as a temporary visitor for $300; and Paul Erdős, at $750 for one term (from Budapest).

Fine Hall, with its library, common room, and massive fireplaces, was both living room, study, and in some cases the only home left in existence to the mathematicians camped out in nearby rooming houses in Princeton's overcrowded downtown. Von Neumann found himself at the center of a thriving mathematical community, assuming the role that Hilbert had played in the Göttingen of 1926. "I would come to Fine Hall in the morning and look for von Neumann's huge car," remembers Israel Halperin, a student in 1933, "and when it was there, in front of Palmer Lab, Fine Hall seemed to be lit up. There was something in there that you might run into that was worth the whole day. But if the car wasn't there, then he wasn't there and the building was dull and dead."[44]

Johnny and Mariette (followed by Johnny and Klári in 1938) sought to reconstruct a fragment of the life they had left behind in Budapest. They entertained lavishly and frequently—in old Princeton style, with domestic servants to help. The household established with Klári on Westcott Road became "an oasis in otherwise somewhat stuffy Princeton," according to Robert Richtmyer, with parties that "were a mixture of intense scientific discussion and then complete irrelevance," according to Oskar Morgenstern. "You went there with a light feeling every time, because there was a spirit of freedom in that house." Adds Richtmyer: "and always something to drink."[45]

Von Neumann did his best to undermine the Institute for Advanced Study's reputation as a refuge where great minds retreated into quiet seclusion to think. "He could not work without some noise or at least the possibility of noise," explains Klári. "Some of his best work was done in crowded railroad stations and airports, trains, planes, ships, hotel lobbies, lively cocktail parties or even among a bunch of shrieking very minor minors whooping it up." At Fine Hall, his office door was always open. "Weyl is happier in a room smaller than yours, and Johnny is productive in a room smaller than Weyl's," Abraham Flexner wrote to Oswald Veblen, arguing against a request for even more luxurious offices in Fuld Hall.[46]

Although happy in small, nondescript offices, von Neumann liked large, fast cars. He bought a new one at least once a year, whether he had wrecked the previous one or not. Asked why he always purchased Cadillacs, he answered, "Because no one would sell me a tank." In 1946 the Diracs were visiting in Princeton, and Mrs. Dirac asked for help in finding an inexpensive used car. "How can I tell her without wounding her feelings," von Neumann wrote to Klári, "that her chances to getting a used car in US '46 are as good as getting a second hand snowball in Hell!"[47]

"I always tried to arrange it so that I could drive," remembers Cuthbert Hurd, who had brought von Neumann in as a consultant, two days a month, for IBM, and often shared the drive up the West Side Elevated Highway to IBM's headquarters in Poughkeepsie, New York. "When the conversation lagged he would sing. The tune was indistinguishable, so he would sway from side to side like this and the car wouldn't go very straight." Von Neumann was regularly ticketed for speeding. "I'd take that ticket and give it to the downtown manager in New York, where the Police court was, and he would go around and pay the fine," says Hurd.[48]

"He drove like mad and only needed to sleep for three or four hours a night," says Marina, recalling an early drive across the United States. "Remember, those were 1930s motels in 1946; nothing had been built during the war. Many of them had no indoor plumbing. I had led a sheltered life, and I had never seen an outhouse, except once at camp." Herman Goldstine, with whom von Neumann occasionally shared hotel rooms while on government assignments, remembers that "he would waken in the night, at two or three in the morning, and would have thought through what he had been working on. He would then write [it] down."[49]

Von Neumann could deliver publishable text, and even mathematical proofs, on the first draft. "I write rather freely and fast if a subject is 'mature' in my mind," he explained in 1945, apologizing for an undelivered manuscript, "but develop the worst traits of pedantism and inefficiency if I attempt to give a preliminary account of a subject which I do not have yet in what I can believe to be in its final form." His handwritten letters sometimes end with an informal "P.S." that continues, for several pages, to explain some new result. "Each day he would start writing before breakfast," says Ulam. "Even at parties in his house, he would occasionally leave the guests to go to his study for half an hour or so to record something that was on his mind." In speech or on paper, every idea expressed was precise. "Von Neumann was one of the greatest of all mathematical artists," says Goldstine. "It was never enough for him merely to establish a result; he had to do it with elegance and grace."[50]

After obtaining U.S. citizenship on January 8, 1937, von Neumann applied for a commission in the army but was rejected for being too old, despite perfect written examination scores. Oswald Veblen arranged for the army to enlist von Neumann as a consultant instead. The U.S. Army Ordnance Department's Proving Ground at Aberdeen had been left in suspended animation at the end of World War I, subsisting on an annual budget of about $6 million until 1937, when funding tripled to $17 million, before jumping to $177 million on the eve of World War II. Von Neumann's involvement with the military increased over the next twenty years. "He seemed to admire generals and admirals and got along well with them," explains Ulam, adding that this "fascination with the military . . . was due more generally to his admiration for people who had power. He admired people who could influence events. In addition, being softhearted, I think he had a hidden admiration for people or organizations that could be tough."[51]

All three military services regarded von Neumann as one of their own. "I think that we have a chance to do some work which is useful, both to the Army and to the mathematical community," von Neumann answered mathematician Saunders Mac Lane, who questioned whether academic mathematicians should take on military work. "We can do this in one particular sector of the Army where the authorities have 'seen the light.' I don't think that we should be influenced too much by the inadequacies in other sectors." According to Ulam, von Neumann was especially favored for his role as chairman of committees, "this peculiar contemporary activity," essential to getting anything done in the United States. "He would press strongly his technical views, but defer rather easily on personal or organizational matters." According to Rear Admiral Lewis Strauss, von Neumann was "able to take the most difficult problem, separate it into its components, whereupon everything looked brilliantly simple, and all of us wondered why we had not been able to see through to the answer as clearly as it was possible for him to do."[52]

While World War I had been a battle for bigger guns, World War II (and the cold war that followed) became a battle for bigger bombs. In 1937, with war looming, it was time to remobilize the scientists, and Veblen was brought back as the army's chief mathematician at the Proving Ground. Von Neumann was appointed, in quick succession, to the Scientific Advisory Committee of the Ballistic Research Laboratory, the War Preparedness Committee of the American Mathematical Society and Mathematical Association of America, and the National Defense Research Committee. "The functions of all these sets and sets of sets are not very well defined as yet, but I suppose they will be, when 'The Day' comes around," he wrote to Ulam in 1940. "So far I have chiefly worried about spherical and Gaussian measures of various functions."[53] This was shorthand for calculating the behavior of high explosives—the surprising thing about large explosions being not how much energy was released, but how unpredictable was the damage produced as a result.

Von Neumann, who believed that mathematics grew best when nourished by "a certain contact with the strivings and problems of the world," became a great friend of the weaponeers. "Physicists— particularly experimental physicists—are more in demand for defense work," he explained to a fellow mathematician once war had been declared, "while we must, so to say, create the demand for our services."[54] Wherever new weaponry went, von Neumann followed—or got there first. The behavior of both high-explosive detonations and supersonic projectiles depended on the effects of shock waves whose behavior was nonlinear and poorly understood. What happens when a discontinuity is propagated faster than the local speed of information ahead of the disturbance (for pressure waves, this being the speed of sound)? What happens when two (or more) shock waves collide?

Shock waves are sudden discontinuities propagated in compressible media—usually air. "Under the conditions in and around an explosion all known substances must be regarded as compressible," von Neumann pointed out.[55] Drawing upon his training in chemical engineering as well as mathematical physics, he took a broad view of weapons design: starting with the chemical energy released by high explosives, through the detonation wave that propagates the explosion, to the blast wave that causes the destructive effects. The resulting insights into shock waves (especially reflected shock waves) contributed to the development of the shaped charges used in antitank weapons, torpedoes, and armor-piercing shells, as well as more effective depth charges against submarines and more effective targeting of conventional bombs. His novel mathematical technique for treating shocks led to the success of the implosion method of initiating a nuclear explosion, and his theory of blast waves helped determine the height at which to explode the resulting weapon for maximum effect. He was one of a handful of scientists who were present at both the conception and the delivery of the atomic bomb.

The amount of fissile material needed to support a self-sustaining chain reaction is a function not only of mass but of density. Squeeze a subcritical mass of plutonium to a high enough density and it becomes critical, and if confined by a surrounding shell of dense, neutron-reflecting material (the "tamper"), it will violently explode. Von Neumann suggested how the requisite high explosive could be formed into implosive lenses that, if arranged like the panels on a soccer ball, and detonated in precise synchronization, would focus the resulting shock wave into a converging front. A far smaller quantity of fissionable material can thereby be made to explode.

Von Neumann's theory of reflected shock waves could then be used to maximize a bomb's effects. "If you had an explosion a little above the ground and you wanted to know how the original wave would hit the ground, form a reflected wave, then combine near the ground with the original wave, and have an extra strong blast wave go out near the ground, that was a problem involving highly non-linear hydrodynamics," recalls Martin Schwarzschild. "At that time it was only just understood descriptively. And that became a problem that I think von Neumann became very much interested in. He wanted a real problem that you really needed computers for."[56]

The results were surprising. In a report to the Navy Bureau of Ordnance in 1943, von Neumann, who normally limited his use of exclamation marks in his mathematical writings to the symbol for factorial ($4! = 1 \times 2 \times 3 \times 4 = 24$), used two exclamation marks, as punctuation, in a row. "Even for a weak shock the reflected shock can be twice as strong as it is head-on, if the angle of incidence is properly chosen!" he reported. "And this happens at a nearly glancing angle, where a weaker reflection would have seemed plausible!"[57]

By the time the United States entered the war (against Japan on December 8 and Germany on December 11, 1941), "Johnny had started on his travels," Klári reported. "Almost continuous: from Princeton to Boston, from Boston to Washington, from Washington to New York—a short stop in Princeton, then to Aberdeen, Maryland, the Army Proving Grounds—back to Washington, maybe one night at home, then starting on the rounds again, not necessarily in the same order, but up and down the Eastern seaboard with occasional forays further inland—not the West yet—that came later."[58]

In February 1943, after a series of false starts, he received orders on behalf of the navy to report to England, the official assignment being to assist with a statistical approach to the problem of mines, submarines, and related countermeasures and counter-countermeasures. Loss of Allied shipping was threatening to turn the tide of the war. What von Neumann actually did during his stay in England remains a mystery, particularly the extent to which he consulted with the British groups who were working, secretly, on code breaking and the feasibility of atomic bombs. We do know that he made a visit in late April of 1943, with John Todd, to Her Majesty's Nautical Almanac Office, one of the largest non-secret computing operations at the time. The office had been evacuated to Bath from Greenwich for safety from German air raids, and on the train ride back to London, having witnessed the capabilities of a six-register National Cash Register Accounting Machine, von Neumann developed a short interpolation routine. He later wrote to Todd that "I received in that period a decisive impulse which determined my interest in computing machines."[59]

Upon his return from England in July of 1943 he was enlisted for "Project Y," as the Manhattan Project was code-named. As a mathematical consultant to the project, he was allowed to travel freely outside Los Alamos, a privilege denied most participants, who were required to bring their families and remain sequestered for the duration of the war. "Progress elsewhere in computing was carried to Los Alamos by von Neumann," says Nicholas Metropolis, "who consulted for several government projects at such a pace that he seemed to be in many places at the same time."[60]

Von Neumann arrived at Los Alamos on September 21, 1943, traveling from Chicago with Isidor Rabi on the Atchison, Topeka and Santa Fe Railway's flagship diesel-electric streamliner *Super Chief.* From the railroad depot at Lamy, New Mexico, they were driven "across a number of good class canyons and mesas" to the new laboratory, which von Neumann described, in a letter to Klári the next day, as "an odd combination of an Army Post, a Western National Park with Lodge, and a few assorted other things." He concluded that the project "is worth meditating about, although one should probably not sell one's soul to it," and added, in a postscript, that "computers are, as you suspected, quite in demand here, too." Two days later he added that "the whole place is queerer than I can describe. And . . . believe me, if I begin to develop a craving for normality and reality, then it's pretty bad."[61]

By "computers" von Neumann meant human computers, the kind that Oswald Veblen had assembled at the Proving Ground during World War I. At the time of von Neumann's arrival at Los Alamos there were about twenty human computers (initially recruited among the wives of physicists, soon aided by reinforcements from the army's Special Engineering Detachment, or SED) equipped with Marchant

10-digit electromechanical desk calculators. The Marchant "Silent Speed" machines, built on San Pablo Avenue in Oakland, California, and requisitioned for the war effort, weighed almost 40 pounds, incorporated 4,000 moving parts, and cycled at 1,300 rpm.

As Nicholas Metropolis, who became director of computing at Los Alamos, put it, "the very nature of the Laboratory's objective—an atomic bomb—precluded extensive field testing."[62] At a time when even one shock wave at a time was poorly understood, predicting the behavior of an implosion weapon accurately enough to build one that stood a reasonable chance of working on the first try was out of reach of the small computing group. To follow the process from start to finish required modeling the initial propagation of a detonation wave through the high explosive, the transmission of the resulting shock wave through the tamper and into the fissile material (including the reflection of that shock wave as it reached the center), the propagation of another shock wave as the core exploded, the passage of that shock wave (followed by an equally violent rarefaction wave) outward through the remnants of the previous explosion and into the atmosphere, and finally the resulting blast wave's reflection if the bomb was at or near the ground. Von Neumann had arrived just in time.

A set of punched-card accounting and tabulating machines were requisitioned from IBM, which could not be told where the machines were going, or why. The machines—three 601 multipliers, a 402 tabulator, a reproducer, a verifier, a sorter, and a collator—arrived, in huge wooden crates, without documentation or an installation crew. IBM was asked for the name of their best technician who had been drafted into the army, who was immediately given a security clearance and reassigned to Los Alamos, but this took time. In the interim, Stanley Frankel, a Berkeley graduate student of Oppenheimer's who had been put in charge of the hand computing group, and Richard Feynman, a graduate student (and amateur safecracker) from Princeton who was game for any unauthorized challenge, managed to uncrate the machines and get them to work.

Feynman and Frankel were hooked. "Mr. Frankel, who started this program, began to suffer from the computer disease that anybody who works with computers now knows about," Feynman later explained. "The trouble with computers is you play with them." Feynman and Frankel, joined by Nicholas Metropolis, adapted the IBM machines to accelerate the work of the hand computing group. "If we got enough of these machines in a room, we could take the cards and put them through a cycle," explained Feynman. "Everybody who does numerical calculations now knows exactly what I'm talking about, but this was kind of a new thing then—mass production with machines."[63]

The strategy was to start from a prescribed initial state and model the progress of the explosion from point to point in space and from step to step in time. A single, initial punched card was established for each point in space, with a deck of these cards representing the state of the explosion at a given instant in time. "Processing a deck of cards through one cycle in the calculation effectively integrated the differential equations ahead one step in the time dimension," explains Metropolis. "This one cycle required processing the cards through about a dozen separate machines with each card spending 1 to 5 seconds at each machine."[64] The result was a new deck of cards used as the input for the next time step. The process was tedious, repetitive, intolerant of error, and quickly bogged down.

"The real trouble was that no one had ever told these fellows anything," explains Feynman. "The Army had selected them from all over the country for a thing called Special Engineer Detachment—clever boys from high school who had engineering ability. They sent them up to Los Alamos. They put them in barracks. And they would tell them nothing." Feynman secured permission from Oppenheimer to give a lecture to the recruits. "They were all excited: 'We're fighting a war! We see what it is!' They knew what the numbers meant. If the pressure came out higher, that meant there was more energy released. Complete transformation! They began to invent ways of doing it better. They improved the scheme. They worked at night."[65] Productivity went up by a factor of ten.

Von Neumann found himself back among the punched cards he remembered from when his father brought parts of a Jacquard loom–control system home from work. "In March or April 1944," says Metropolis, "von Neumann spent two weeks working in the punched-card-machine operation, pushing cards through the various machines, learning how to wire plugboards and design card layouts, and becoming thoroughly familiar with the machine operations."[66]

There were fewer than two years from the first, tentative theoretical models to the successful test, code-named Trinity, of an implosion weapon at the northern extremity of the Alamogordo Bombing Range on July 16, 1945. Despite the pressure to complete the job, the physicists found time to relax. "We used to go for walks on Sunday," remembers Feynman. "We'd walk in the canyons, and we'd often walk with Bethe, and Von Neumann, and Bacher. It was a great pleasure. And Von Neumann gave me an interesting idea; that you don't have to be responsible for the world that you're in. So I have developed a very powerful sense of social irresponsibility as a result of Von Neumann's advice. It's made me a very happy man ever since."[67]

With von Neumann, the veil was rarely lifted. "One time, in early 1945, he came back from Los Alamos and proceeded to behave in the most unusual 'Johnnyesque' manner," Klári recounts. "He arrived home sometime mid-morning, immediately went to bed and slept twelve hours. Nothing he could have done would have had me more worried than Johnny skipping two meals, not to speak of the fact that I had never known him to sleep that long in one stretch. Sometime late that night he woke up and started talking at a speed which, even for him, was extraordinarily fast."

"What we are creating now is a monster whose influence is going to change history, provided there is any history left," he said, in Klári's account, "yet it would be impossible not to see it through, not only for the military reasons, but it would also be unethical from the point of view of the scientists not to do what they know is feasible, no matter what terrible consequences it may have. And this is only the beginning!"

The concerns von Neumann voiced that night were less about nuclear weapons, and more about the growing powers of machines. "While speculating about the details of future technical possibilities," Klári continues, "he gradually got himself into such a dither that I finally suggested a couple of sleeping pills and a very strong drink to bring him back to the present and make him relax a little about his own predictions of inevitable doom."

"From here on, Johnny's fascination and preoccupation with the shape of things to come never ceased," concludes Klári's account. For the next seven years he neglected mathematics and devoted himself to the advance of technology in all forms. "It was almost as if he knew that there was not very much time left."[68]

We have only hints of von Neumann's final thoughts. "As he came more and more to realize that the control over the physical forces of nature which he and his co-workers had placed in the hands of their fellow men could be used for evil as well as for good, he felt with steadily increasing intensity the moral problems bound up with the greatest of modern scientific triumphs," said Father Anselm Stritt-matter, the Benedictine monk who spent many hours at von Neumann's bedside during his final months and delivered the last rites at his death. "As for his own role in this complex situation, in spite of the dismal possibilities he envisioned, he knew no hesitation, he had no regrets."[69]

"There is a unifying force behind all manifestations of nature, which we cannot fully comprehend, but we can try to explain it with the means at our disposal," says Nicholas Vonneumann, summing up his brother's life. "It was in this spirit that John tried to comprehend . . . the mysteries of atomic and subatomic particles through quantum mechanics, the mysteries of weather . . . through hydrodynamics and statistics, the mysteries of the central nervous system through . . . artificial computers, the mysteries of genetics and inheritance through his theory of self-reproducing automata."[70]

Even Klári, closer to von Neumann than anyone else, was never fully able to understand this "strange, contradictory, and controversial person; childish and good-humored, sophisticated and savage, brilliantly clever yet with very limited, almost primitive lack of ability to handle his emotions—an enigma of nature that will have to remain unsolved."[71]

"No matter which way you looked he always seemed to belong somewhere else," explains Klári. "The pure mathematicians claimed that he had become a theoretical physicist; the theoretical physicists looked at him as a great help and advisor in applied mathematics; the applied mathematician was awed that such a pure and ivory-towerish mathematician would show so much interest in his applied problems and, I suspect, in certain government circles they may have thought of him as an experimental physicist, or even an engineer."[72]

On August 6, 1945, a uranium-fueled atomic bomb yielding 13 kilotons was dropped on Hiroshima, followed by a plutonium-fueled bomb yielding 20 kilotons on Nagasaki on August 9. The Japanese surrendered on August 15. "Isn't it wonderful that the war is over?" Marina wrote to Klári on August 28. "Is Daddy still going to travel so much now the war is over? I hope not."[73] Von Neumann's travels—between Princeton, Aberdeen, Los Alamos, Santa Monica, Chicago, Oak Ridge, and Washington, D.C.—continued.

World War II was over, but the cold war had begun.

# FIVE

# MANIAC

*Let the whole outside world consist of a long*
*paper tape.*

—John von Neumann, 1948

On Monday, November 12, 1945, at 12:45 p.m., six people, led by John von Neumann, gathered in Vladimir Zworykin's office at RCA's research laboratories in Princeton, New Jersey. Vladimir Kosma Zworykin was a pioneer of television (and the last entry in many encyclopedias) who would live to regret that his invention's capacity for transmission of intelligence had become a channel for so much noise. Captain Herman Goldstine (on loan from U.S. Army Ordnance and the Aberdeen Proving Ground) was one of the principal organizers of the army's Electronic Numerical Integrator and Computer, or ENIAC, whose existence would not be made public until February 1946. Statistician John Tukey (of Princeton University and Bell Laboratories) provided a direct link to Claude Shannon, whose mathematical theory of communication showed how a computer built from unreliable components could be made to function reliably from one cycle to the next. Jan Rajchman and Arthur Vance were engineers, and George Brown a statistician, from RCA. This first meeting of the Institute for Advanced Study's Electronic Computer Project established principles that would guide the destiny of computing for the next sixty years.

"The heart of the system is a central clock, carrying an enormous load," the minutes report. The circuitry would be modular, because "this sort of design is favorable for mass production," explained the engineers. "'Words' coding the orders are handled in the memory just like numbers," explained von Neumann, breaking the distinction between numbers that *mean* things and numbers that *do* things. Software was born. Numerical codes would be granted full control—including the power to modify themselves.[1]

The age of electronics began in 1906 with Lee De Forest's invention of the vacuum tube, or, as the British (led by John Ambrose Fleming, whose work preceded De Forest's) described it, the thermionic valve. Within an evacuated glass envelope, a charged cathode was heated to a temperature high enough to boil off electrons, whose flow to the anode (or plate) could be controlled by a secondary current applied to a very thin filament (or filaments) known as the grid. Switching (and signal amplification) was now possible at radio frequencies, rather than the speed of relays and Morse code.

Zworykin, the youngest of seven children, was born in 1889 to a family of steamship owners on the Oka River in Russia. He was seventeen years old and a student at the Petrograd (St. Petersburg) Institute of Technology when he was discovered using the instruments in the physics laboratory for unauthorized experiments going beyond the problems that were assigned in class. Professor Boris Rosing took Zworykin aside and, instead of reprimanding him, offered him a position in his own private lab. Rosing was producing his own electron tubes, which at that time required building his own vacuum pumps and formulating his own glass. He introduced Zworykin not only to the behavior of electrons within the evacuated glass envelope, but also to how these captive electrons could be coaxed into communication with the world of light outside.

"I found that he was working on the problem of television about which I had never heard before," remembered Zworykin sixty years later. "This was my first introduction to the problem which eventually occupied most of my life." By the time Zworykin graduated with a degree in electrical engineering in 1912, "Rosing had a workable system consisting of rotating mirrors and a photocell on the pickup end, and a cathode ray tube with partial vacuum which reproduced very crude images over the wire across the bench."[2] Much of Zworykin's subsequent career would be devoted to inventing better ways to translate, in both directions, between photons and electrons—commercial television being the way to make this pay.

Rosing secured an appointment for Zworykin to work on X-ray diffraction with Paul Langevin in Paris, until interrupted by the outbreak of World War I. Returning to Russia, Zworykin was inducted into the army and rose to become an officer in the Signal Corps, where his knowledge of radio and ability to fix machinery, from generators to machine guns, enabled him to move freely through the ranks and escape execution by a succession of captors as the war drew to a close. During the Bolshevik Revolution and counterrevolution, wireless was the only way to determine, in the more remote parts of Russia, who was in power at any given time. Zworykin eventually escaped down the River Ob, through country whose residents, lacking telecommunication, were unaware there had even been a revolution, to the Russian Arctic, and then, with stops at Novaya Zemlya, Tromsø, Copenhagen, and London, he arrived in New York City on New Year's Eve 1919.

Zworykin presented himself to Boris Bachmeteff, the Russian ambassador in Washington, and secured a job as an adding machine operator at the Russian Purchasing Commission in New York. His wife, Tatiana, whom he had left behind in Russia, soon followed him to the United States. In 1920, after the birth of their first child, Zworykin joined a small group of fellow Russian émigrés at the Westinghouse laboratories in East Pittsburgh, where he was able to return to work on television in his spare time. He faced a series of obstacles, including the implosion of a prototype picture tube that slid off the back seat of his car as he stopped for a red light. The noise, mistaken for a gunshot, attracted the attention of a police officer, who grew suspicious at Zworykin's attempted explanation, in broken English, of how pictures could be transmitted by radio waves to the device that lay shattered in the back of his car. "So you see pictures on the radio now? Sure . . . buddy!" the officer muttered, before taking Zworykin to jail until the facts were sorted out.[3]

After failing to interest Westinghouse—then engaged in a bitter struggle against General Electric—in the commercialization of television, Zworykin transferred to RCA (successor to the American Marconi Company and progenitor of NBC), where David Sarnoff, a fellow Russian expatriate, placed the resources of RCA at his disposal. Sarnoff would end up sinking $50 million into the development of broadcast television—and fought a protracted patent-interference case against the American inventor Philo Farnsworth, who had independently developed an improved charge-storing camera tube that, in the opinion of the courts, predated Zworykin's iconoscope, upon which the RCA television system, adopting Farnsworth's improvements, was based.

In 1941, Zworykin was appointed director of RCA's new research laboratories in Princeton, situated adjacent to the Rockefeller Institute for Medical Research, two miles from the Institute for Advanced Study on the west side of the former Trenton–New Brunswick Turnpike, now Route 1. In addition to commercial television Zworykin helped bring the world the photomultiplier tube (for seeing in the dark) and the electron microscope (for seeing beyond the resolution of visible light). He devoted his later years to applying electronics to medical and biological research. "One cannot stumble on an idea unless one is running," Zworykin advised those who joined his lab.[4]

In October 1945, on the recommendation of von Neumann, Zworykin, who had been leasing a house next door to the Veblens, secured an exception to Institute housing policy that allowed him to purchase a home in the faculty enclave at the end of Battle Road, paying $30,000 in cash to paleographer Elias Lowe. Herbert Maass objected—not to Zworykin, but to the "rather substantial profit to Professor Lowe."[5]

Zworykin's close relationship with Theodor von Kármán, who granted him access to secret military installations to work on electronic weapon systems, was viewed with suspicion by the FBI. Despite his anti-Soviet record, and contributions to the American defense effort that included night-vision gunsights and television-guided bombs, he was denied permission to travel to Moscow with a delegation of American technologists in 1945. J. Edgar Hoover personally labeled him a subversive, and his activities (including visits with his Philadelphia mistress) were monitored until 1975. In 1956 he refused to cooperate with an FBI interviewer, saying, "I left Russia to get away from state police."[6]

The development of electronics, according to Zworykin, could be divided into three epochs. "In the first, beginning with DeForest's invention of the audion in 1906 and ending with the First World War, electron currents were controlled in vacuum tubes in much the same manner as a steam valve controls the flow of steam in a pipe," he explained. "No more attention was paid to the behavior of the individual electrons in the tube than is customarily expended on the motion of the individual steam molecules in the valve."

"In the second period," beginning in the 1920s, Zworykin continued, "the directed, rather than random, character of electron motion in vacuum was applied in the cathode-ray tube." In the third period, beginning in the 1930s, beams of electrons were further subdivided into groups. "This subdivision was either on the basis of time, the electrons being bunched at certain phases of an applied high-frequency field as in the klystron or magnetron, or of space, as in image-forming devices," Zworykin explained. "The electron microscope and the image tube are typical representatives of this group."[7]

During World War II, Zworykin and his protégé Jan Rajchman, an expatriate Pole educated in Zurich who had joined Zworykin's group on New Year's Day of 1936, sought to launch a fourth epoch in the evolution of vacuum tubes. In 1939, as Germany invaded Poland, Colonel Leslie Simon of the U.S. Army's Ballistic Research Laboratory approached RCA about how to improve an antiaircraft gunner's chances of shooting down enemy planes. To hit targets on the ground, a gunner could make use of firing tables prepared in advance. To place a shell in the path of a moving airplane required on-the-spot computation, including a last-minute estimate of time-of-flight to set a timed fuse so that the shell exploded as close as possible to the plane. "The Germans had a great dominance in the air and the Allies were very poor at antiaircraft fire control," Rajchman explains. "Colonel Simon had the foresight to believe that electronics could provide the required speed."[8]

With Zworykin's encouragement, Rajchman developed a series of digital processing and storage tubes: switching, gating, and storing pulses of electrons within single envelopes at megacycle speeds. The Computron and the Selectron, antediluvian ancestors of solid-state integrated circuits, were vacuum-tube versions of the microprocessor and memory chip. "The idea was to make a single tube which could multiply two numbers and add a third number to the product, the numbers being expressed in digital binary code," Rajchman explains. "A number of electron beams emanating from a single central cathode were deflected each by three electrodes, corresponding respectively to a digit of the multiplier, a digit of the multiplicand, and a 'carry-over' digit. . . . In effect, the tube was made by 'integrated vacuum technology,' as we would describe it today."[9]

The Computron, invented by Rajchman and Richard L. Snyder, was a 64-pin, 14-bit arithmetical processing tube containing 737 separate parts. No adjustments were possible once the envelope was sealed. "Numbers may be added or multiplied quickly and without the complication of timing impulses, clearing impulses or the like," wrote Rajchman and Snyder in their patent application for a "Calculating Device" of July 30, 1943. By the time a proof-of-principle prototype was demonstrated, however, "it became clear that our pioneer work could not lead to an anti-aircraft director that could be used in actual combat soon enough."[10]

The Selectron was an all-digital, random-access, 4,096-bit electrostatic storage tube, built with vacuum-tube technology but functionally equivalent to modern silicon-based memory chips. "One should be able to go to any element without having to go to all the oth-

ers [and] it should remember indefinitely without rejuvenation . . .
just memorize forever until we want the information," Rajchman
explained in 1946.[11] It was the prospect of the Selectron that convinced
von Neumann that the path to digital computing lay through RCA.
"John von Neumann came to see us frequently," says Rajchman, "and
became very familiar with our research." Along with the Computron
and the Selectron, Rajchman also developed the resistor-matrix function table, providing read-only memory, or ROM. "We made fairly
large matrix arrays which had about a hundred and fifty thousand
resistors in them," he says. On October 30, 1943, he filed a patent for an
all-digital "Electronic Computing Device" that would perform binary
arithmetic at electronic speeds, using resistor matrices to store both
invariant function tables as well as variable data to be operated upon.
"The whole computation is made in the binary system of numeration so that any number is expressed as a sum of powers of two."[12]
The proposed computer, both parallel and asynchronous, would have
been exceptionally fast. It had no moving parts. The resistor matrices
could be initialized with different functions and data as needed, tailored to different guns.

The elements of universal computation were falling into place.
"Exactly when we started to forget the problems for which the
machine was supposed to be built and started to work in earnest on
a universal computer for all problems," Rajchman recalled in 1970, "I
can't say."[13]

As the United States prepared to enter the war, there was a shortage of human computers at the Aberdeen Proving Ground, and an
auxiliary computing section was established at the Moore School
of Electrical Engineering at the University of Pennsylvania, where
human computers could be recruited among the student population,
augmented as needed from neighboring colleges and schools.

Shells and targets were moving at ever-higher speeds, and the combined efforts of both computing sections were unable to keep up. A
human computer working with a desk calculator took about twelve
hours to calculate a single trajectory, with hundreds of trajectories
required to produce a firing table for any particular combination of
shell and gun. The electromechanical differential analyzer at the Ballistic Research Laboratory (a ten-integrator version of the analog
computer that Vannevar Bush had developed at MIT) took ten or
twenty minutes. To complete a single firing table still required about
a month of uninterrupted work. Even with double shifts (and a sec-

ond, fourteen-integrator differential analyzer) working at the Moore School, the army was falling behind. "The number of tables for which work has not been started because of lack of computational facilities far exceeds the number in progress," reported Herman Goldstine in August 1944. "Requests for the preparation of new tables are being currently received at the rate of six per day."[14]

Herman Heine Goldstine had been teaching introductory classes in exterior ballistics for Gilbert A. Bliss at the University of Chicago in July 1942 when he was called up to join the army, assigned to the air force (still under the wing of the army at the time), and sent to Fort Stockton, California, to prepare for deployment in the Pacific Theater against Japan. Alerted by Gilbert Bliss, Oswald Veblen "started the wheels moving," according to Goldstine. "It was touch-and-go whether I would go overseas or Veblen would get there first." On the same day that Goldstine received orders to ship out for the Pacific, he also received orders, thanks to Veblen, to report to Aberdeen. He telephoned the commanding general, who advised him, "Son, if I were you, I would get out of the camp. If you've got an auto, I'd get in the auto, and start driving." Goldstine headed east.[15]

Upon arrival at Aberdeen, Lieutenant Goldstine was assigned to Colonel Paul N. Gillon, who was responsible for the Ballistic Research Laboratory's computational substation at the Moore School. The situation did not look good. "No amount of augmentation of the staff of human computers—then around 200—would suffice," wrote (now Captain) Goldstine in a postwar report. "It was accordingly decided . . . to sponsor the development of a radically new machine, the ENIAC, which if it were successful would reduce the computational time consumed in preparing a firing table from a few months to a few days."[16]

The ENIAC was built by a team led by John W. Mauchly and J. Presper Eckert, thirty-six and twenty-four years of age when their project was launched in 1943. Mauchly had been teaching physics at Ursinus College, on the outskirts of Philadelphia, and attempting to demonstrate a statistical correlation between sunspot activity and climate variation in his spare time, when he took an introductory training course in defense electronics at the Moore School. Before he had completed his training, he was asked to join the faculty, and did. Eckert was a native Philadelphian whose first job, while he was still in high school, had been in Philo Farnsworth's television research laboratory, leaving him with a deep understanding of electronics—and a lingering mistrust of Zworykin and RCA.

The Philadelphia area was home to Philco, RCA, and numerous smaller electronic research laboratories, including the Franklin Institute, whose origins went back to Benjamin Franklin's efforts to bring experimental philosophy to the New World. Eckert and Mauchly were American entrepreneurs, their backgrounds far removed from Zworykin's prerevolutionary St. Petersburg or von Neumann's 1920s Budapest. According to classmate Willis Ware, "Pres Eckert was the kid who always had everything. . . . His father was very wealthy—real estate—in Philadelphia, so Pres always had the biggest and bestest and the finest." In Eckert's opinion, many of the scientists who left academia for the computer industry, with no prior experience of American business life, "were seeing the world the way they thought it ought to be and not the way it really was."[17]

Mauchly and Eckert began by attempting to improve the precision of the Moore School's differential analyzer, substituting electronic circuits for electromechanical linkages whose accuracy deteriorated unpredictably, especially on problems with many steps. In August 1942, Mauchly formalized their convictions that an all-digital, all-electronic computer (which Mauchly spelled *computor,* to distinguish the proposed machine from the human computers that already populated the Moore School) was the way to move ahead.[18]

The official proposal for an "Electronic Diff. Analyzer" was submitted on April 2, 1943. The intentionally ambiguous "Diff." signified a transition from the analog "differential" to the digital "difference"—with implications that have reverberated ever since. "Mauchly, familiar with Geiger counters in physics laboratories, had realized that if electronic circuits could count, then they could do arithmetic and hence solve, inter alia, difference equations—at almost incredible speeds!" explains Nicholas Metropolis, who brought the initial "Los Alamos problem" to run on the new machine. Eckert, the lab instructor for Mauchly's training course, became the project's chief engineer.[19]

Facing wartime deadlines and a series of six-month funding horizons, workers ranging from "factory girls to moonlighting telephone linemen" were enlisted to construct the 30-ton machine, whose modular design would facilitate moving it to the Proving Ground when complete.[20] Twenty separate, communicating processors (or "accumulators") were assisted by a multiplier and combined divider/

square-rooter, with input and output via IBM punched card machines. Programming was locally distributed among the individual processors, coordinated by a "master programmer" with a storage capacity of 60 bits.

Ten two-state flip-flops (two vacuum tubes toggled together so that one tube or the other was always in an energized, conducting state) were formed into ten-stage ring counters representing each decimal digit in the ten-digit accumulators, forming, in effect, the electronic equivalent of a Marchant adding machine, but running at 300,000 rpm. An initiating and cycling unit provided a 5-kilocycle central clock, while a constant transmitter translated punched card data, buffered by banks of relays, into electrical signals that were intelligible to the machine. Three resistor-matrix function tables, whose design was contributed by Jan Rajchman at RCA, stored 104 12-digit numbers each.

The ENIAC, incorporating 17,468 vacuum tubes and 1,500 relays, consumed 174 kilowatts of power and occupied a 33-by-55-foot room. There were 500,000 hand-soldered joints. "The ENIAC itself, strangely, was a very personal computer," remembers mathematician Harry Reed, who arrived at Aberdeen in 1950. "Now we think of a personal computer as one which you carry around with you. The ENIAC was actually one that you kind of lived inside." The army had initially offered the contract for its construction to RCA, but according to Rajchman, "Zworykin . . . estimated it would take about 20,000 tubes . . . and that the mean free path, the time between failure, would therefore be ten minutes or so. . . . He didn't want to be involved in something as massive and unreliable as that." RCA declined the contract but freely contributed their expertise. "We were asked to tell everything we knew to the Moore School," remembers Rajchman. "It was all an atmosphere, of course, of great fervor for the war, and nobody worried about patents or priorities."[21]

In the opinion of Minneapolis district judge Earle R. Larson, who presided over six years and 34,426 exhibits submitted as evidence in the *Honeywell Inc. v. Sperry Rand Corp.* patent dispute, key elements of the ENIAC (whose patent rights, originally claimed by Eckert and Mauchly, had descended to Sperry Rand) were anticipated by John Vincent Atanasoff of Ames, Iowa, who had demonstrated an electronic digital computer to Mauchly in June of 1941. Atanasoff's computer had an all-electronic central processor, with a memory consisting of 3,000 individual capacitors distributed in 30-bit tracks on the surface of two rotating drums. Mauchly went to the grave claiming he had learned little from Atanasoff, while Atanasoff abandoned his own project when enlisted to help in the wartime effort at anti-aircraft fire control. When the war was over, Atanasoff launched a computer project under the sponsorship of the Naval Ordnance Laboratory, but when von Neumann's project took the lead his funding was withdrawn.

It was no accident, given the mission to speed up the production of firing tables, that the ENIAC had the computational architecture of a roomful of twenty human computers working with ten-place desk calculators and passing results back and forth. The accumulators operated in parallel, similar to the multiple-core processors of today. "The ENIAC had some very modern features—it's just that we did not then use the modern terminology to describe them," Eckert explains. Converted to serial, stored-program control in 1947, and upgraded with 100 words of magnetic-core memory in 1953, the ENIAC logged a total of 80,223 hours of operation until it was shut down for the last time on October 2, 1955, at 11:45 p.m.[22]

"The ENIAC" was "an absolutely pioneer venture, the first complete automatic, all-purpose digital electronic computer," according to von Neumann's assessment in 1945, although he warned Nicholas Metropolis and other early programmers to "watch it like a hawk, and trust it only as far as one can throw it." The project was revolutionary not for the technology used to build it, but for its scale. "All electronic tubes, resistors, and diodes used in the ENIAC were joint Army-Navy rejects," explains Metropolis, "so that in principle the ENIAC could have been built before the war."[23]

The ENIAC proposal was not an easy sell. "When as a matter of precaution the proposal was sent to several well-known persons for review, the resulting recommendations were somewhat uniformly negative," J. Grist Brainerd, who supervised the ENIAC contract for the Moore School, recalls. It was Oswald Veblen, chairman of the scientific committee of the Ballistics Research Laboratory at Aberdeen, who gave the order to move ahead. On April 9, 1943, Herman Goldstine gave a briefing to the director of the laboratory, Colonel Leslie E. Simon, "at which Veblen after listening for a short while to my presentation and teetering on the back legs of his chair brought the chair down with a crash, arose, and said 'Simon, give Goldstine the money.' He thereupon left the room and the meeting ended on this happy note."[24]

On June 5, 1943, a six-month contract for "research and development of an electronic numerical integrator and computer" in the amount of $61,700 was signed. Colonel Gillon took on the job of "fighting off the competition of apparently more urgent but actually less important conflicting projects through the war." After a visit on October 18, 1945, Samuel H. Caldwell of Harvard's Computation Laboratory reported to Warren Weaver that "the boys were engulfed in engineering difficulties. That machine contains more assumptions concerning the reliability of cheap, mass-production, radio parts than anything I have laid eyes on, and I wouldn't give you a nickel for its chances of successful operation until after extensive rebuilding and replacement of poor parts."[25] His report was dated January 16, 1946—at which point the ENIAC had already been running the first Los Alamos hydrogen bomb problem for over a month.

Von Neumann first visited the ENIAC in either August (according to Goldstine) or September (according to Eckert and Mauchly) of 1944. "That moment," says Goldstine, "changed his life for the rest of his days." Goldstine was on his way back to the Moore School from a meeting at the Aberdeen Proving Ground when, as he remembers, "I saw Professor von Neumann standing on the railroad platform, all alone, and decided that I would go and talk to this famous man . . . but he was totally uninterested. Then, gradually we began to talk about more things. Pretty soon he learned that we were building a machine that would do 300 multiplications a second, and suddenly he changed."[26]

Von Neumann, a member of the Ballistic Research Laboratory's Scientific Advisory Board, was cleared to examine the ENIAC, and shown the first two accumulators undergoing initial tests: computing the solution to a differential equation, and exchanging coded pulses at a 5-kilocycle pace. "If he had first asked questions like 'How fast does it work?' we would have been disappointed," says Eckert. "Because he asked about the control logic, there was an immediate rapport."[27]

The ENIAC was aimed at the backlog of firing tables, but Eckert, Mauchly, Goldstine, and then-twenty-eight-year-old Arthur Burks (a logician and philosopher turned electronic engineer for the duration of the war) had begun thinking of other applications from the start. "A sufficiently approximate solution of many differential equations can be had simply by solving an associated difference equation," Mauchly had written in August of 1942. This was true whether calculating firing tables, predicting weather, or solving the implosion problems that would soon consume the Los Alamos computing group. "The entire economy of computing changed overnight," Goldstine explained. "Instead of being in a world of expensive multiplication and cheap storage, we were thrown into one in which the former was very cheap and the latter very expensive. Virtually all the algorithms that humans had devised for carrying out calculations needed reexamination."[28]

The ENIAC was programmed by setting banks of 10-position switches and connecting thousands of cables by hand. Hours, sometimes days, were required to execute a programming change. "Programming steps were very expensive to come by," says Eckert. "It took boxes and cables and things. Doing something a second time, or reiterating something—we were 100,000 times faster than a human being—was very cheap."[29]

Data and instructions were intermingled within the machine. "A pulse, no matter for what use, had the same physical definition in almost all situations throughout the ENIAC," Mauchly explains. "Some pulses were used to control operations and others to signify data . . . but a pulse representing an algebraic sign for some data, or one representing a digit value, could be fed into a control circuit and expected to function just as any control pulse might function." As to when the concept of stored programming first took form, Mauchly adds that "there were less than 700 bits of high-speed storage in the ENIAC," while in the master programmer and individual program counters "about 150 bits of fast electronic storage were available for 'program control.'"[30] More than 20 percent of the ENIAC's original high-speed storage was used for storing program information that could be modified while computations were under way.

The ENIAC was limited by storage, not by speed. "Imagine that you take 20 people, lock them up in a room for 3 years, provide them with 20 desk multipliers, and institute this rule: during the entire proceedings all of them together may never have more than one page written full," Von Neumann observed. "They can erase any amount, put it back again, but they are entitled at any given time only to one page. It's clear where the bottleneck . . . lies."[31]

Punched cards could be used to store intermediate results, but the process was error-prone and slow. The shakedown run on the ENIAC—a hydrogen bomb calculation that Stan Frankel and Nick Metropolis brought from Chicago and Los Alamos in December of 1945—consumed nearly one million cards, most of them for temporary storage of intermediate results. "I can remember Metropolis and

Frankel coming and explaining to us their task," says Arthur Burks. "They made it very clear that they couldn't tell us what the equations were."[32]

"The Los Alamos calculations which commenced December 10, 1945, were . . . the first time that the machine as a whole was being used . . . [and] employed 99 percent of the capacity of the ENIAC machine," concluded Judge Larson in his findings of fact in the *Honeywell v. Sperry Rand* patent dispute. The calculations continued for well over a month, into January 1946. "The difficulties encountered were not with the machine," testified Presper Eckert, singling out mathematics rather than physics, "but with the mathematical nature of the problem and mistakes of the mathematicians who had designed the problem for the machine."[33]

"I have often been asked, 'How big was the ENIAC storage?'" says Mauchly. "The answer is, infinite. The punched-card output was not fast, but it was as big as you wished. Every card punched out could be read into the input again, and indeed that is how Metropolis and Frankel managed to handle cycle after cycle of the big problem from Los Alamos." The difficulty, as Mauchly described it, was that "the fast memory was not cheap and the cheap memory was not fast."[34] A vacuum-tube flip-flop had a response time in the order of a microsecond, whereas it took on the order of one second to read or write an IBM card. A gap of six orders of magnitude lay in between.

Eckert proposed a way to bridge this gap with memory that was both fast *and* cheap. "Mr. J. P. Eckert, Jr., then of the Moore School, conceived the idea that the acoustic tank, then used for Moving Target Indicator equipment, could easily be made the basis for a dynamic form of memory," von Neumann and Goldstine later explained. "The use of such a device would enable one to store 1,000 binary digits at the cost of 5 to 10 vacuum tubes as compared to 1000 flip-flops in the ENIAC scheme."[35] Acoustic delay lines, developed at MIT's wartime Radiation Laboratory, took advantage of how slowly sound waves travel through liquid compared to radar waves traveling at the speed of light. An incoming radar signal was converted into an acoustic signal through a crystal transducer at one end of a tube filled with liquid—mercury being ideal—and when the wave train reached the far end of the tube it was converted back into an electrical signal by a second transducer, delayed but otherwise unchanged. By inverting the delayed signal, and synchronizing it with the next radar echo that returned, it was possible to subtract background clutter to distinguish objects (such as enemy aircraft) that had moved between one sweep of the radar beam and the next.

One thousand pulses, about a microsecond apart, could be stored in the millisecond it took an acoustic signal to travel the length of a 5-foot "tank." By regenerating the pulse train, and listening to the data stream as it passed by, it was possible to read and write data with millisecond access times. "After the central control organ listens to all 32 words of one column it passes to the next column," von Neumann explained to Warren Weaver in 1945, referring to 30-bit code segments for the first time as "words." Acoustic delay-line memory was used in many first-generation stored-program computers, although, as British topologist Max Newman complained, "its programming was like catching mice just as they were entering a hole in the wall."[36]

At the time von Neumann began collaborating with the ENIAC group, planning for ENIAC's delay-line successor was already under way. The Electronic Discrete Variable Automatic Computer, or EDVAC, "would be quite flexible in its control facilities, would have about 50 times as large a memory, i.e., be able to store about 1000 ten decimal digit numbers, and contain only about $\frac{1}{10}$ as many tubes," Goldstine and von Neumann reported.[37] The machine would be programmed by loading coded sequences into high-speed memory rather than by setting cables and switches by hand.

"The idea of the stored program, as we know it now, and which is a clear-cut way of achieving a universal computer, wasn't invented overnight," explains Rajchman. "Rather it evolved gradually. First came manually changeable plug-ins, relays, and finally the modifying contacts themselves became electronic switches. Next came the idea of storing the state of those switches in an electronic memory. Finally this resulted in the idea of the modern stored program in which 'instructions' and 'data' are stored in a common memory."[38]

Even before the advent of the ENIAC, the elements of stored program computing were falling into place. In July of 1944, on behalf of the implosion effort at Los Alamos, von Neumann and Stan Frankel were briefed on the series of Bell Telephone relay computers being built by Samuel B. Williams and George R. Stibitz in New York. The new machines were controlled by punched paper tape, and von Neumann reported to Oppenheimer on August 1 that the "problem tape carries numerical data, and operational instructions." As he described it to Oppenheimer, "an instruction on the control-tape therefore looks like this: 'Take the contents of register a, also the contents of register b, add (or subtract, or multiply, etc.), and put the result into register c.'" Not only were data and instructions intermingled, but also the computer could, in principle, modify its own instructions. "The machine can use in (a) a tape which comes from its own reperforator—i.e., which it has punched itself."[39]

Eckert and Mauchly were thinking along similar lines. "All through 1944, and in 1945 as well, we were leading a 'double life,'" Mauchly recalls. "For much of two shifts, 8 AM to Midnight, both ENIAC construction and testing needed supervision. Then as hourly workers went home and project engineers 'thinned out,' Eckert and I were left time to consider that 'next machine.' Naturally, 'architecture' or 'logical organization' was the first thing to attend to. Eckert and I spent a great deal of thought on that, combining a serial delay line storage with the idea of a single storage for data and program."[40]

In the closing months of World War II, von Neumann circulated between Princeton, Los Alamos, Washington, Philadelphia, and Aberdeen, conveying a stream of new ideas. "None of us was important enough to have persuaded people to accept this kind of thing," says Goldstine. "In the first place, von Neumann had a real built-in need at Los Alamos. . . . They had an enormous IBM punched card installation out there doing implosion calculations. I just don't believe any of us could have gone and persuaded somebody like Fermi of the importance of numerical calculation the way von Neumann could."[41]

In early 1945, during the final push to finish and test the atomic bomb, von Neumann's notes on the EDVAC project were typed up under Goldstine's supervision and distilled into a 105-page report. The "First Draft of a Report on the EDVAC," reproduced by mimeograph and released into limited distribution by the Moore School on June 30, 1945, outlined the design of a high-speed stored-program electronic digital computer, including the requisite formulation and interpretation of coded instructions—"which must be given to the device in absolutely exhaustive detail."[42]

The functional elements of the computer were separated into a hierarchical memory, a control organ, a central arithmetic unit, and input/output channels, making distinctions still known as the "von Neumann architecture" today. A fast internal memory, coupled to a larger secondary memory, and linked in turn to an unlimited supply of punched cards or paper tape, gave the unbounded storage that Turing had prescribed. The impediment of a single channel between memory and processor is memorialized as the "von Neumann bottleneck," although its namesake attempted, unsuccessfully, to nip this in the bud. "The whole system will be well balanced, so, that if it is properly and intelligently used, there will be no bottlenecks," he explained to Max Newman, "even not at the outputs and inputs of the human intellects with which it has to be matched."[43]

When a subject captured von Neumann's attention, he reconstituted it on his own terms from the bottom up. Digital computing required no such process of reduction; it was all axioms from the start. In 1945 the ENIAC and EDVAC were still classified military projects. Von Neumann could speak freely in logical abstractions, but not in specific electronic circuits. So he did. He was also, as Julian Bigelow put it, "clever enough to know that his forte was not in experimental work or in making things function in the real world."[44]

During the war, both open publication and individual credit had been suspended—for both computers and bombs. After the war, it was decided that bombs would be kept secret and computers would be made public, with a scramble for credit as a result. The EDVAC report engendered widespread controversy, despite the small number of copies that were distributed before the mimeograph stencils gave out. Von Neumann was listed as the sole author, without any acknowledgment of the contributions made by other members of the EDVAC group. Eckert and Mauchly, who had been pledged to silence about the ENIAC and EDVAC, felt slighted by a publication that was based on their own unpublished work. "Johnny was rephrasing our logic, but it was still the SAME logic," says Mauchly.[45] Adding injury to insult, the EDVAC report would be deemed to constitute a legal publication invalidating any patents not filed within a year.

"It wasn't even a draft when he wrote it," explains Eckert. "He wrote these letters to Goldstine, and when we asked what he was doing this for at the time, Goldstine said, 'He's just trying to get these things clear in his own mind and he's done it by writing me letters so that we can write back if he hasn't understood it properly.'" The report, compiled by Goldstine and accompanied by rough hand-stenciled sketches, contained blank spaces where references were to be inserted. The word EDVAC never appears in the text of the report. "He grasped what we were doing quite quickly," adds Eckert. "I didn't know he was going to go out and more or less claim it as his own."[46]

"I certainly intend to do my part to keep as much of this field 'in the public domain' (from the patent point of view) as I can," von Neumann explained to Stan Frankel, arguing for an open systems approach at IAS.[47] "The primary purpose of this report was to contribute to clarifying and coordinating the thinking of the group working on the EDVAC," he testified in 1947, when questions as to the disposition of patent rights first arose. The secondary purpose was to publish the preliminary results as soon as possible, "in order to further the development of the art of building high speed computers," he explained, concluding that "my personal opinion was at all times, and is now, that this was perfectly proper and in the best interests of the United States."[48]

With the war over, individual interests eclipsed the interests of the United States. The Moore School was too academic for Eckert and Mauchly, and not academic enough for von Neumann. Eckert and Mauchly left to form the Electronic Control Company and build commercial computers—first BINAC and then UNIVAC, a brand synonymous with computing for a time. Von Neumann decided to go build his own computer, as a scientific instrument, somewhere else. Spare time on the ENIAC and even the EDVAC would not be enough. "It was, therefore, the most natural thing that von Neumann felt that he would like to have at his own disposal such a machine," says Willis Ware. "If he really wanted a computer, the thing to do was to build it," adds Arthur Burks.[49]

Von Neumann's initial thinking was to transplant the entire core of the ENIAC group. "At the end of the war, he had a whole new set of ideas that were not incorporated in the ENIAC," Ware explains. "I can imagine Johnny thinking to himself, 'Well, here's myself and Herman and Eckert and Mauchly, and Burks. What a team to go do this thing that I want to do!' "[50]

Eckert declined von Neumann's invitation to lead the IAS engineering team, going into business with Mauchly for himself, while von Neumann entered into a series of lucrative personal consulting contracts with IBM. "Von Neumann agrees to assign to IBM, with the exception of the inventions specified below, the entire rights to any and all improvements and inventions made by him," reads the draft of a retainer agreement with IBM, dated May 1, 1945. As Eckert later complained, "he sold all our ideas through the back door to IBM."[51]

Von Neumann's customary good nature appeared to be breaking down. "Eckert and Mauchly are a commercial group with a commercial patent policy," he explained to Stanley Frankel, who remained on good terms with Eckert and Mauchly ever since the weeks of all-night troubleshooting in late 1945 and early 1946. "We cannot work with them directly or indirectly, in the same open manner in which we would work with an academic group," he warned. "If you wish to maintain the same type of close contact with the Eckert-Mauchly group—which is for you and you alone to decide—then you should not put yourself into an incompatible position by communicating with us too."[52]

At the Moore School, the EDVAC was orphaned, superseded by the Eckert-Mauchly enterprise on the one hand and the Institute for Advanced Study project on the other. By the time the EDVAC was finished, in 1951, its mercury delay-line memory and serial architecture had been overtaken by advances sparked by its own draft report. To achieve a "practicable" and fully random-access memory, it was explained near the end of the EDVAC report, the Farnsworth-Zworykin iconoscope, rather than the Eckert delay line, might be the "more natural" approach. "This device in its developed form remembers the state of $400 \times 500 = 200,000$ separate points," it was noted. "These memories are placed on it by a light beam, and subsequently sensed by an electron beam, but it is easy to see that small changes would make it possible to do the placing of the memories by an electron beam also." The iconoscope "acts in this case like 200,000 independent memory units," with the switching between individual capacitors being performed "by a single electron beam—the switching action proper being the steering (deflecting) of this beam so as to hit the desired point on the plate."[53]

Sixty years later, most primary computer memory is embodied, in silicon, as dynamically refreshed arrays of capacitors—the current implementation of Farnsworth's, Zworykin's, and Rajchman's original translation between coded sequences in time and arrays of charge in space. Ten million capacitors now cost less than one cent. Memory locations are addressed directly by digital switching rather than indirectly by the deflection of an electron beam, but the underlying principle and logical architecture remain unchanged. Our ever-expanding digital universe is directly descended from the image tube that imploded in the back seat of Zworykin's car.

Where to build the new computer? The Institute did not even have a workbench where you could plug in a soldering gun. "One could hardly imagine a more improbable environment," Julian Bigelow adds. "How does all of this fit in with the Princetitute?" asked Norbert Wiener, in March of 1945. "You are going to run into a situation where you will need a lab at your fingertips, and labs don't grow in ivory towers."[54] Wiener helped arrange for an invitation, at department-head level, to MIT, with the assurance that all the resources of MIT would be at von Neumann's disposal to build the envisioned computing machine.

Competing offers were made by Harvard, the University of Chicago, and IBM. "We are all very much interested in your man, von Neumann," Harvard president James Conant wrote to Frank Aydelotte. "The question is, could we get him." Von Neumann played the invitations against one another—and against those who resisted the construction of a computer at the Institute—until he got his way. "The question of how to hold on to John von Neumann is growing more urgent from day to day. . . . It would be a tragedy if we lost him," James Alexander warned Frank Aydelotte. "I doubt whether he would be willing to stay with us if this meant giving up entirely the work on high speed mathematical machines."[55]

Loaning von Neumann to Los Alamos during the wartime emergency was one thing, but losing him to a rival institution would be a serious blow to the IAS. Von Neumann was quick to take advantage of this. "These negotiations," in Bigelow's opinion, "were a part of a bargaining way of life which von Neumann, so to speak, managed with his left little finger, while the other fingers on his hands were doing more effective and important work." Aydelotte countered the rival offers, reassuring Alexander that "you may feel free to say to von Neumann that I have every confidence in finding the funds from one source or another to enable him to carry out his plans." There was little time. Not only was von Neumann impatient to get started on the computer, but Aydelotte was about to leave for Palestine, as a member of the Joint Anglo-American Commission that concluded, unanimously, in April 1946, that "Palestine must ultimately become a state which guards the rights and interests of Moslems, Jews and Christians alike."[56]

At the Institute for Advanced Study in early 1946, even applied mathematics was out of bounds. Mathematicians who had worked on applications during the war were expected to leave them behind. Von Neumann, however, was hooked. "When the war was over, and scientists were migrating back to their respective Universities or research institutions, Johnny returned to the Institute in Princeton," Klári recalls. "There he clearly stunned, or even horrified, some of his mathematical colleagues of the most erudite abstraction, by openly professing his great interest in other mathematical tools than the blackboard and chalk or pencil and paper. His proposal to build an electronic computing machine under the sacred dome of the Institute, was not received with applause to say the least." It wasn't just the pure mathematicians who were disturbed by the prospect of the computer. The humanists had been holding their ground against the mathematicians as best they could, and von Neumann's project, set to triple the budget of the School of Mathematics, was suspect on that count alone. "Mathematicians in our wing? Over my dead body! and yours?" Aydelotte was cabled by paleographer Elias Lowe.[57]

Aydelotte, however, was ready to do anything to retain von Neumann, and supported the Institute's taking an active role in experimental research. The scientists who had been sequestered at Los Alamos during the war, with an unlimited research budget and no teaching obligations, were now returning in large numbers to their positions on the East Coast. A consortium of thirteen institutions petitioned General Leslie Groves, former commander of the Manhattan Project, to establish a new nuclear research laboratory that would be the Los Alamos of the East. Aydelotte supported the proposal and even suggested building the new laboratory in the Institute Woods. "We would have an ideal location for it and I could hardly think of any place in the east that would be more convenient," Aydelotte, en route to Palestine, cabled von Neumann from on board the *Queen Elizabeth*. At a meeting of the School of Mathematics called to discuss the proposal, the strongest dissenting voice was Albert Einstein, who, the minutes record, "emphasizes the dangers of secret war work" and "fears the emphasis on such projects will further ideas of 'preventive' wars."[58] Aydelotte and von Neumann hoped the computer project would get the Institute's foot in the door for lucrative government contract work—just what Einstein feared.

Aydelotte pressed for a proposed budget, and von Neumann answered "about $100,000 per year for three years for the construction of an all-purpose, automatic, electronic computing machine." He argued that "it is most important that a purely scientific organization should undertake such a project," since the government laboratories were only building devices for "definite, often very specialized purposes," and "any industrial company, on the other hand, which undertakes such a venture would be influenced by its own past procedures and routines, and it would therefore not be able to make as fresh a start."[59]

Aydelotte first sought funding from philanthropist Samuel Fels, emphasizing the "contributions to mathematics, physics, biology, economics and statistics which might be made by an electronic computer," and promising that the new device would open up new areas of knowledge "in the same remarkable way that the two hundred inch telescope promises to bring under observation universes which are at the present moment entirely outside the range of any instrument now existing."[60] Despite being offered a private audience with Einstein, Fels declined to lend support.

Aydelotte then approached the Rockefeller Foundation's Warren Weaver, who was familiar with the other laboratories working on computers, and in a unique position to evaluate the proposal from IAS. "I am somewhat surprised that von Neumann is himself interested in the actual construction and operation of a great new computing engine," Weaver answered on October 1, 1945. "The device we are all dreaming about is something very much more than a computing device. . . . It is a device wherewith one carries out, accurately and rapidly, certain electrical and mechanical processes which are isomorphic with certain important mathematical processes." Weaver needed no persuasion as to the importance of von Neumann's project, but he explained to Aydelotte that "I start out with the idea that the Institute is not a natural physical setting for such a development. I would, however, love to have you change my mind."[61]

Von Neumann did his best. "I propose to store everything that has to be remembered by the machine, in these memory organs," he explained to Weaver in an eleven-page letter in early November 1945. "This includes the . . . numerical information which defines the problem, including . . . intermediate results, produced by the machine while it works . . . [and] the coded, logical instructions which define the problem and control the functioning of the machine." He described how "a very simple code of orders is adequate to handle everything" and "can be used to route the central control through subroutines, which may be organized into hierarchies of any desired structure." This intermingling of data and instructions, he noted, "permits modification of orders in dependence upon numerical results of calculations which are carried out in the course of the process." Finally, he explained how coded instructions, stored in internal memory, "endow the machine with 'virtual organs,' i.e. they make it behave as if it possessed certain organs which do actually not exist in the physical sense."[62]

Weaver lent his personal influence and support, but hesitated to commit the Rockefeller Foundation to financing a venture in direct partnership with both the military and RCA. He decided to wait and see how "this rather novel combination of Institute-University-industrial laboratory-Army-Navy" played out before adding Rockefeller to the mix. During Aydelotte's mission to Palestine, a letter from Weaver asking for an update on the computer project was intercepted by Marston Morse, to whom Aydelotte had delegated responsibilities while away. "I want to say to you confidentially that it would be a great service to the Institute if you would continue your insistence on a budget for the new project with details for the future," answered Morse. "A few underestimates and the whole character of the Institute might be changed to follow through. The bigger it gets the more ambiguous it is."[63] Morse's fear was not that the computer project would be a failure but that it would be too much of a success.

Others, asked by Weaver to review the von Neumann proposal, were less kind. "Von Neumann shows some tendency to regard the problem as one which begins in the scientific stratosphere and works its way down, instead of one which begins on the ground and works up," answered Samuel Caldwell, of the Harvard Computation Laboratory. "The relay computer contains '5000 to 15000 relays each.' So what? Does von Neumann think the electronic machine will not contain thousands of something?"[64]

Von Neumann found an unwavering ally in Lewis Strauss, the IAS trustee, merchant prince, and rear admiral who wielded power over the Office of Naval Research. Strauss saw the merit in von Neumann's no-strings-attached approach. After what Los Alamos had accomplished with desk calculators, what might be next? "If we devote in this manner several years to experimentation with such a machine, without a need for immediate applications, we shall be much better off at the end of that period in every respect, including the applications," argued von Neumann. "The importance of accelerating approximating and computing mathematics by factors like 10,000 or more, lies not only in that one might thereby do in 10,000 times less time problems which one is now doing, or say 100 times more of them in 100 times less time—but rather in that one will be able to handle problems which are considered completely unassailable at present."[65]

Strauss took the lure. "The projected device, or rather the species of devices of which it is to be the first representative, is so radically new that many of its uses will become clear only after it has been put into operation," von Neumann assured him. "These uses which are not, or not easily, predictable now, are likely to be the most important ones. Indeed they are by definition those which we do not recognize at present because they are farthest removed from . . . our present sphere."[66]

Strauss, who was preparing to leave the navy, promised not only to make sure the computer project was funded before his departure, but, as reported by Marston Morse on Christmas Eve 1945, "to obtain a Quonset hut without cost." The funding was delivered, as promised, but not the Quonset hut. The paperwork was minimal, and a one-page budget was sufficient to secure the needed funds. When the navy raised questions over title to the computer and associated patent rights, the contract was shifted to the army instead. "Professor von Neumann and I believe," Goldstine wrote in 1951, "that the Institute has an almost unique contract with the Ordnance Department in that the Government has, in fact, given us a grant to build a machine for ourselves."[67]

"We would have, once a year, kind of a pass-the-hat session in which we would sit up in the board room in the Institute with these representatives of all these government agencies," remembers James Pomerene, "and they would say, 'Well, I can put in $10,000' and another guy would say 'I can put in $20,000.' And one would say 'How about you, Joe? You're good for $30,000, aren't you?' We would have our $200,000 put together, and it all worked fine."[68]

Half the first contingent of engineers, as well as the name MANIAC, were imported from the Moore School. "Originally we called ENIAC the 'MANIAC' when it didn't work right," J. Presper Eckert remembers. "And later they borrowed that name."[69] The IAS project combined the practical experience derived from the ENIAC with the theoretical possibilities of Turing's Universal Machine. There was regular contact between the IAS group and their British counterparts, although the British were constrained by the Official Secrets Act, which prevented them from confirming the existence of the code-breaking computers that had been built during the war.

"Von Neumann was well aware of the fundamental importance of Turing's paper of 1936 'On computable numbers . . .' which describes in principle the 'Universal Computer' of which every modern computer (perhaps not ENIAC as first completed but certainly all later ones) is a realization," Stanley Frankel explains. "Von Neumann introduced me to that paper and at his urging I studied it with care. . . . He firmly emphasized to me, and to others I am sure, that the fundamental conception is owing to Turing."[70] Von Neumann knew that the real challenge would be not building the computer, but asking the right questions, in language intelligible to the machine. For this, if not for its machine shops and laboratories, the Institute for Advanced Study, Oppenheimer's "intellectual hotel," was ideal.

"Johnny had by then a very definite idea of how and why he wanted this machine to function with the emphasis on the why," remembers Klári. "He wanted to build a fast, electronic, completely automatic all purpose computing machine which could answer as many questions as there were people who could think of asking them."[71]

SIX

# Fuld 219

*We have been trying to see how far it is pos-*
*sible to eliminate intuition, and leave only*
*ingenuity. We do not mind how much inge-*
*nuity is required, and therefore assume it to*
*be available in unlimited supply.*

—Alan Turing, 1939

"THE PROSPECT of a visit from an architect usually costs Professor Veblen a day's work and a night's sleep," noted Abraham Flexner, when the construction of a headquarters for the Institute for Advanced Study was first announced. Veblen and Flexner had been at odds, from the beginning, over the question of building buildings and buying land. "The way to reform higher education in the United States is to pay generous salaries and then use any sort of makeshift in the way of buildings," Flexner had argued when the Institute first opened for business—with Veblen, Einstein, Alexander, von Neumann, and little else.[1]

The Institute operated out of an assortment of temporary facilities for its first nine years. "Everybody was working somewhere else," Klári von Neumann observed upon her arrival in Princeton in 1938. "Flexner had his office in one of the buildings along Nassau Street; the mathematicians had rooms in Fine Hall, which was the University's mathematical building; the economists had some kind of an office in the basement of the Princeton Inn; and the few archeologists who were members, essentially worked in their own homes when in Princeton and then went out 'on location' to dig."[2]

When Flexner conceded to the construction of Fuld Hall, he registered one last complaint, to Veblen, that "I would far rather rent additional floor space in 20 Nassau Street and get our minds so full of the purposes for which we exist that we will all become relatively indifferent to buildings and grounds." He then warned Aydelotte that

88

"I am still a little uneasy in mind about Veblen, for I think he has a hankering for rooms that are unduly large."[3]

Fuld Hall, named after Louis Bamberger's sister, Carrie, and her husband Felix Fuld, was constructed in 1939. The architect was Jens Frederick Larson, who had made a name for himself designing additions to university campuses—including Swarthmore and Dartmouth—and was drawn to the challenge of designing a new institution from the ground up. Born in Boston in 1891, Larson had enlisted in the First Canadian Overseas Contingent in 1915, heading to France as a member of the infantry and working his way up to lieutenant in the artillery until he signed up for flight training, transfixed by the air battles overhead. In 1917 he became a founding member of the 84 Squadron of the Royal Flying Corps, piloting a Royal Aircraft Factory SE-5A, an experimental design that not only outperformed the better known Sopwith Camel but was easier to fly.

According to British records, Larson, known as "the Swede," scored at least eight aerial victories between November 1917 and April 1918. According to Canadian records, the score was nine. On April 3, 1918, he downed two opponents on a single day, having met "two formations of Pfalz and V-Strutters in the clouds at 7,000 feet."[4] He led his four companions up into the clouds and then dove back down upon the Germans, who were unable to escape. After scoring one more kill, on April 6, he retired from combat and became a flight instructor in England, before returning to the United States.

The earliest floor plan of Fuld Hall, with a central common room and offices in the wings (accompanied by notes suggesting that one wing be allocated to women) was sketched in pencil on the back of a menu from the City Mid-Day Club, 25 Broad Street, New York City, for Thursday, October 21, 1937. (The lunch choices offered to the Committee on Buildings and Grounds included Blue Point oysters on the half shell for forty cents, or Cape Cods for forty-five cents.) The result, two years later, was an imposing redbrick Georgian edifice with white trim and a copper roof, its lateral symmetry culminating in a clock tower that dominated the otherwise nondescript landscape of Olden Farm. According to Robert Oppenheimer, two young boys were once overheard in conversation on the private road, now named Einstein Drive, in front of Fuld Hall:

"What's that? Is it a church?"

"It's the Institute."

"What's the Institute?"

"It's a place to eat."

In addition to the dining room on the top floor, Fuld Hall housed the Institute's administrative and faculty offices, and, in the center of the ground floor, a common room with a large fireplace surrounded by leather-bound armchairs and presided over by a grandfather clock from the Bambergers' South Orange estate. A chessboard (and later a Go board, favored by Oppenheimer's young particle physicists) sat near the windows overlooking the Institute Woods. Fresh newspapers, including the Air Edition of the *Times* of London, were skewered every morning on a polished wooden rack. Afternoon tea—a ritual introduced at Fine Hall by Oswald Veblen, who, according to Herman Goldstine, "tried awfully hard to be an Englishman"—was served on real china daily at exactly three o'clock. According to Oppenheimer, "tea is where we explain to each other what we do not understand."

Surrounded by woodlands, open fields, and private drives, Fuld Hall resembled a private sanatorium or a large European country estate. The facilities were maintained by a dedicated staff of maids, groundskeepers, and janitors, including the extended Rockafellow family, who occupied one of the former farmworkers' houses at the end of Olden Lane. Many of the staff remained with the Institute for life. On February 16, 1946, Mrs. Alice Rockafellow, described as "reliable and dependable in every emergency of which there are many in the cafeteria," was given a raise from seventy to eighty dollars per month. "Her salary is less than other maids as the Rockafellows get very low rent," the faculty minutes report.[5] In one of the far corners of the Woods, beyond a stand of primeval beeches, a family of subsistence farmers who had been there since the land was acquired by the Institute were allowed to remain. During the war, the Institute's fields were planted in alfalfa, rotated with corn and other crops. The Woods were declared a game reserve in 1945, although a limited bow hunting season for deer continued, and hunting platforms can still be seen in the trees if you look up.

Lunch and dinner were served in the fourth-floor dining room at below-market cost. The menu for October 14, 1946, included "Creamed Halibut with Eggs on Potatoes," for twenty-five cents, or "Fresh Boiled Salmon, Parsley Sauce, Potatoes," for fifty cents. Coffee was five cents. A note in the kitchen reminded staff of the "Einstein diet: No fat; no vegetable of cabbage family or bean; nothing ice cold." Einstein preferred his eggs boiled for four minutes, and a baked apple for dessert.

The cafeteria was managed by Alice Rockafellow. The menus (for the Einstein diet, too) were produced, on a manual typewriter, by one of Larson's fellow World War I–era aviators, Bernetta Miller, born in Canton, Ohio, in 1884 and the fifth woman to obtain a pilot's license in the United States. In 1912 she had demonstrated the new Blériot monoplane—being built under license by Moisant Aviation on Long Island—for the U.S. Army. "Of course, I had no illusions as to why I was sent to College Park to demonstrate the monoplane to the U.S. government officials who were exclusively devoted to the idea of the biplane," Miller later explained. "If a mere woman could learn to fly one, so surely could a man." She served as a volunteer on the ground in World War I, receiving the Croix de Guerre from the French government for "helping the injured in the advanced aid stations" in the Tours, Toul, and Argonne sectors. A letter of commendation from the commander of the U.S. Army's Eighty-second Division, dated January 13, 1919, cites how "under enemy fire, she visited the front lines, carrying a supply of cigarettes and other comforts to the men."[6] She was wounded at least once.

After the war, Miller became bursar of the American Girls School in Istanbul, and then returned to the United States, serving as bursar of St. Mary's Hall, a girl's school in Burlington, New Jersey, until signing on as Frank Aydelotte's personal assistant and bookkeeper in 1941. Her memoranda are strongly worded, with capital letters for emphasis, and signed with a bold, firm, hand. "No BREAD will be consumed on THURSDAYS, No FOOD STUFFS will be fried on Wednesdays, and No PIES OR CAKES will be served on MONDAYS OR FRIDAYS," she announced in May of 1946, when food conservation due to postwar shortages was put into effect.[7]

"I cannot urge too strongly the urgent necessity of the Computer people taking over ENTIRELY their accounting," she noted in a memorandum to the director on September 13, 1946. "It is swamping my office to the point where we cannot give even fair attention to Institute matters." Miller kept meticulous records of the tea service in Fuld Hall, reporting that during the six months of the 1941–1942 academic term, 9,605 servings were consumed, at a cost of 5.2 cents in tea, sugar, cookies, and labor each. She personally led the delegation of mothers who petitioned Oppenheimer and Aydelotte to start a nursery school at the Institute. "There are now 34 children on the project of whom 15 are of nursery school age. . . . The matter is urgent if the parents are to have reasonable quiet at home," she reported in September 1947, requesting permission to turn one of the visitor apartments into a school. "The confusion in a small apartment with children about is considerable as you know."[8] Crossroads Nursery School opened in 1947 and has been filled to capacity ever since. Mathematicians produce their best work at about the same time that they produce their children, and the nursery school helped keep the two apart.

Miller, who had no children of her own, "was obviously interested in women, not men," adds geneticist Joseph Felsenstein, whose maternal grandmother was Miller's first cousin and who was taken on family visits to Bernetta and her companion, Betty Faville, by then retired in New Hope, Pennsylvania, when he was growing up. At the Institute, he explains, "she became one of the people who had to stand between Albert Einstein and the world." She held Aydelotte in high regard, but disliked Oppenheimer, who fired her in 1948. "I think the man was a complete snake," she later claimed. "But I would *never* say he was disloyal."[9] Einstein gave her a personal letter of recommendation when she left.

Arriving at the main entrance to Fuld Hall in 1946—when the overcrowded offices were numbered differently than they are today—the telephone switchboard was on the left. From the ground floor, in the center, a half flight of stairs led down to the common room, with high French doors looking out onto the open field that extended as far as the old Princeton-Trenton trolley line marking the edge of the Institute Woods. The main column of George Washington's army, under General John Sullivan, had been traversing the edge of this field on the morning of January 3, 1777, when General Mercer's party engaged the British, just on the other side of where the Institute's Social Science Library and new dining hall now stand.

On the second and third floors of Fuld Hall, in the center, with tall ceilings, was the library, with adjacent reading and talking rooms. On the fourth and top floors were the dining room, kitchen, and board room, with a balcony and terrace overlooking the lower part of Olden Farm, toward Stony Brook. (It is from this balcony that a love-struck Meg Ryan and Tim Robbins, their paths crossed via Einstein, gaze up at the stars in the 1994 Walter Matthau film *I.Q.*) Offices extended into the wings on both sides of the common room and library, with the humanists (and the director) occupying the right side of the building, and the mathematicians (and Einstein) occupying the left.

At the time the computer project was launched in 1946, Veblen occupied room 124 on the ground floor, with bay windows looking out toward the end of Olden Lane. Einstein was directly above, in room 225. Von Neumann was in room 120 on the ground floor, adjacent to the common room, and flanked to the left by secretaries Betty Delsasso in 121 and Gwen Blake in 122. On the second floor, the right wing was occupied by economists Walter W. Stewart in 212, Winfield Riefler in 210, and Robert B. Warren in 213, with Judy Sachs, the librarian, in 215. Stewart, Riefler, and Warren, respectively of the Bank of England, the U.S. Federal Reserve, and the U.S. Treasury, jointly constituted the School of Economics and Politics whose establishment, without the approval of the other faculty, led to Flexner's resignation in 1939. In the left wing, adjacent to Weyl and Einstein, and above von Neumann, next to the library, was Kurt Gödel, in Fuld 217.

"Formal logic has to be taken over by mathematicians," Veblen had announced on New Year's Eve 1924, when the plans for what would become the Institute for Advanced Study were first taking form in his mind. "There does not exist an adequate logic at the present time, and unless the mathematicians create one, no one else is likely to do so."[10] It was Gödel, above anyone else—and now directly above von Neumann—who proved Veblen's instincts correct.

In 1924 both von Neumann and Gödel were working on the logical foundations of mathematics, before Gödel's incompleteness theorems brought the Hilbert program to a close. Von Neumann "believed in Hilbert's goal of a final and conclusive axiomatization of mathematics," according to Stan Ulam, "and yet, in a 1925 paper, in a mysterious flash of intuition, he pointed out the limits of any axiomatic formulation of set theory. That was perhaps a sort of vague forecast of Gödel's result."[11] The seeds of doubt were sown.

In September of 1930, at the Königsberg conference on the epistemology of the exact sciences, Gödel made the first, tentative announcement of his incompleteness results. Von Neumann immediately saw the implications, and, as he wrote to Gödel on November 30, 1930, "using the methods you employed so successfully . . . I achieved a result that seems to me to be remarkable, namely, I was able to show that the consistency of mathematics is unprovable," only to find out, by return mail, that Gödel had got there first.[12] "He was disappointed that he had not first discovered Gödel's undecidability theorems," explains Ulam. "He was more than capable of this, had he admitted to himself the possibility that Hilbert was wrong in his program. But it would have meant going against the prevailing thinking of the time."[13]

Von Neumann remained a vocal supporter of Gödel—whose results he recognized as applying to "all systems which permit a formalization"—and never worked on the foundations of mathematics again. "Gödel's achievement in modern logic is singular and monumental . . . a landmark which will remain visible far in space and time," he noted. "The result is remarkable in its quasi-paradoxical 'self-denial': It will never be possible to acquire with mathematical means the certainty that mathematics does not contain contradictions. . . . The subject of logic will never again be the same."[14]

Gödel set the stage for the digital revolution, not only by redefining the powers of formal systems—and lining things up for their physical embodiment by Alan Turing—but by steering von Neumann's interests from pure logic to applied. It was while attempting to extend Gödel's results to a more general solution of Hilbert's *Entscheidungsproblem*—the "decision problem" of whether provable statements can be distinguished from disprovable statements by strictly mechanical procedures in a finite amount of time—that Turing invented his Universal Machine. All the powers—and limits to those powers—that Gödel's theorems assigned to formal systems also applied to Turing's universal machine, including the version that von Neumann, from his office directly below Gödel's, was now attempting to build.

Gödel assigned all expressions within the language of the given formal system unique identity numbers—or numerical addresses—forcing them into correspondence with a numerical bureaucracy from which it was impossible to escape. The Gödel numbering is based on an alphabet of primes, with an explicit coding mechanism governing translation between compound expressions and their Gödel numbers—similar to, but without the ambiguity that characterizes the translations from nucleotides to amino acids upon which protein synthesis is based. This representation of all possible concepts by numerical codes seemed to be a purely theoretical construct in 1931.

"Metamathematical notions (propositions) thus become notions (propositions) about natural numbers or sequences of them; therefore they can (at least in part) be expressed by the symbols of the system . . . itself," wrote Gödel in the introduction to his proof.[15] Gödel constructed a formula, the Gödel sentence (G) saying, in effect, "The sentence with Gödel number g cannot be proved," where the details of the system are manipulated so that the Gödel number of G is g. G cannot be proved within the specified system, and so it is true. Since, assuming consistency, its negation cannot be proved, the Gödel sentence is therefore formally undecidable, rendering the system incomplete. Thus Gödel brought Hilbert's dream of a universal, all-encompassing formalization to a close.

Gödel arrived at the Institute in the fall of 1933 but, suffering depression, returned to Vienna in May of 1934. After retreating to the sanatorium at Purkersdorf, where he was diagnosed with nervous exhaustion, he returned to Princeton in September of 1935, where he fell into an even more severe depression, resigning his position and returning to Austria at the end of November. He admitted himself to the sanatorium in Rekawinkel, and then recovered sufficiently to spend several weeks with his future wife, Adele Nimbursky (née Porkert), a Viennese cabaret dancer, at the spa in Aflenz.

Veblen, Marston Morse, and von Neumann (who visited Gödel in Vienna) were determined to bring him back to the Institute, although Aydelotte, who later confided to Gödel's psychiatrist that "I have always been a little worried by the fact that he does not take more recreation," maintained his reservations, and, in 1950, when Gödel was finally offered a faculty appointment, "took the point of view that Gödel is not the type of person to be appointed full professor."[16] Nonetheless, Aydelotte supported bringing Gödel back to the United States

After his September marriage to Adele, Gödel returned to Princeton in late 1938, but after a semester at Notre Dame, he returned once again to Vienna, in June of 1939, just as war was breaking out. Gödel now found himself caught by the same quasi-paradoxical self-contradiction that had characterized his recent mathematical results. He was born in Brünn, Czechoslovakia, and had naturalized as an Austrian citizen in 1928. After Austria was annexed by the Hitler government in 1938, he lost his teaching position in Vienna and, although not Jewish, was accused of "having travelled in liberal-Jewish circles"; his application to be appointed *Dozent neuer Ordnung* (Lecturer of the New Order) was declined. With Austria officially nonexistent, he was forced to acquire a German passport, even for a temporary visit to the United States. With the issuance of a German passport, however, came eligibility for German military service, and without fulfilling his military obligation, any request for an exit visa would be denied.

The German authorities would not grant an exception without a visa from the Americans, and the Americans would not grant a visa without an exception from the Germans. "You will appreciate of course that if Professor Gödel's difficulty arises from some question relating to military or other matters within the jurisdiction of a foreign government, our consular officer at Vienna would be unable to intervene on his behalf since Professor Gödel is not an American citizen," the chief of the U.S. Visa Division wrote to Abraham Flexner in October 1939.[17]

"Gödel is absolutely irreplaceable; he is the only mathematician alive about whom I would dare to make this statement," appealed von Neumann, in a letter circulated at the highest diplomatic levels available through Flexner's Rockefeller Foundation connections at the time. "Salvaging him from the wreck of Europe is one of the great single contributions anyone could make." The Visa Division countered that to be admitted to the United States under a non-quota visa, the applicant had to both hold a current teaching position in the country of residency and be offered a teaching position in the United States. Flexner, Aydelotte, and von Neumann confirmed that Gödel would be "teaching" (even though the Institute had neither students nor classes), but that was not enough. "The objection made against Gödel," von Neumann explained to Flexner, "is that the two years teaching in the country of origin have to be immediately preceding their application; whereas Gödel was suspended from his position by the Nazis after the *Anschluss* in 1938. This requirement I think is altogether illogical."[18]

Diplomacy succeeded where logic failed. The German authorities in Vienna granted the Gödels permission to leave, and the American authorities in Washington granted them permission to enter the United States. On January 2, 1940, Gödel cabled von Neumann with the news. "The only complication which remains," Gödel reported to Aydelotte, "is that I shall have to take the route through Russia and Japan."[19] With visas issued by the Americans on January 8, the Gödels left Berlin for Moscow on January 15, taking the Trans-Siberian Railway to Vladivostok, where they transferred by ship to Yokohama, arriving on February 2, and just missing their intended passage to San Francisco on the *Taft*, which had sailed on February 1. Aydelotte came to the rescue, wiring $200 to the Gödels at the Yokohama New Grand hotel and booking passage for them on the *Cleveland* via Honolulu (where Gödel requested another $300) to San Francisco, where they arrived on March 4. They finally arrived in Princeton, by train, on March 9.

They had escaped just in time. By June, Paris was occupied and Italy had declared war on Britain and France. "My worst premonitions became true," Stan Ulam wrote to von Neumann on June 18. "My faith in America has almost completely disappeared."[20] The United States did not declare war until December 8, 1941, but many at the Institute were either already displaced by the war or engaged in preparing for it. Von Neumann was already immersed in weapons research; there was talk of the "uranium problem"; Veblen and Morse were both preparing to return to positions with the Army Proving Ground.

While von Neumann was looking for targets that should be bombed, the Institute's humanists were enlisted (by the American Commission for the Protection and Salvage of Artistic and Historic Monuments in War Areas) to help identify targets that should *not* be bombed. Erwin Panofsky, the art historian, was responsible for identifying culturally important resources in Germany, while the Institute's classicists and archaeologists helped supply similar intelligence for the Mediterranean and Middle East. Even Einstein was debriefed.

As the war dragged on, the Institute battened down the hatches, conserving fuel by "trying to heat the common room from the fireplace," and otherwise attempting to keep spirits up. Supplies and materials grew scarce, and purchases were postponed, while the Institute community continued to expand. "Will you please advise us if there is any law against using a trailer for passengers," Bernetta Miller wrote to the Department of Motor Vehicles, when the station wagon that shuttled between the Institute and the train station became inadequate for the passenger load.[21]

The Gödels, holding German passports, were required to register as enemy aliens, and could not leave Princeton without written permission from the Department of Justice in Trenton, even for routine visits to their doctors in New York. "I continue to be a little troubled about the idea of our so-called enemy aliens traveling too far afield," wrote Aydelotte in December 1941, who had to intermediate with local authorities for the release of individuals picked up on suspicion when they left the IAS.[22]

"I have never taken an oath of allegiance to Germany. My wife . . .

has never taken an oath of allegiance to Germany," Gödel wrote to the Department of Justice in Washington, D.C., requesting to amend their status under the Alien Registration Act. "Since we came to this country on German passports and were under the impression that Austrian citizenship was no longer recognized in this country, and were not advised to the contrary when we questioned the officials on this point, we felt that we had no choice but to register as Germans."[23]

"The procedure for such amendment or correction has not, as yet, been set up but in all probability it will be soon," Earl G. Harrison, special assistant to the attorney general, answered. "In the meantime your letter will be filed appropriately with your record in the Alien Registration Division." Aydelotte stepped in to help. "When Dr. and Mrs. Gödel filed their declaration of intention they were put down as of German nationality, he being listed as born in Brünn, Germany, and she as being born in Vienna, Germany. These cities were, of course, not German at the time that Dr. and Mrs. Gödel were born and these statements on the declaration of intention should, it seems to me, be corrected," he wrote to the U.S. District Court. "I am at a loss to know just how to go about it to get this correction made."[24]

"As Mr. Gödel is a naturalized Austrian citizen and Mrs. Gödel an Austrian Citizen through birth, their nationality, as far as the declaration of intention is concerned, will have to remain German due to the fact that this Country recognized Germany's conquest of Austria thereby making it a part of the German Reich," came the answer from the court. "This is borne out by the issuance of a German passport. However, when Mr. and Mrs. Gödel file their petitions for citizenship, this status will be changed in accordance with the modified rule regarding Austrians."[25]

Despite these obstacles, Gödel produced his third landmark work, a monograph on the consistency of the continuum hypothesis, published in 1941. "Gödel obtained this result by a very ingenious construction which uses the tricks of his proofs in formal logics! Did you hear about this?" von Neumann wrote to Stan Ulam in May of 1941. "Please send Gödel continuum hypothesis notes," Alan Turing cabled from King's College, Cambridge, on December 16.[26] Proposed by George Cantor in 1877, and presented in 1900 as the first of Hilbert's twenty-three unsolved problems, the continuum hypothesis states that the set of real numbers (the continuum) is the smallest infinity whose size is larger than the set of integers, and that no intermediate-sized infinities lie in between. Gödel proved that within a strictly defined system it was impossible to disprove the hypothesis—a result that has been strengthened in recent years.

Unable to return to Austria, Gödel grew more anxious. "The evidence we have had here of Dr. Gödel's difficulties comes from the fact that he thinks the radiators and ice box in his apartment give off some kind of poison gas," Frank Aydelotte wrote to Max Gruenthal, Gödel's psychiatrist, in December of 1941. "He has accordingly had them removed, which makes the apartment a pretty uncomfortable place in the winter time. Dr. Gödel seems to have no such distrust of the heating plant at the Institute and he carries on his work here very successfully."[27]

Aydelotte requested a prognosis, and finally got to the point. "I should like also especially to know," he asked, "whether you consider that there is any danger of his malady taking a violent form." Dr. Gruenthal wrote back, politely but tersely, refusing to discuss Gödel's condition without his patient's permission, but willing "to reassure you insofar of his malady taking a violent form."[28]

With no hope of returning to Europe, the Gödels settled down in Princeton and applied for permanent residency in the United States. One more obstacle remained. Once Gödel's status was recognized as formerly Austrian, not German—allowing him some security on the path to U.S. citizenship  he became eligible for the draft, and was classified 1A. In April of 1943 he was ordered to report to the Trenton Army Induction Center for examination.

"Dr. Gödel, like most refugees from Nazi Germany, is eager to do anything he can in support of the American war effort," Aydelotte answered the draft board on Gödel's behalf, "but under the circumstances I think I ought to inform the Selective Service Board that Dr. Gödel has twice since he has been in Princeton shown such signs of mental and nervous instability as to cause the doctors who were consulted to diagnose him as a psychopathic case." Aydelotte went on to extol Gödel's genius, while asking the Draft Board to consider that "this ability, however, is unfortunately accompanied by certain mental symptoms which, while they do not prevent active work in mathematics, might prove serious from the standpoint of the Army."[29]

"Although the Board is in sympathy with your knowledge of Mr. Gödel's condition, we are unable to effect a disqualification for this man at the local board," the draft board answered. "It will be necessary for him to be forwarded to the induction station for the Army examination."[30] The army had its own psychiatrists, and they would make the decision concerning Dr. Gödel for themselves.

"I have secured a certain amount of additional evidence concerning him which I believe the Selective Service Board would be glad to have," Aydelotte replied, explaining that during Gödel's convalescence in Austria he "had the idea that all the sanitarium food was poisoned and he would eat only things that were prepared and brought to him by a young woman friend of the family (whom he later married) and then only on condition that she eat with him from the same plate and with the same spoon," and that his mother "was so frightened concerning his condition that she slept always in a locked room at night."[31] With this statement, Gödel's Selective Service file comes to an end.

Gödel, who had first been brought to the Institute on a stipend of $200 per month, received a raise to $4,000 per year upon his return in 1940. His salary was paid by the Rockefeller Foundation, under an arrangement negotiated one year at a time. With the war over, von Neumann campaigned for a permanent appointment. "Gödel did some of his best work (continuum hypothesis) at the Institute—actually at a time when he was less normal than now," von Neumann argued. "The Institute is clearly committed to support him, and it is ungracious and undignified to continue a man of Gödel's merit in the present arrangement forever." As to the argument that his best work was already behind him, "he may easily do more work in mathematics proper," von Neumann continued. "His probability of doing some is no worse than that of most mathematicians past 35."[32]

On December 19, 1945, Gödel was made a permanent member, with a stipend of $6,000 and, in an evident concession to those who opposed his appointment, "it was resolved that Professor Gödel's stipend as a permanent member should not be drawn from the Mathematics budget but should rather come from general Institute funds."[33]

The Gödels lived in a series of rented apartments and, in 1949, purchased a house on Linden Lane for $12,500, taking advantage of an Institute mortgage at 4 percent. "We have found a place which is very convenient for us and which (so we both think) is exceptionally pretty," Gödel wrote to Oppenheimer. "We hope we shall soon have the pleasure of seeing you and Mrs. Oppenheimer in our new home and letting you judge for yourself."[34]

After completing his work on the continuum hypothesis, Gödel became preoccupied with two areas of research: cosmology, as a result of having discovered a solution to Einstein's equations that implied a rotating universe; and the legacy of Gottfried Wilhelm Leibniz, the seventeenth-century pioneer of calculus, binary arithmetic, universal language, the Monadology, and much else. "He seemed to believe," said Stan Ulam, "that much of Leibniz's work including mathematical logic and computing has been lost or concealed." Critics dismissed Gödel's study of the Leibniz manuscripts as a waste of his mathematical talents, verging on the occult, but according to von Neumann, "a man of his caliber and record ought to be the sole judge of what he does."[35]

In early 1946, when von Neumann was authorized to proceed with building a computer at the Institute, Herman Goldstine and Arthur Burks, both from the ENIAC project, showed up for work at Fuld Hall. Burks commuted from Swarthmore, Pennsylvania, catching the same train from the Thirtieth Street Station in Philadelphia as Goldstine, who, still on active duty, "had obtained an Army car which sat at the Princeton railroad station . . . and drove us out to the Institute and then we went home the same way."[36]

There was no room at the inn, either in the town of Princeton or in Fuld Hall. Those who had been away for the duration of the war had now returned, and in their absence all nonessential construction had come to a halt. The shortage of building materials remained as severe as it had been during the war, and new construction or even remodeling required authorization from the Civilian Production Administration. Adding to the overcrowding at the Institute, the entire staff of the Economic, Financial and Transit Department of the League of Nations, granted refuge by Aydelotte after the disbanding of their headquarters in Geneva, were crammed into all available space, including the board room, on the third and fourth floors of Fuld Hall. A staff of thirty-six individuals, from eight countries, remained encamped for almost five years, despite Veblen, among others, advocating "a firm refusal to let the other work of the Institute be further hampered by our hospitality to the league."[37]

World government, advocated by Albert Einstein and Edward Teller alike, was spilling out into the halls. "Until the nations of the world can combine into some kind of effective political organization, sacrificing some part of their national sovereignty and delegating to this supra-national organization the power to enforce peace and to settle international controversies by political and judicial processes, the world will continue to live in danger of war," Aydelotte had warned in February of 1941.[38] When the war ended, the debate over the next one had already begun. In adjacent offices two floors below those of the League, von Neumann was arguing for preventive war against the Soviet Union to be followed by a Pax Americana, while Albert Einstein was contributing his call to global disarmament, "One Way Out," to the Federation of American Scientists' manifesto *One World or None.*

All Institute members except permanent faculty were either sharing offices or relegated to temporary desks in the library. Some were even sleeping in Fuld Hall. One room, however, remained unoccupied: Fuld 219. This was a small office reserved for the secretary to the occupant of Fuld 217. "It was agreed that the room connected with Dr. Gödel's office might be used for people working on the computing machine," the School of Mathematics reported on February 13, 1946. "Kurt Gödel didn't have a secretary, didn't want one, I assume," says Arthur Burks. "So for that summer, when of course we didn't yet have a building for the computer, Herman and I occupied the secretary's office next to Gödel's office. It had a blackboard on the wall."[39]

"We spent most of our time the first few months planning this new machine, working out the structure and the instructions, and we would consult periodically with von Neumann," Burks recalls. "After we'd done a certain amount of planning, we decided we'd better write it up now. Which was fine with me. So Herman and I wrote the first draft and I don't remember how we divided it but we both worked writing it. Then we'd show it to von Neumann and he would revise, or we'd discuss it, and so forth, and that was issued as a report at the end of June."[40]

"Preliminary Discussion of the Logical Design of an Electronic Computing Instrument" was issued on June 28, 1946. In fifty-four pages, opening with a discussion of the "principal components of the machine," comprising "certain main organs relating to arithmetic, memory-storage, control and connection with the human operator," and closing with a list of the order codes, the report specified the logical architecture, if not the physical embodiment, of the new machine. Because, as the authors put it, "the moment one chooses a given component as the elementary memory unit, one has also more or less determined upon much of the balance of the machine," a full five pages are devoted to the memory architecture, envisioned as 40

Selectron tubes each storing 4,096 bits—reduced to 1,024 as engineering considerations began to set in.

The report went on to describe—in enough detail to begin planning and coding of actual problems—a complete set of order codes. There were twenty-one instructions, supplemented by a number of input/output orders discussed at the end of the report. Finally, "there is one further order that the control needs to execute," the authors concluded. "There should be some means by which the computer can signal to the operator when a computation has been concluded, or when the computation has reached a previously determined point. Hence an order is needed which will tell the computer to stop and to flash a light or ring a bell."[41]

Some 175 copies of the report were distributed and then, in May 1947, Goldstine reported that the stencils "are practically worn out and so it is probably not possible to make more copies using them."[42] A second edition, reset on the Institute's new Varityper, was issued on September 2, 1947. Few technical documents have had as great an impact, in the long run. "Preliminary Discussion of the Logical Design of an Electronic Computing Instrument," drafted in the annex to Gödel's office, would end up fulfilling Leibniz's dreams of digital computing and universal language that Gödel believed had been overlooked.

Gottfried Wilhelm Leibniz, born in Leipzig in 1646, enrolled in the University of Leipzig as a law student at age fifteen. Our universe, Leibniz theorized, was selected from an infinity of possible universes, optimized so that a minimum of laws would lead to a maximum diversity of results. Leibniz's reflections on the nature of mind culminated in his *Monadology* of 1714, a short text describing a universe of elementary mental particles that he called monads, or "little minds." These entelechies (the local actualization of a universal mind) reflected in their own inner state the state of the universe as a whole. According to Leibniz, relation gave rise to substance, not, as Newton had it, the other way around. "Back to Leibniz!" is how Norbert Wiener titled an article on quantum mechanics in 1932. "I can see no essential difference between the materialism which includes soul as a complicated type of material particle and a spiritualism which includes material particles as a primitive type of soul," Wiener added in 1934.[43]

Leibniz believed, following Hobbes and in advance of Hilbert, that a consistent system of logic, language, and mathematics could be formalized by means of an alphabet of unambiguous symbols manipulated according to mechanical rules. In 1675 he wrote to Henry Oldenburg, secretary of the Royal Society and his go-between with Isaac Newton, that "the time will come, and come soon, in which we shall have a knowledge of God and mind that is not less certain than that of figures and numbers, and in which the invention of machines will be no more difficult than the construction of problems in geometry." Envisioning what we now term software, he saw that the correspondence between logic and mechanism worked both ways. To his "Studies in a Geometry of Situation," sent to Christiaan Huygens in 1679, he appended the observation that "one could carry out the description of a machine, no matter how complicated, in characters which would be merely the letters of the alphabet, and so provide the mind with a method of knowing the machine and all its parts."[44]

With his logical calculus, or calculus ratiocinator, Leibniz took the first steps toward his vision of a "universal symbolistic in which all truths of reason would be reduced to a kind of calculus." Believing that "a kind of alphabet of human thoughts can be worked out and that everything can be discovered and judged by a comparison of the letters of this alphabet and an analysis of the words made from them," he proposed a universal coding in which primary concepts would be represented by prime numbers—an all-encompassing mapping between numbers and ideas.[45]

"I think that a few selected men could finish the matter in five years," Leibniz claimed. "It would take them only two, however, to work out, by an infallible calculus, the doctrines most useful for life, that is, those of morality and metaphysics." Anticipating Gödel and Turing, Leibniz promised that through digital computing "the human race will have a new kind of instrument which will increase the power of the mind much more than optical lenses strengthen the eyes. . . . Reason will be right beyond all doubt only when it is everywhere as clear and certain as only arithmetic has been until now."[46]

Leibniz saw binary coding as the key to a universal language and credited its invention to the Chinese, seeing in the hexagrams of the *I Ching* the remnants of "a Binary Arithmetic . . . which I have rediscovered some thousands of years later." Leibniz's notes show the development of simple algorithms for translating between decimal and binary notation and for performing the basic functions of arithmetic as mechanically iterated operations on strings of zeros and ones. "In Binary Arithmetic there are only two signs, 0 and 1, with which we can write all numbers," he explained. "I have since found that it further expresses the logic of dichotomies which is of the greatest use."[47]

In 1679, Leibniz imagined a digital computer in which binary numbers were represented by spherical tokens, governed by gates under mechanical control. "This [binary] calculus could be implemented by a machine (without wheels)," he wrote, "in the following manner, easily to be sure and without effort. A container shall be provided with holes in such a way that they can be opened and closed. They are to be open at those places that correspond to a 1 and remain closed at those that correspond to a 0. Through the opened gates small cubes or marbles are to fall into tracks, through the others nothing. It [the gate array] is to be shifted from column to column as required."[48]

Leibniz had invented the shift register—270 years ahead of its time. In the shift registers at the heart of the Institute for Advanced Study computer (and all processors and microprocessors since), voltage gradients and pulses of electrons have taken the place of gravity and marbles, but otherwise they operate as Leibniz envisioned in 1679. With nothing more than binary tokens, and the ability to shift right and left, it is possible to perform all the functions of arithmetic. But to do anything with that arithmetic, you have to be able to store and recall the results.

"There are two possible means for storing a particular word in the Selectron memory," Burks, Goldstine, and von Neumann explained. "One method is to store the entire word in a given tube and . . . the other method is to store in corresponding places in each of the 40 tubes one digit of the word." This was the origin of the metaphor of handing out similar room numbers to 40 people staying in a 40-floor hotel. "To get a word from the memory in this scheme requires, then, one switching mechanism to which all 40 tubes are connected in parallel," their "Preliminary Discussion" continues. "Such a switching scheme seems to us to be simpler than the technique needed in the serial system and is, of course, 40 times faster. The essential difference between these two systems lies in the method of performing an addition; in a parallel machine all corresponding pairs of digits are added simultaneously, whereas in a serial one these pairs are added serially in time."[49]

The 40 Selectron tubes constituted a 32-by-32-by-40-bit matrix containing 1,024 40-bit strings of code, with each string assigned a unique identity number, or numerical address, in a manner reminiscent of how Gödel had assigned what are now called Gödel numbers to logical statements in 1931. By manipulating the 10-bit addresses, it was possible to manipulate the underlying 40-bit strings—containing any desired combination of data, instructions, or additional addresses, all modifiable by the progress of the program being executed at the time. "This ability of the machine to modify its own orders is one of the things which makes coding the non-trivial operation which we have to view it as," von Neumann explained to his navy sponsors in May of 1946.[50]

"The kind of thinking that Gödel was doing, things like Gödel numbering systems—ways of getting access to codified information and such—enables you to keep track of the parcels of information as they are formed, and . . . you can then deduce certain important consequences," says Bigelow. "I think those ideas were very well known to von Neumann [who] spent a fair amount of his time trying to do mathematical logic, and he worked on the same problem that Gödel solved."[51]

The logical architecture of the IAS computer, foreshadowed by Gödel, was formulated in Fuld 219. "In the 1930s the realization of an actual physical device that could function as a general-purpose information-processing programmable computer was still decades in the future, yet someone knowledgeable about modern programming languages today looking at Gödel's paper on undecidability written that year will see a sequence of forty-five numbered formulas that looks very much like a computer program," says Martin Davis, who arrived at the Institute under the sponsorship of the Office of Naval Research in September of 1952. "In demonstrating that the property of being the code of a proof in PM [*Principia Mathematica*] is expressible inside PM, Gödel had to deal with many of the same issues that those designing programming languages and those writing programs in those languages would be facing," Davis notes.[52]

Gödel had demonstrated, in 1931, the powers of numerical addressing and self-reference. In a stored-program computer, one of the rules is that you can change the rules. Gödel was well aware that Turing's Universal Machine and von Neumann's implementation of it were demonstrations, if not the direct offspring, of his, Gödel's, ideas. "What von Neumann perhaps had in mind appears more clearly from the universal Turing machine," he later explained to Arthur Burks. "There it might be said that the complete description of its behavior is infinite because, in view of the non existence of a decision procedure predicting its behavior, the complete description could be given only by an enumeration of all instances. The universal Turing machine, where the ratio of the two complexities is infinity, might then be considered to be a limiting case."[53]

Leibniz's belief in a universal digital coding embodied his principle of maximum diversity: infinite complexity from finite rules. "Nothing is a better analogy to, or even demonstration of such creation than the origin of numbers as here represented, using only unity and zero or nothing," he wrote to the Duke of Brunswick in 1697, urging that a silver medallion be struck (with a portrait of the duke on the reverse) to help bring the powers of binary arithmetic, and "the creation of all things out of nothing through God's omnipotence," to the attention of the world.[54]

Where does meaning come in? If everything is assigned a number, does this diminish the meaning in the world? What Gödel (and Turing) proved is that formal systems will, sooner or later, produce meaningful statements whose truth can be proved only outside the system itself. This limitation does not confine us to a world with any less meaning. It proves, on the contrary, that we live in a world where higher meaning exists.

"Our earthly existence, since it in itself has a very doubtful meaning, can only be a means toward the goal of another existence," Gödel wrote to his mother in 1961. "The idea that everything in the world has meaning is, after all, precisely analogous to the principle that everything has a cause, on which the whole of science rests."[55]

# SEVEN

# 6J6

*Absence of a signal should never be used as a signal.*

—Julian Bigelow, 1947

JULIAN HIMELY BIGELOW, the fourth of five siblings, was born in Nutley, New Jersey—forty-two miles from Princeton—on March 19, 1913. At the age of three, while staying with an aunt, "he found a screw driver, and removed all the door knobs and put them in a big pile, and it took him a really long time to put all these door knobs back."[1] His father, Richard Bigelow, gave up a teaching career at Wellesley to raise his family—and outwait the Depression—by retreating to Millis, Massachusetts, in search of a self-sufficient rural life.

The Bigelows lived in a hand-hewn eighteenth-century farmhouse with no electricity except for one circuit in the basement that powered a water pump. Julian surreptitiously installed an additional circuit that terminated, in his bedroom, at a single electric light. He entered the Massachusetts Institute of Technology at the age of seventeen, delivering milk in a Model T Ford to pay for his tuition and graduating with a master's degree in electrical engineering in 1936. "When I was at MIT," he remembers, "the electronics and radio work which had been going on was considered a rather suspect and maybe a frivolous thing. One should really be designing large generators, or at least a large arc discharge thyrotron or something."[2]

Bigelow's first job was with the Sperry Corporation, in Brooklyn, New York, building navigational gyroscopes and machinery for automatic detection of flaws in railroad tracks. Sperry later merged with office equipment manufacturer Remington Rand to become Sperry Rand, the early computer conglomerate that acquired the Eckert-Mauchly Electronic Control Company and unsuccessfully defended the ENIAC patents against Honeywell—after implementing a cross-licensing agreement with IBM. At the end of 1938, Bigelow left Sperry and joined IBM in Endicott, New York, as their first employee with the job title of electronic engineer. "At that time, IBM was a very mechanically oriented company, and the notion of electronic computing was almost repugnant," he recalls.[3]

At the onset of World War II, Bigelow, an amateur aviator for most of his life, returned to MIT to retrieve his academic records and enlist in the navy as an aviation cadet. "But when I got up there," he explains, "I had to go see my department head, and he grabbed me and said, 'We can't let you go, we need you. We've got this fellow, Norbert Wiener, going around saying he knows how to win the war singlehandedly, so to speak, with his intellectual ideas. Nobody can find out what he's talking about, so we need you to work with him to see what there is to it.'"[4]

At the close of World War I, after leaving Oswald Veblen's group at the Aberdeen Proving Ground, Wiener had secured a job at the *Boston Herald,* where his career as a reporter and features writer was short-lived. The problem, as he described it, was that "I had not learned to write with enthusiasm of a cause in which I did not believe." After being fired from the *Herald,* he was hired as an instructor at MIT, his home for the next forty-five years. Wiener was "daring and uncautious, instinctive as often as logical, and utterly unsuited to meticulous step by step analytical-experimental staircase procedures," Bigelow reported to von Neumann in 1946. "He has had sad experiences trying to work with large groups such as might be expected to carry out reliable experimental programs; he finds himself obliged to work with little funds and a few enthusiastic individual supporters."[5] Bigelow served as Wiener's assistant from 1940 to 1943.

Wiener, whose nearsightedness had disqualified him from the infantry in World War I, decided to take on the antiaircraft fire-control problem: the most intractable targeting challenge of World War II. In 1940, German bombers were raining high explosives on Great Britain. U.S. targets might be next. The newly established National Defense Research Committee (NDRC) of the Office of Scientific Research and Development (OSRD) was fielding a wide range of proposals—with the Wiener-Bigelow collaboration being one of the longest shots. Wiener approached the problem from mathematical first principles, while Bigelow attempted to embody Wiener's mathematics in an automatic antiaircraft fire director—dubbed the "debomber"—that was never built.

Wiener's first suggestion, made to Vannevar Bush in September of 1940, was to circumvent the need for precision by "bursting in the air containers of liqu[e]fied ethylene or propane or acetylene gases so that an appreciable region will be filled with an explosive mixture [and] interdicted to enemy aircraft."[6] This unsportsmanlike proposal got no response from Vannevar Bush.

Wiener then approached Warren Weaver, who had been assigned responsibility for the antiaircraft effort in the United States, proposing to investigate "the design of a lead or prediction apparatus" which "anticipates where the airplane is to be after a fixed lapse of time."[7] Weaver, taking a wartime break from serving, as he described it, as the Rockefeller Foundation's chief "philanthropoid," awarded the requested $2,325 in December 1940, and the NDRC's D.I.C. (Detection, Instruments, Controls) Project 5980 was launched. In 1940 an antiaircraft gunner facing high-altitude bombers had roughly 10 seconds to observe an approaching target before estimating its range, setting a timed fuse, and firing a 90-mm shell that would spend up to 20 seconds in flight. The job of the gunner was to guess where the airplane would be at the designated instant, while the job of the pilot was to guess where the shell would be at that instant—and to be somewhere else.

Wiener and Bigelow considered the observer, gun, airplane, and pilot as an integrated, probabilistic system. The odds favored the pilot: in 1940 only one out of about 2,500 antiaircraft shells scored a hit. In a preliminary report, they explained how they intended "to place the analysis of the problem of prediction upon a purely statistical basis, by determining to what extent the motion of a target is predictable on the basis of known facts and history, and to what extent the motion of the target is not predictable."[8]

The predictable elements would indicate the most likely future position of the target, with the unpredictable elements determining the optimum "spread"—the degree to which the gunner would scatter fire because the exact position of the target was unknown. This distinction was equivalent to the distinction, in communications theory, between signal and noise. Similar ideas were formalized by Claude Shannon (working in consultation with Wiener) and Andrey Kolmogorov (working independently in the Soviet Union) at about the same time. "The transmission of a single fixed item of information is of no communicative value," Wiener explained in his report to Weaver in 1942. "We must have a repertory of possible messages, and over this repertory a measure determining the probability of these messages."[9]

Wiener had launched his mathematical career with a theory of Brownian motion—the random trajectory followed by a microscopic particle in response to background thermodynamic noise. He was thus prepared for the worst case possible: an aircraft that changes course at random from one moment to the next. Wiener's theory, strengthened by Bigelow's experience as a pilot, held that the space of possible trajectories (equivalent to the space of possible messages in communications theory) was constrained by the performance envelope of the aircraft and the physical limitations of the human being at the controls. Almost all combat flying, Bigelow observed, was composed of curves, not straight lines. Straight-line extrapolation of a flight path was a reliable predictor only of where the airplane would *not* be at any given future time.

Wiener was strictly a theoretician. Bigelow, "a quiet, thorough New Englander, whose only scientific vice is an excess of scientific virtue," in Wiener's assessment, was an engineer at home with machines. "For many years, Bigelow nursed a series of old and decrepit cars," Wiener explained, "which, by all the canons of the motorist, should have been consigned to the junk heap years ago." Alice Bigelow, Julian's daughter, remembers "learning how to jump start a car while driving it, as soon as I was big enough to see, from age nine or something—because the cars were always broke. Dad would push it: 'It's fine, it's fine, just a little push, and it will go.' And in Princeton that was just unthinkable."[10]

It was equally unthinkable, in Princeton, to live in anything except a conventional house. Bigelow, however, purchased a former blacksmith shop on Clay Street, in central Princeton, and in 1952 moved it to a vacant lot on Mercer Street, between the Battlefield and the Friends meetinghouse at Stony Brook. When negotiations with the Township and Borough of Princeton over the cost of moving overhead wires out of the way to allow passage of the building broke down, "he cut the thing in half like a layer cake, and then bolted it back together," remembers Jule Charney, leader of the IAS meteorological group.[11]

Bigelow maintained a succession of small aircraft, including a Cessna he purchased damaged in Wyoming, restored to airworthiness, and flew home. A dismantled aircraft engine once occupied the Bigelows' Princeton living room, concealed by a tablecloth when there were guests. Wiener, who had been unable to stay upright on a horse, was afraid of flying, but "was willing to take the chance and fly with me," Bigelow recalls. "We flew from Framingham to Providence and back again. Inside the airplane there were some steel tubes which brace the windshield and his hands left fingerprints on them."[12]

Asked why Wiener—who had the resources of MIT at his disposal, and whose interest in digital computing preceded von Neumann's—did not build his own computer, Bigelow answered, "He wasn't a man to do something that was practical. A computer has to work."[13] So did an antiaircraft director, and on October 28, 1941, Warren Weaver sent Wiener and Bigelow a list of questions centered on one uncertainty: Was Wiener's theory going to amount to anything that could affect the outcome of the war?

On December 2, 1941, five days before the Japanese attack on Pearl Harbor, Bigelow responded with a fifty-nine-page letter to Weaver, marked "to be destroyed after reading," reporting progress on the "debomber" so far. The goal, as Bigelow described it, was an antiaircraft director that kept the signal (the aircraft's flight path) separate from the noise: noise introduced both by a pilot attempting to behave unpredictably, and by observation and processing errors along the way. "To re-separate the signal from the soup in these last two terms is no cinch, and in the case of random or Brownian noise with no simple spectrum it is quite impossible to do the filtering perfectly," noted Bigelow. "Result: lost ground."[14]

Bigelow compiled a list of fourteen "Maxims for Ideal Prognosticators" starting with MIP 1: "Make all observations in same coordinate system as will finally be used by the gun-pointer." Maxims 2–4 advised separating the available information into that needed immediately and that needed later, while Maxim 5 added that "if noise is ever to be filtered from signal, it must be done at the earliest possible stage rather than after the two are tangled with other noises and signals, for the same reason that repeater stations are used on a signal line rather than filters and amplifiers at the ends." Maxim 7 advised "Never estimate what may be accurately computed"; Maxim 8 advised "Never guess what may be estimated"; and, if a guess was absolutely necessary, "Never guess blindly" was Maxim 9.

Maxims 10 through 14 specified how to implement optimal prediction when the target "has the character of a Brownian motion impressed upon a resonator system." Existing methods of tracking a target's changing position "of necessity refers it to an irrelevant point of observation thus destroying its fundamental symmetry," while an ideal predictor should assume that the target obeys the conservation laws of physics "upon which is superimposed a random modulation symmetrical in time."[15] The Wiener-Bigelow debomber would model the behavior of the airplane within the frame of reference belonging to the airplane, rather than referring it to that of the observer on the ground.

"We should clear any fog surrounding the notion of 'prediction,'" Bigelow confessed. "Strictly and absolutely, no network operator—or human operator—can predict the future of a function of time. . . . So-called 'leads' evaluated by networks or any other means are actually 'lags' (functions of the known past) artificially reversed and added to the present value of the function."[16] Nonetheless, Bigelow's strategy paid off. A proof-of-principle model was constructed, allowing an operator controlling a white spot of light to follow a red spot of light, driven by a modified phonograph turntable and representing an evasive target, around a darkened room. Wiener "was excited by the thought that his calculations were relevant and serviceable," Bigelow recalls. "He showed it by puffing on his cigar in a violent way. The room would be full of smoke. He'd sort of jump up and down. He was a little bit too eager to accept the demonstration I produced as a proof that it would work."[17]

Wiener "was really flying on Cloud 9," adds Bigelow. "I did not want to put down the mathematical ideas he was talking about, but simply to realize those in a continuous and effective way in time for this war was probably wildly impossible."[18] As the chances of putting his ideas into practice diminished, Wiener pushed harder on the theoretical side. "I tried to work against time," he explained. "More than once I computed all through the night to meet some imaginary deadline which wasn't there. I was not fully aware of the dangers of Benzedrine, and I am afraid I used it to the serious detriment of my health."[19]

On July 1, 1942, George Stibitz, chairman of the Anti-Aircraft Director Division of the NDRC, spent the day with Bigelow and Wiener, noting in his diary that "their statistical predictor accomplishes miracles. . . . For a one-second lead the behavior of their instrument is positively uncanny. Warren Weaver threatens to bring along a hacksaw on the next visit and cut through the legs of the table to see if they do not have some hidden wires somewhere."[20]

The Wiener-Bigelow collaboration, although failing to produce a working debomber, was influential on other fronts. With neurophysiologist Arturo Rosenblueth, Bigelow and Wiener coauthored a 1943 paper, "Behavior, Purpose and Teleology," that suggested unifying principles underlying purposeful behavior among living organisms and machines. "Teleology has been interpreted in the past to imply purpose and the vague concept of a 'final cause,'" they noted, explaining that "we have restricted the connotation of teleological behavior by applying this designation only to purposeful reactions which are controlled by the difference between the state of the behaving object at any time and the final state interpreted as the purpose." Teleology was thus identified with the Bigelow-Wiener definition of negative feedback, where "the signals from the goal are used to restrict outputs which would otherwise go beyond the goal."[21]

This paper served as the namesake for the informal Teleological Society, whose inaugural meeting, hosted by von Neumann, was held at the Institute for Advanced Study on January 4–6, 1945. With the sponsorship of the Josiah Macy Jr. Foundation, a series of more formal conferences followed, and what came to be known as the Cybernetics movement took form. "Cybernetics came into its own," explained neurophysiologist Warren McCulloch, "when Julian Bigelow pointed out the fact that it was only information concerning the outcome of the previous act that had to return."[22]

In 1943, Bigelow left MIT, reassigned by Warren Weaver to the NDRC Applied Mathematics Panel's Statistical Research Group. Under the auspices of Columbia University, eighteen mathematicians and statisticians—including Jacob Wolfowitz, Harold Hotelling, George Stigler, Abraham Wald, and the future economist Milton Friedman—tackled a wide range of wartime problems, starting with the question of "whether it would be better to have eight 50 caliber machine guns on a fighter plane or four 20 millimeter guns."[23] Bigelow was brought in to help with an automatic bomb sight being developed for high-speed dive bombers trying to hit fixed targets on the ground: the debomber problem upside down. He was promoted to associate director, and remained with the group for thirty-one months.

Back in Princeton, von Neumann was trying to get the Electronic Computer Project off the ground. Presper Eckert, who was expected to lead the engineering team, was reluctant to leave the Moore School for the uncertainties of the Institute, and sent his brother-in-law, mechanical engineer John Sims, instead. Sims was hired on January 18, 1946, and assigned to begin searching for tools, electronic components, and materials, becoming the project's first employee. Herman H. Goldstine, awaiting release from the army, became the second employee, accepting a position (first offered on November 27, 1945) as associate director on February 25, 1946. His salary was set at $5,500—lower than an Institute professor, but higher than an Institute visitor, upsetting the distinction that had been in place since 1933.

As negotiations with Eckert stalled—grinding to a halt once Eckert and Mauchly decided to go into business for themselves—von Neumann began searching for an alternate chief engineer. Wiener, asked for his recommendation, put Bigelow at the top of the list. "We telephoned from Princeton to New York, and Bigelow agreed to come down in his car," Wiener recalls. "We waited till the appointed hour and no Bigelow was there. He hadn't come an hour later. Just as we were about to give up hope, we heard the puffing of a very decrepit vehicle. It was on the last possible explosion of a cylinder that he finally turned up with a car that would have died months ago in the hands of anything but so competent an engineer."[24]

Bigelow was hired on March 7, 1946, at a salary of $6,000, effective June 1, with an interim $25 per diem as a consultant until he could move to Princeton from New York. The Aydelottes offered Julian and Mary Bigelow temporary accommodation in Olden Manor, including "use of the kitchen for as many meals as Mrs. Bigelow feels equal to preparing."[25] There were several months of commuting while Julian completed his obligations to the Statistical Research Group and Mary, a psychologist, arranged to move her practice to Princeton from New York.

The Bigelows became pillars of the close-knit Institute community. Mary was a gifted therapist, and Julian was fluent not only in mathematics and physics, but in the undocumented practices that were required, in postwar New Jersey, to get anything built—or fixed. "I came to Princeton in fall 1948 with three year old Katharina, both of us quite lost in that vast new country," Verena Haefeli, now Verena Huber-Dyson, remembers. "Everybody was friendly, something that did not just happen to you in Switzerland without the appropriate preamble of introductions. It was Mary Bigelow that managed to put me at ease, with her warm, naturally outgoing ways and her sensitive understanding of the human psyche. I remember Julian's handsome good looks, imposing figure and especially his clear blue eyes. To me, fresh from mixed up Europe, he was the prototype of American uprightness and purposefulness."[26]

Arthur Burks, Goldstine's colleague from the ENIAC project, was hired on March 8, 1946 (at a salary of $4,800), and electrical engineer James Pomerene was hired (at a salary of $4,500) on March 9. Twenty-six years old and recently married, Pomerene showed up for work on April 1, and was soon joined by Hazeltine colleague (and Moore School alumnus) Willis Ware, who accepted a position on May 13 and began work on June 1. Pomerene and Ware took the train down to Princeton to meet Bigelow. "We drove back to New York City with Julian in his little old green Austin," Ware recalls. "By the time Pom and I got back to New York City—we were still at Hazeltine—we were absolutely enthralled. He just bubbles with ideas all the time."[27] Pomerene and Ware were able to trade their New York apartments with two Princeton residents who were working for the United Nations in Manhattan, exchanging a long commute by train for a short commute by bicycle down Nassau Street and Olden Lane.

Pomerene and Ware had both worked on pulse-coded IFF (Identification Friend or Foe) radar systems during the war. As soon as radar had made it possible to hit targets at night or beyond visible range, otherwise adversarial air forces agreed on a system of coded signals identifying their aircraft as friend or foe. In contrast to the work of wartime cryptographers, whose job was to design codes that were as difficult to understand as possible, the goal of IFF was to develop codes that were as difficult to *mis*understand as possible. Pomerene and Ware (and their counterparts Frederic C. Williams and Tom Kilburn on the British side), who had developed circuitry to communicate coded pulses, at high speeds across noisy channels between aircraft, now faced the same problem in building electronic digital computers: how to transmit coded pulses, thousands of times per second, from one machine cycle to the next. We owe the existence of high-speed digital computers to pilots who preferred to be shot down intentionally by their enemies rather than accidentally by their friends.

A small team began to coalesce. Richard W. Melville, a seaman first class technical specialist in radar, "came in with a sailor's hat on, and asked for a job," according to Bigelow, "and I liked him and he seemed to be on the ball."[28] Melville turned out to be "a wizard," in Pomerene's assessment, at improvising the requisite laboratory facilities, while supervising the mechanical engineering of the machine. He kept everything running smoothly in tight quarters, worked miracles at finding war surplus materials and parts, and hired mechanically inclined high school students once designs were finalized and forty-stage copies of the prototype shift registers and accumulators had to be produced. His wife, Claire, taking over a vacant apartment, opened a nursery school for Institute children too young to attend the Princeton public schools.

William S. Robinson, mechanic, was hired on March 21, 1946, and was joined in the machine shop—once there was one—by Winfield T. Lacey and Frank E. Fell. Ralph Slutz, a graduate student in physics at Princeton, accepted an offer on April 5 to start full-time work on July 1. "I went and knocked on John von Neumann's door," Slutz remembers, "and said 'I hear you're going to build a computer, how about a chance to work on it?' He said yes." Slutz had met von Neumann in the course of wartime work on blast waves, and heard about the prospects for doing computation with vacuum tubes. "I remember sitting in my classes," he recalls, "sketching out adders rather than paying any attention to quantum mechanics."[29]

Robert F. Shaw, one of the ENIAC veterans still at the Moore School, accepted a position as a member of the engineering staff on May 13. John (Jack) Davis, also at the Moore School (and former neighbor and high school classmate of Willis Ware), accepted an offer on April 13 to report to work on June 1. "I used to sit on Jack Davis' bed and listen to the short wave radio on receivers that we'd built on our mothers' pie pans," remembers Ware.[30] Ames Bliss, son of ballistician Gilbert Bliss, accepted a position as contracts administrator on May 14, at a salary of $4,000 per year. Akrevoe Kondopria of Philadelphia, Goldstine's secretary at the Moore School, transferred to the IAS and reported for work on June 3, 1946.

"I was sixteen and came from a Greek immigrant family. My father was from a poor Ionian island and could barely read and write. It was made quite clear to me that I couldn't go to college, much as I wanted to," she remembers. Her guidance counselor advised forgoing college and applying for a job as a secretary at the Moore School of the University of Pennsylvania. "There I met Captain Goldstine, with his two gold bars, slim and elegant in his uniform, and Mrs. Goldstine, Adele, who dressed casually and smoked a lot. For some reason they hired me. I, who had never taken algebra, was thrown into a world completely foreign to me, and it changed my life."

"The Goldstines were wonderfully encouraging, and as the time grew nearer to move to Princeton, they asked me to come along with them," she explains. She commuted by train from Philadelphia until offered a room in mathematician Salomon Bochner's house, at the end of Springdale Road. Using a manual typewriter at first, and the Institute's new Varityper later, she helped produce the progress reports that were being generated even before the construction of the machine got off the ground. "I remember how tedious it was, because you had to change the type font from text to the mathematical symbol disc, and of course, you had to be very, very accurate," she recalls. She had no inkling, at first, of the hydrogen bomb calculations that were driving the project at IAS. "Nicholas Metropolis was always sending letters to P.O. Box 1663, Santa Fe, New Mexico, and was getting letters from the same address," she remembers. "My guess was that he must have had a girlfriend there."

Akrevoe remained in Princeton until August 1949, when her mother, who "thought I was getting beyond my station," insisted she move back to Philadelphia, presenting the ultimatum as "It's time to come back; you've gotten too many big ideas." It was hard to leave. "The Goldstines and the young engineers treated me like a kid sister," she remembers, "and probably taught me more than I would have learned in college." The presence of a bright, red-haired seventeen-year-old was just what the computing group needed. "Some of these guys were so helpless," she adds. "They didn't have very many social skills."[31]

Between the stuffiness of Princeton and the rarefied atmosphere at the Institute, it was not easy for the engineers to fit in. Goldstine, a mathematician underneath his army uniform, and Burks, a logician finishing up his wartime service as an electronic engineer, had been granted sanctuary in the annex to Gödel's office on the second floor, from where they assimilated easily into the culture of Fuld Hall. When engineers began showing up, the welcome cooled. The ivory tower was full. "I have thought very carefully over the problem of disposing of these fifteen workers who are to arrive the middle of June," wrote Aydelotte to von Neumann. "The only really useable space in our basement is that adjoining the men's lavatory, to which you are most heartily welcome."[32]

"There was no space for us, and so for the first five or six months, we were crowded into the boiler room with a few work benches we set out," Bigelow explains. "There was not even an office for me to go to and hide, and think about circuit logic, without having people walking over my desk and crawling all over me." All purchases of building materials, down to a single two-by-four, had to be reviewed by the Civilian Production Board. "Finished lumber was rationed because of housing shortages," adds Bigelow, "so we purchased rough-sawn oak from a local purveyor of fireplace wood."[33]

The engineers were shunned by the scholars upstairs. "The prevailing attitude among the humanists toward the notion of a laboratory at the Institute was one of undisguised horror," remembers Bigelow. "The attitude of the mathematicians ranged from about the same extreme to some instances of mild interest; however, the extremists among them tempered their objections in accord with the universal respect and esteem enjoyed by von Neumann."[34]

"We were doing things with our hands and building dirty old equipment. That wasn't the Institute," remembers Ware. "The coming of six engineers with their assortment of oscilloscopes, soldering irons, and shop machinery was something of a shock."[35] When the project first started, "we were given temporary space in the second basement, surrounding the boilers," he adds. "It wasn't bad since it was summer and they were turned off." The basement storage room was bare. "Our first job was to build work tables for us to work on," remembers Slutz. "We asked von Neumann if he would pay for the paint if we painted the walls a more reasonable color than they were when we moved in. This he did." The engineers also had to install their own wiring, and, reminiscent of Julian Bigelow's installation of a single circuit to his bedroom, the first engineering expenditure by the Electronic Computer Project was four dollars, recorded by Bernetta Miller in April 1946, for "Electrical work."[36]

"Our work benches surrounded the boilers and miscellaneous shop and laboratory equipment was stuffed in any available corner," explains Ware. "With the coming of autumn, the situation warmed up quite considerably; so much so in fact that the group enjoyed an improvement in social status and moved to some unoccupied storerooms in the first basement."[37] As they moved up from the boiler room level to the first basement directly below the ground floor of Fuld Hall, there were vocal protests on all fronts. To the humanists, the computer people were mathematicians, and to the mathematicians, they were engineers. "Even the curiosity so natural to all scientific people was overruled by the passionate distance toward anything that might conceivably deviate from pure and theoretical thinking," Klári von Neumann explained. Bigelow describes the situation as one of people "who had to think about what they were trying to do" objecting to people "who seemed to know what they were trying to do."[38]

The mathematicians and the humanists occupied opposite wings of Fuld Hall. Their defenses were entrenched. "I have learned with some dismay that a group of electronic experts has moved into half of the basement of our wing at the Institute," classicist Benjamin Merritt complained to Aydelotte in 1946.[39] The engineers and the scholars were forced to coexist. "There would be social gatherings from time to time," remembers Ware, "and they'd say 'Well, I'm in mathematics' or 'I'm in physics,' or 'I'm in . . . what are you in?' And then when one answered, it became clear that you were a social outcast. We were sort of fifth class citizens around there."[40]

Salary disparities did not help. The engineers received $5,000 or $6,000 a year, barely enough to keep them from leaving for better-paying jobs in industry, but more than what was paid to visiting scholars with advanced degrees. "Many of these people, who were good electro-technologists, had, for example, only a bachelor's degree," explains Bigelow, "whereas the Institute itself had visiting members with Ph.Ds from four or five important universities in the world, who came over here for fellowship stipends which amounted to $2,500 or $3,000. So there were really substantive kinds of jealousy arising from this."[41] The budget for the Computer Project—funded entirely by the government—soon grew larger than that of any of the existing schools.

The computer group needed to move out of Fuld Hall, and soon. With the postwar shortage of building materials, and the resistance of both the Institute's old guard and the neighboring residents to anything resembling a laboratory, this was not easy to do. "At that time, building materials were under rationing," explains Bigelow. "You could not go out and build yourself a house or a garage or something. If you wanted a few feet of wood, you had to have a certificate for it, much less things like workbenches and hardware and tools."[42] It turned out to be more difficult to obtain things under the Civilian Production Administration than it had been under the War Production Board. A lot of bargaining went on behind the scenes. "I just finished talking to Sam Feltman (at the Ordnance Department) regarding our contract," Goldstine explained to von Neumann, when money was running low. "He says he received his allotment of money just this morning and hopes for War Dept. approval within 2 or 3 days at which time he will authorize Phila Ord. Dist. to give me the dough, Gott sei Dank!! In turn for this he has two favors: First, he wants a copy of the

*Theory of Games,* which I'll try and send him. Second, he wants help in placing his son in medical school."[43]

The other problem was where to put the building. "There is a distinct feeling in the faculty that the computer building should not be on the same plot of land with Fuld Hall," Aydelotte wrote to Herbert Maass, suggesting a location on the other side of Olden Lane, near the old barn. "Most of the Institute buildings are placed close together, connected with a path or archways," noted Klári von Neumann. "This one was put way across a big empty field edged with tall bushes, beyond it a road and then the building; one could easily pretend that it did not belong with the rest at all."[44]

Arthur Burks remembers helping Herman Goldstine and Oswald Veblen choose the site. "And we walked through the woods, but it was clear that Veblen didn't want any trees to be cut down for the building. In the end, he picked a site which was low down, and not too far away. . . . He wanted the building to be one story only, so that this would not be a conspicuous building." The site was too swampy to build on as is. "The site we have selected needs a little fill, and I notice that you people are excavating immense quantities of earth in preparation for building the stacks of the new Princeton library," Aydelotte wrote to the university. "As far as I can see your trucks seem to be hauling the earth a considerable distance away, and I am wondering if we could buy a few truckloads of this earth."[45]

The Electronic Computer Project's contract with its government sponsors allowed $23,000 for the construction of a "temporary structure" to house the computer, whereas the bids for a building whose appearance was acceptable to the Institute community came in at $70,000. After much negotiation, it was decided that a flat-roofed, cement-block building would be constructed for $51,000, with the Institute paying an additional $9,000 for a cosmetic brick veneer, and reserving the right to add a gabled roof at a later time. "The exterior walls of the building are being veneered with brick in order to insure a weather-tight job, since we were advised that the present grade of stucco was of questionable weatherproofing quality," Aydelotte wrote to Colonel Powell, chief of army ordnance, who questioned the cost of the brick veneer. "However, in order that there be no possible criticism of us in this connection, we had always planned to pay the $9,000 for this adjunct to the building out of Institute funds."[46]

The new building would not be ready to move into until Christmas of 1946. In the interim, the engineering team prepared for the construction of the machine. A small machine shop (equipped with a lathe, drill press, and planer) was established in the boiler room, and the engineers began accumulating electronic components, instruments, and tools. "We obtained surplus components and built our own power supplies," remembers Bigelow. "We really worked from the ground up." Electronic components were still restricted for civilian use. "Whenever we wanted anything, we tried to get the Army Materiel Command to find it for us, and at that time there were itinerant salesmen who would buy war surplus and then go around and try to peddle it, and so we got a lot of stuff that way," adds Willis Ware. "The Princeton machine was built through war surplus. We used whatever the army could find for us, so in a subtle sort of way that pushed the design."[47]

The engineers used their personal connections and personal equipment to help build the computer, and used computer project resources for personal projects in their spare time. Everything from Einstein's hi-fi to Herman Goldstine's TV antenna was made in the ECP shop. "Princeton has the unique characteristic that it's almost exactly halfway between Philadelphia and New York, so RCA designed an antenna that, with the flip of a switch, electronically pointed to New York or electronically pointed to Philadelphia," says Willis Ware. "So we all, in our shop, with Melville's aid, tooled up and made those antennas for a lot of people."[48] Jack Rosenberg took things one step further and made his own *recordings*. "He used to listen on WQXR every Saturday to Toscanini playing in New York and he would record it with tremendous high fidelity," explains Morris Rubinoff. "Now tremendous high fidelity considering the antenna he had meant that you heard every bit of static, crackle and noise that came through, and you were proud that you could hear it all, it certainly was hi-fi—from 15 cycles to 20,000 or 30,000 or something."[49] With Aydelotte's replacement by Oppenheimer, the freewheeling atmosphere began to change. "It has proven convenient, especially for those engaged in work on the computer, to avail themselves of the discount rates of the computer contract in making essentially personal purchases of radio equipment and parts," Oppenheimer wrote to von Neumann in 1949. "However, it appears that in a number of cases we have carried the items so charged on our books for a quite improper length of time."[50]

The design of the computer was determined partly by the commandments that Burks, Goldstine, and von Neumann dispensed from above ("They were our bible," says Rosenberg)[51] and partly by what the availability of war surplus components dictated from below. "We bought quantities of surplus electrical and electronic components, electron tubes, etc., rather indiscriminately," says Bigelow. "We read through *Preliminary Discussion . . .* often and discussed the engineering task it presented with each other and with Johnny and Herman, who were already trying out exploratory coding procedures on paper."[52]

From the beginning, there was tension between Goldstine, the associate director of the project, and Bigelow, the chief engineer. On almost every important question, from circuit design (with Bigelow clinging to "a weird and inchoate idea as how to build an adder," as Goldstine complained to von Neumann) to the disposition of patent rights, the two disagreed. The chain of command was often in dispute. Only von Neumann was strong enough to intercede. "He kept Herman and I from fighting by some marvelous technique," says Bigelow. "We got along like oil and water, or cat and dog; and von Neumann would keep this here, and this there, and smooth things over."[53]

"I didn't talk to von Neumann very often," says Ralph Slutz. "It was more like I talked to Bigelow and Bigelow talked to von Neumann. You know, the Cabots speak to the Lodges and the Lodges speak to God." Conversations with von Neumann were often long-distance calls. "He had a habit of telephoning at any hour of the day or night," says Goldstine. "Even at 2 in the morning he might telephone and say 'I see how to do this.' And then tell you. The main problem with working long distance with von Neumann was that telephone connections weren't that good in those days, and von Neumann spent most of his time saying 'Hello!' So that whenever the line was clear, all we were doing was being busy saying 'Hello.' But in spite of these things, we got a lot done by these means."[54]

Von Neumann wanted to know *how* everything worked, but he left it to the engineers to *make* it work. "The experimental business wasn't really for von Neumann," Goldstine explains. "Once he understood the principle of it, the ghastly details like the fact that you'd have to put by-pass condensers on things, and all sorts of dirty engineering things—that didn't really interest him. He recognized that these were essential, but it wasn't his thing. He would not have had the patience to sit there and do it; he would have made a lousy engineer."[55]

According to Bigelow, "Von Neumann had one piece of advice for us: not to originate anything." This helped put the IAS project in the lead. "One of the reasons our group was successful, and got a big jump on others, was that we set up certain limited objectives, namely that we would not produce any new elementary components," adds Bigelow. "We would try and use the ones which were available for standard communications purposes. We chose vacuum tubes which were in mass production, and very common types, so that we could hope to get reliable components, and not have to go into component research."[56]

The fundamental, indivisible unit of information is the bit. The fundamental, indivisible unit of digital computation is the transformation of a bit between its two possible forms of existence: as structure (memory) or as sequence (code). This is what a Turing Machine does when reading a mark (or the absence of a mark) on a square of tape, changing its state of mind accordingly, and making (or erasing) a mark somewhere else. To do this at electronic speed requires a binary element that can preserve a given state over time, until, in response to an electronic pulse or some other form of stimulus, it either changes or communicates that state. "Most of the essential elements or 'Cells' in the machine are of a binary, or 'on-off' nature," Bigelow and his colleagues explained in their first interim progress report. "Those whose state is determined by their history and are time-stable are memory elements. Elements of which the state is determined essentially by the existing amplitude of a voltage or signal are called 'gates.'"[57]

In 1946, on the eve of the transistor, it was uncertain whether the non-zero probability of error in any individual digital transformation would bring a computation involving millions of transformations to a halt. The ENIAC was the only large-scale precedent. "The mere fact that the ENIAC was and that the ENIAC ran gave me tremendously more confidence that something could be done than would have been the case if there hadn't been that demonstration of such a large machine running," says Ralph Slutz.[58] But the new machine would be to the ENIAC as the ENIAC was to a desktop calculator. What available fundamental computational element was reliable enough to work?

The answer was the 6J6, a miniature twin-triode vacuum tube that was produced in enormous numbers during and after World War II. Three-quarters of an inch in diameter and 2 inches in length, with a 7-pin base, the 6J6 drove military communications during the war and the consumer electronics industry that followed. Effectively two tubes in one envelope, a common cathode (pin 7) served two separate plates (pins 1 and 2) and grids (pins 5 and 6). The twin-triode architecture allowed the tube to be used as a "toggle," with one side or the other in a conducting state, and less than a microsecond required to make the switch. "And that was the word insisted on by Julian Bigelow as being the more accurate word for what the flip-flop does. And he's right," says Pomerene. "Flip-flop is not the right word for a bi-stable circuit which stays in whatever state you put it in." This constituted a far more secure representation of binary data than an element whose state is represented by simply being on or off—where failure is indistinguishable from one of the operational states. As Bigelow later described it, "A binary counter is simply a pair of bistable cells communicating by gates having the connectivity of a Möbius strip."[59]

"If the 6J6, which was the twin triode, had not existed during the war and had not been widely used, I don't know what we would have used for a tube," says Willis Ware. Not only did the widespread use of the 6J6 mean that it was available inexpensively, but it was found to be more reliable as well. One of Bigelow's last assignments at the Statistical Research Group at Columbia had involved the reliability of munitions. "There had been a lot of accidental explosions of rocket propellant units on airplanes in which the explosion would take the wing off a plane," he explains. "And this would happen in a very rare and erratic fashion. So we had some excellent people in statistics there, including no less than Abraham Wald, who founded sequential analysis while working with our group. Statistical thinking had become a part of my way of thinking about life." It turned out that the most reliable tubes were those produced in the largest quantities—such as the 6J6. As Bigelow described it, "We learned that tube types sold at premium prices, and claimed to be especially made for long life, were often less reliable in regard to structural failures than ordinary tube types manufactured in larger production lots."[60]

That higher quality did not require higher cost was not readily accepted, especially since IBM, who had used the 6J6 as the computing element in its popular model 604 electronic calculator, had recently established its own experimental tube production plant in Poughkeepsie, New York, to develop special computer-quality tubes at a much higher cost. There was intense debate over whether the choice of the mass-market 6J6 was a mistake. Of the final total of 3,474 tubes in the IAS computer, 1,979 were 6J6s. "The entire computer can be viewed as a big tube test rack," Bigelow observed.[61]

"It was considered essential to know whether such miniature tubes as the 6J6 have radically inferior lives compared to other types, to an extent rendering their use in design a major blunder; and accordingly a crude life-test set up was devised and operated to get some sort of a statistical bound on their reliability," Bigelow reported at the end of 1946. Four banks of 6J6 tubes, twenty in each bank, for a total of eighty tubes, were installed in a test rack so they were oriented up, down, and in the two horizontal positions (cathode edge-wise and cathode flat). The entire rack was mounted on a vibrating aluminum plate, and the tubes left to run for three thousand hours. "A total of six failed, four within the first few hours, one about 3 days and one after 10 days," was the final report. "There were four heater failures, one grid short and one seal failure."[62]

The problem wasn't tubes that failed completely—built-in self-diagnostic routines made these easy to identify and replace—it was tubes that either were not up to specification in the first place, or that drifted off specification with age. How could you count on getting correct results? While von Neumann was beginning to formulate, from the top down, the ideas that would develop into his 1951 "Reliable Organizations of Unreliable Elements" and 1952 "Probabilistic Logics and the Synthesis of Reliable Organisms from Unreliable Components," the technicians at the Institute were facing the same problem, from the bottom up.

It took an engineer fluent in wartime electronics and wartime ingenuity to solve the problem of building a reliable computer from unreliable war surplus parts. Jack Rosenberg, from New Brunswick, New Jersey, was the first in his family to attend college, entering MIT in 1934 at age sixteen. As a senior in high school he had attended the Century of Progress exhibit in Chicago, and "spent nearly the whole week in what was called the Hall of Science. And I saw there a booth by MIT, and I talked to the man at the booth, and he said MIT is probably the toughest school to get into. So I applied to MIT."

Rosenberg started out in mathematics but switched to electrical engineering, graduating at the top of his class with two degrees. "When I went around for interviews in 1939, I saw many of my classmates already working," he says. "I knew that I was more intelligent than them, but that's the way it went." So he took a job as a civilian engineer for the U.S. Army Signal Corps, becoming an officer when the United States joined the war.

July of 1945 found Rosenberg on board an army troop ship sailing at eight knots across the Pacific to the Philippines, to prepare for the invasion of Japan. "As a radio ham, I spent most of my waking time in the radio room listening to the short wave," he says. Since the slow transport was a sitting duck, transmission was not allowed. On August 6, 1945, he heard news of the atomic bombing of Hiroshima, followed by news of the bombing of Nagasaki on August 9. "The ship's troop commander was as startled as I had been," he says. "He told me to keep listening to the radio. His orders for the invasion had not been changed." Then the news of Japan's unconditional surrender arrived. "The bombs had saved our lives," says Rosenberg, and however difficult von Neumann (and Oppenheimer) proved to be as his employers, he never forgot that.[63]

Rosenberg remained in the southern Philippines until April 1946. In the post exchange, he found a copy of *Atomic Energy for Military Purposes*, a swiftly declassified nontechnical account of the Manhattan Project by Henry Smyth, chairman of the physics department at Princeton University. Upon his discharge from the army, at Fort Dix, New Jersey, in July of 1946—having returned across the Pacific on a turbine-driven steamship at thirty knots—Rosenberg went to Princeton to seek a job in nuclear energy research. He was hired by the Physics Department to work on the instrumentation for the university's new cyclotron, but, he says, "my enthusiasm lasted about a month."

"Early in 1947," he continues, "I was informed that at the Institute for Advanced Study, a famous scientist was looking for an engineer to develop an electronic machine of a sort no one but he understood." Rosenberg interviewed with Bigelow and von Neumann, and started work in July. "There was a lot of anti-Semitism in the army. But there wasn't anti-Semitism with Johnny," he says.

"Johnny used to meet with each of us individually about once a week, asking what we had built, how it worked, what problems we had, what symptoms we observed, what causes we had diagnosed," says Rosenberg. "Each question was precisely the best one based on the information he had uncovered so far. His logic was faultless—he never asked a question that was irrelevant or erroneous. His questions came in rapid-fire order, revealing a mind that was lightning-fast and error-free. In about an hour he led each of us to understand what we had done, what we had encountered, and where to search for the problem's cause. It was like looking into a very accurate mirror with all unnecessary images eliminated, only the important details left."[64]

When Rosenberg arrived, the problem was how to build a forty-stage shift register, this being at the heart of the machine's ability to compute. "It was easy to build a two-stage register that worked reliably," says Rosenberg. "When a third stage was added occasional errors crept in. Adding a fourth stage made the register useless. We discovered that the electrical characteristics of the vacuum tubes were very different from the specifications published for them in tube handbooks, even when the tubes were new."

According to Rosenberg, after further extensive testing, and consultation with the major tube manufacturers, whose response was that "no one else had ever complained about their product, and they had enough customers without us," von Neumann was informed that "there were no reliable tubes, and no reliable resistors." The response from him was that "we would have to learn how to design a reliable 40-stage machine with thousands of unreliable components," said Rosenberg. And they did.[65]

They switched from designing according to the published tube specifications to what is now called "worst-case design"—which, "in recognition of the then current new fashion in women's clothing, Bigelow called 'The New Look.'" As Ralph Slutz explains it, "we tested a batch of a thousand tubes and took the weakest tube we found and the strongest tube we found, and then allowed an extra 50% safety factor over that."[66]

The new design parameters were extended from individual tubes to toggles, gates, standard circuit modules, and finally to full-scale forty-stage registers, which, after tedious debugging, worked. Bigelow also argued, counterintuitively, that the machine's overall reliability could be improved by speeding it up, noting that "increasing speed may actually increase certainty rather than the reverse." Unlike mechanical devices, vacuum tubes are weakened by age, not use, and "suffer accidental failures in proportion to their population," not their operating speed. Optimum reliability can therefore be achieved by operating as few tubes as possible, at maximum speed. "Finally," noted Bigelow, "intermittent errors are most embarrassing and difficult to detect when the intermittency corresponds roughly to the operating rate."[67]

The IAS engineers coaxed collectively acceptable digital behavior out of tubes whose performance would have been largely unacceptable if tested against the specifications for individual tubes. This was achieved by following the same principle that Bigelow and Wiener had developed in their work on the debomber: separate signal from noise at every stage of the process—in this case, at the transfer of every single bit—rather than allowing noise to accumulate along the

way. This, as much as the miracle of silicon, is why we have microprocessors that work so well today. The entire digital universe still bears
the imprint of the 6J6.

"One time we thought it would be a good idea to take the vacuum
tubes out of the machine and just run them through a regular testing
routine," remembers James Pomerene, "and you never saw a crum-
mier bunch of tubes in your life!"[68]

# EIGHT

# V-40

*Far as all such engines must ever be placed at an immeasurable interval below the simplest of Nature's works, yet, from the vastness of those cycles which even human contrivance in some cases unfolds to our view, we may perhaps be enabled to form a faint estimate of the magnitude of that lowest step in the chain of reasoning, which leads us up to Nature's God.*

—Charles Babbage, 1837

PENTAERYTHRITOL TETRANITRATE (PETN) was an important high explosive in World War I, yet its molecular structure remained unknown in World War II. Andrew Booth, whose father invented, among other things, the automotive ignition advance and a stovetop thermocouple designed to power a radio in a household without electricity, was a graduate student at Birmingham University when he was assigned the job of using X-ray crystallography to find this out.

Booth, who had "alarmed his mother by mending fuses at the age of two," won an undergraduate mathematical scholarship to Cambridge University in 1937, where he was assigned to pure mathematician G. H. Hardy, a match doomed from the start. Given an ultimatum by Hardy either to quit wasting time on other subjects or to lose his scholarship, Booth, who believed that "if mathematics isn't useful, it's not worth doing," left Cambridge to pursue physics, engineering, and chemistry on his own terms. During an apprenticeship at the Armstrong Siddeley aircraft engine works in Coventry, he improved the design of searchlights and established an X-ray facility for the inspection of engine parts. This led to a graduate scholarship, sponsored by the British Rubber Producers Research Association, to determine the molecular structure of PETN, although this was not made explicit at the time. "I was just given some stuff and asked to find out what the structure was," he explains. "Which we did, and of course we realized it was what it was by then."[1] In addition to PETN, Booth's group derived the structure of RDX, a new plastic explosive that would play a key role in the development of atomic bombs.

By recording the diffraction patterns produced by X rays scattered from a sample of crystalline material, it is possible, if difficult, to infer the pattern of electron densities, and thus the molecular structure, that produced the observed diffraction. Given a known molecular structure, it is easy to predict the pattern of diffraction, but given an observed diffraction pattern, there is no easy way to determine the molecular structure. You start by taking a guess, and then perform the calculations to see if the guess was anywhere near correct. After many iterations, a plausible structure may—or may not—emerge.

"What people did previous to the time when I came along with the computing suggestions was to jiggle things about and hope for the best," remembers Booth. "And it took a hell of a long time, and usually gave a very bad result." The physics behind the scattering of X rays was straightforward, but to work backward took brute force. "There is a lot of very nasty calculation," he says. "For a typical structure like the one I did, there are about four thousand reflections . . . and you have to calculate the sum over all of the wretched atoms in your structure to calculate this wretched phase angle. It takes a hell of a long time by hand. I had a boy and one girl who were helping me with the calculations. It took us three years."[2]

With the structure of PETN determined, and his PhD in hand, Booth transferred to the main BRPRA laboratories at Welwyn Garden near London. Under the supervision of John W. Wilson, he began to build a series of mechanical and electromechanical calculators to speed up the X-ray analysis work. This attracted the attention of crystallographer Desmond Bernal, who was launching a "bio molecular" laboratory at Birkbeck College to tackle more complex organic molecules that had been resistant to structural analysis so far. "I was building this special-purpose crystallographic digital computer," says Booth. "That's what Bernal was interested in, and that's why he wanted me to come work with him." One of the members of Bernal's group was Rosalind Franklin, later to play a key role in determining the helical structure of DNA.

In 1946, Bernal sent Booth on a mission to survey the state of computing in the United States. Warren Weaver, now back at the Rockefeller Foundation, agreed to sponsor Booth's initial visit, and to follow up with a Rockefeller fellowship at the laboratory of Booth's choice. Booth made the rounds: visiting George Stibitz at Bell Labs, Howard Aiken at Harvard, Jay Forrester at MIT's Servomechanism Laboratory, the EDVAC group at the Moore School, von Neumann and Bigelow at the Institute in Princeton, and finally Eckert and Mauchly at the Electronic Control Company (who "were totally hostile," he says). He even visited an abacus-based computing center at the Bank of Hong Kong in San Francisco, and gave talks to audiences ranging from ladies' clubs to Linus Pauling's research group.

He also "made some side trips," including staying "for a couple of nights" with Irving Langmuir of General Electric, whose interests ranged from weather modification to protein structure, but who was most excited, at that moment, by his invention of the garbage disposal, which he termed the "electric pig." The device was demonstrated to Booth. "I had just eaten a banana," says Booth, "and I just threw the banana skin in. There was a horrible scrunch and the thing jammed solid. So it finished up with this Nobel Prize winner crawling about on his tummy on the floor, taking the bottom off this thing and getting the banana skin out."[3]

"Then I went back to New York, and I met with Weaver," remembers Booth. "And he said, 'What do you want to do?' I was quite clear by then, and I said, 'Well the only group that is worth talking to is the lot at Princeton.' They were the only people who had really got themselves out of the business of just waving their hands in the air and doing nothing."[4]

Familiar with both the Analytical Engine of Charles Babbage and with Turing's Universal Machine, Booth saw the IAS project as the practical implementation of these ideas. Booth, who knew Turing from Cambridge, was later asked by the National Research Council to review some of the circuits Turing had designed for the Automatic Computing Engine (ACE) being built at the National Physical Laboratory in London. "They were incredibly complicated," he says. "And for each of these circuits, I drew an equivalent one the way I would have designed it. And they were about a quarter as expensive in components. I think some of Turing's would not have worked."[5]

Bigelow's approach was minimalist. "Julian's fixed principle was that you had to build things with no capacitors," Booth explains. "If you have capacitors, then you have a limiting speed. If you don't have capacitors, then if you can put enough juice into it you can get any speed you like."[6] The speed of the IAS computer could also be slowed down and, while debugging programs, even "stepped," one instruction at a time. Much of the useful computation was done at 8 kilocycles, or about half throttle. There was no fixed "clock speed." As soon as one instruction was executed, the computer proceeded to the next.

In early 1947, Booth sailed for New York on the *Queen Mary*, accompanied by his assistant, Kathleen Britten, later the author of an early textbook on computer programming, who had been spearheading the X-ray calculator work. Britten's passage (and salary) was being paid by John Wilson of the BRPRA, who booked her a cabin in first class. Booth, whose passage (and fellowship at the Institute) was being paid by the Rockefeller Foundation, was booked into steerage. Bernal registered a complaint. "Eventually Johnny [Wilson] put out the money for me to go first class as well," adds Booth.

When Booth and Britten arrived in Princeton at the end of February 1947, the question facing all Institute visitors was, where could they live? "I think there will be no problem in taking care of Miss Britten as well as of Booth," Goldstine wrote to von Neumann. "Miss [Bernetta] Miller is trying to hire Miss [Hetty] Goldman's housekeeper to move into one of the apartments and fix meals. In this case presumably both Booth and Miss Britten could move in with Miss Goldman's housekeeper and observe all the necessary proprieties." Booth and Britten—who were married in 1950, "after Kathleen had received her Ph.D. in trans sonic aerodynamics"—were among the first group of Institute visitors to set up housekeeping in the new housing project across Olden Lane from Fuld Hall.[7]

In March of 1946, Veblen had proposed constructing emergency housing for the computer project personnel. "Professors Panofsky and Morse were not in favor of this proposal," the minutes report. "They did not think it wise that the Institute divert its funds for a project of this sort for the almost exclusive benefit of the computing group."[8] In June of 1946, as engineers began showing up for work, the situation became acute. New hires were forced to commute from as far as Philadelphia and New York. "The best solution would be to rent a block of apartments from the New York Life Insurance Company," Aydelotte reported to an emergency meeting of the trustees. "However, the New York Life has been hesitant about leasing these apartments to the Institute; and Dr. Aydelotte suspects that they prefer not to rent to Jews, although they have said nothing to this effect. The Committee decided to submit the New York Life a tentative list of tenants, on which Jews would be included, but no Hindus or Chinese."[9]

This approach failed, and in desperation the surrounding community was canvassed for help. The Lawrenceville School, whose headmaster "liked the idea that these men would mingle with the boys, make speeches in chapel, etc.," agreed to take in two or three Institute scholars, while a chicken farm near Rosedale offered four apartments, with central heating, "especially suitable for couples with children." Marston Morse proposed that "as a last resort, cots be put in some of the rooms at Fuld Hall." At the end of 1946, a family of four was still camped out in Fuld Hall.[10]

The breakthrough came in August 1946, when a cluster of wood-frame apartment houses, built to house a wartime influx of workers at the Republic Steel Company's iron mines in Mineville, upstate New York, were put up for sale. "I sent an engineer, Mr. Bigelow, of the computer project, to Mineville the same day," Aydelotte reported to the trustees. "He found representatives of two other universities on the spot eager to secure the houses available. Thanks to the enterprise of Mr. Bigelow, we were able to buy eleven buildings, containing thirty-eight apartments of two and three bedrooms each. These apartments are substantially built, they have hardwood floors, are insulated with rock wool against the Adirondack winters, and are fitted with storm windows, fly screens, clothes lines and garbage pails."[11]

There was only one problem: Mineville and Princeton were three hundred miles apart. Under Bigelow's supervision, the buildings were dismantled into sections, transported by rail to Princeton, and reassembled, with poured-concrete foundations, on Institute property between the Springdale golf course and Fuld Hall. The entire project was completed by January 1947, at a cost of $30,000 for the houses and $212,693.06 for the site preparation and the move—despite the complaints of nearby Princetonians who sought to halt the project "because of its deleterious effects upon the fashionable housing area which it will invade."[12]

The Mineville houses were built in the same wartime style as the government housing at Los Alamos, and some who had stayed in the Los Alamos housing project during the war, under Oppenheimer's directorship, found themselves staying in the Institute housing project after the war, under Oppenheimer's directorship once again. By February 1947 the first seventeen families, including the Bigelows, were occupying the new apartments, and more were moving in. "Since we have been here," Bigelow reported to Aydelotte, "we have come to know many of our neighbors quite well, not only those working in mathematics and physics, with whom we have much in common, but what is often more stimulating, we have met people working in other fields with experience and outlook different from our own."[13]

In April 1947, the Institute awarded Bigelow an honorarium of $1,000 for his efforts, and Bernetta Miller reported, in September, that "there are now 30 odd children and more coming," while "tenants are grateful for the prospect of grass now beginning to show." Informal gatherings among the Mineville houses soon became a fixture of Institute life. "We had a tendency to congregate during the evening, and we got to know one another extremely well," remembers Morris Rubinoff, one of the second contingent of engineers who arrived in June of 1948.[14]

"Julian and Mary were the heart and soul of the housing project," remembers Freeman Dyson, who also arrived in 1948. "If you had any personal problems, you went to Mary. She would give you what you needed: moral support, good advice, and just the warmth of her character. And if you had practical problems: a car that needed fixing, or a rat in the basement, or a problem to get the coal furnace either to work at all or not to work too much, Julian would always be able to fix it. Those were wonderful years."[15]

With the housing crisis solved, and the computer building completed, the engineers could start building the computer—along with its power supplies and cooling equipment—rather than working on one component at a time. The computer building was adjacent to the housing project, and, says Rubinoff, "you could get to work and get home to lunch and back to work in less time than you can do it anywhere else." It was Los Alamos all over again. "They would work till eight or nine o'clock, go out to dinner for two hours and then go back to work again," remembers Thelma Estrin, an electrical engineer who arrived, with her husband, engineer Gerald Estrin, in June 1950, in the middle of the final push to finish the machine. "Sometimes they would work all night."[16]

"I had just completed my doctorate," adds Gerald. "I had never heard of a computer; I didn't know anything about it," but while searching for a job, he was told that "there's an interesting project going on at the Institute of Advanced Study." Von Neumann invited the Estrins for a visit, and hired them on the spot. "Von Neumann liked to be in the middle of stuff," says Gerald. "We stepped off on the grounds and fell in love with the place. There were little signs on the grass, tiny ones, that said, 'Please Don't Walk on the Grass.'" The

Estrins spent the next three years at the Institute, before moving to Israel to build a copy of the IAS machine. "They were very concentrated years. Being in such a small group, I really learned about every part of the computer, helping in every way that I could."[17]

In an age when most microprocessors require only a single voltage—somewhere between 1 and 5 volts—it is difficult to comprehend how many different voltages it took to run a vacuum tube computer. There were seven main branches leading from the 120-volt, three-phase power lines entering the building from outside. First there were three branch circuits that supplied three-phase power to the vacuum tube heaters: about 6.5 kw to the arithmetic unit, and 1.5 kw to the memory. Second, DC power was supplied to the core of the computer through four separate rectifiers, subdivided further into twenty-six different voltages, ranging between –300 and +380 volts. Finally, the Williams tube deflection circuits required a regulated 1075, 1220, and 1306 volts. A useful current in one circuit could easily induce noise somewhere else, not to mention noise introduced by transients in the incoming power lines. Initially, there was so much trouble with noise in the DC power supplies that a 300-volt, 180-ampere-hour battery house was built outside the computer building, to supply clean DC power until the memory could be better shielded and more stable power supplies designed.

None of these voltages meant anything without reference to a common ground. This resulted in a landmark known, briefly, as the Rosenberg ground. "We had developed the machine in two sections," James Pomerene explains. "And at some point, due to things which I don't remember anymore, when we came to put them together, what I called 'ground' in my design was at a different voltage level than what Rosenberg called 'ground.' And so for a while we had batteries that made the adjustment between my ground and the Rosenberg ground."[18]

The computer consisted of four "organs": input/output, arithmetic, memory, and control. The choice of memory drove the design, yet was the last element to be resolved. "Once the form of the high speed memory has been decided most of the other components of an electronic computer become semi-invariant," Booth and Britten observed in their report on their stay at IAS.[19] The anticipated all-digital Selectron memory tube from RCA, if not yet in existence, was specified accurately enough that the rest of the computer could be

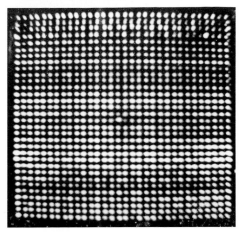

The digital universe in 1953. A 32-by-32 array of charged spots—serving as working memory, not display—is visible on the face of a Williams cathode-ray storage tube (stage thirty-six) in this diagnostic photograph from the maintenance logs of the Institute for Advanced Study Electronic Computer Project, February 11, 1953. (*Shelby White and Leon Levy Archives Center, Institute for Advanced Study*)

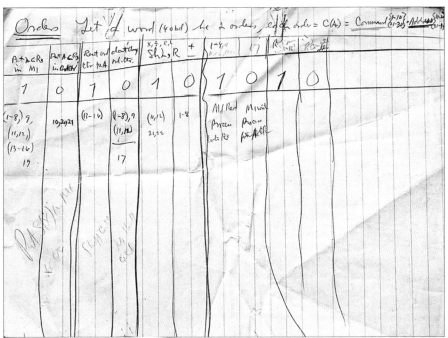

In the beginning was the command line. "Orders: Let a word (40bd) be 2 orders, each order − C(A) = Command (1–10, 21–30) • Address (11–20, 31–40)," reads the top line of this undated note, saved by Julian Bigelow and evidently written, given the use of the abbreviation *bd* for *binary digit*, in late 1945 or early 1946 before the introduction of the term *bit*. (*Bigelow family*)

An IAS General Arithmetic Operating Log entry for March 4, 1953, notes "over to" a thermonuclear weapons design code, immediately after Nils Barricelli's numerical evolution code (terminating at memory location 18,8) is run for the first time. (*Shelby White and Leon Levy Archives Center, Institute for Advanced Study*)

Alan Turing at age five. *(King's College Archive, Cambridge; courtesy of the Turing family)*

John von Neumann at age seven.
*(Nicholas Vonneumann and Marina von Neumann Whitman)*

Alan Turing's "On Computable Numbers, with an Application to the Entscheidungsproblem" was published in the *Proceedings of the London Mathematical Society* shortly after Turing's arrival in Princeton in 1936. The Institute for Advanced Study's copy was consulted so frequently it became unbound. (*Institute for Advanced Study*)

PROCEEDINGS

OF

THE LONDON MATHEMATICAL SOCIETY

INSTITUTE
FOR ADVANCED STUDY

SECOND SERIES

Colossus at Bletchley Park in 1943. To help decipher digitally encrypted enemy telecommunications during World War II, British cryptanalysts built a series of versatile, if not yet universal, logical computing machines. Supervised by Dorothy Du Boisson and Elsie Booker, "Colossus" compares a coded sequence stored in an internal vacuum-tube memory with a sequence stored on external punched paper tape by scanning at high speed with photoelectric reading heads. (*National Archives Image Library, Kew, U.K.*)

Alan Turing (far left) in 1946. With the war over, Turing began designing the Automatic Computing Engine (ACE) to be constructed at the National Physical Laboratory in London, while von Neumann began designing the Mathematical and Numerical Integrator and Computer (MANIAC) to be constructed at the IAS. Turing's design was influenced by von Neumann's implementation, and von Neumann's implementation was influenced by Turing's ideas. *(King's College Library, Cambridge)*

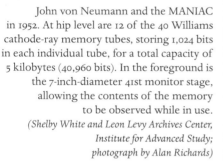

John von Neumann and the MANIAC in 1952. At hip level are 12 of the 40 Williams cathode-ray memory tubes, storing 1,024 bits in each individual tube, for a total capacity of 5 kilobytes (40,960 bits). In the foreground is the 7-inch-diameter 41st monitor stage, allowing the contents of the memory to be observed while in use. *(Shelby White and Leon Levy Archives Center, Institute for Advanced Study; photograph by Alan Richards)*

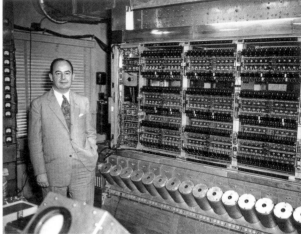

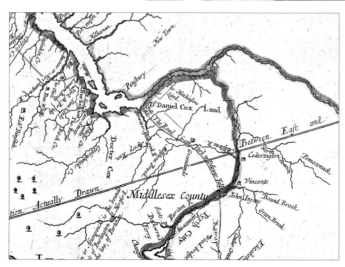

"The Road from York to Delaware Falls," formerly a Lenni Lenape footpath across the "waist" of New Jersey between the Raritan and Delaware estuaries, with Greenland's Tavern at the halfway point. The diagonal line marks the division between East and West New Jersey, decided by a meeting at the tavern in 1683. The site of the Stony Brook Quaker settlement and future town of Princeton is near the center of the illustration, just below the *f* in "from," in this detail from "A new mapp of East and West New Jarsey: being an exact survey taken by Mr. John Worlidge," London, 1706. *(Library of Congress, Geography and Map Division)*

Fuld Hall, headquarters of the Institute for Advanced Study, was constructed in 1939 at Olden Farm in Princeton, New Jersey, on land that had changed hands only twice since the ownership of William Penn. (*Abraham Flexner,* I Remember *[New York: Simon & Schuster, 1940]*)

Oswald Veblen, nephew of Thorstein Veblen (who coined the phrase "conspicuous consumption" in his 1899 *The Theory of the Leisure Class*), was a topologist, geometer, ballistician, and outdoorsman who, as a student, earned one prize in sharpshooting and another in math. The first professor hired by the Institute for Advanced Study, in 1932, it was Veblen who had suggested the idea of an autonomous mathematical institute to Simon Flexner, at the Rockefeller Foundation, in 1923. (*Shelby White and Leon Levy Archives Center, Institute for Advanced Study; photograph by Wilhelm J. E. Blaschke, Oslo, 1936*)

Norbert Wiener (far right), with U.S. Army mathematicians at the Aberdeen Proving Ground, 1918, worked on ballistics with Oswald Veblen in World War I and founded the field of cybernetics based on his work on antiaircraft fire control with Julian Bigelow in World War II. *(MIT Museum)*

Abraham Flexner, who began his career as a high school teacher in Louisville, Kentucky, envisioned the Institute for Advanced Study as a refuge from "dull and increasingly frequent meetings of committees, groups, or the faculty itself. Once started, this tendency toward organization and formal consultation could never be stopped." *(Shelby White and Leon Levy Archives Center, Institute for Advanced Study)*

## THE USEFULNESS OF USELESS KNOWLEDGE

### BY ABRAHAM FLEXNER

On the eve of war in Europe, in October 1939, Abraham Flexner announced in *Harper's Magazine* that, "among the most striking and immediate consequences of foreign intolerance I may, I think, fairly cite the rapid development of the Institute for Advanced Study . . . a paradise for scholars who, like poets and musicians, have won the right to do as they please." (Harper's Magazine)

Founder's Rock, at the entrance to the six-hundred-acre Institute Woods. The Bambergers, Newark dry goods merchants who endowed both Flexner's educational experiment and Veblen's land acquisitions, urged that less attention be devoted to land and buildings, and more attention "to the cause of social justice which we have deeply at heart." *(Courtesy of the author)*

IAS School of Mathematics, meeting in Fuld Hall, 1940s. Left to right: James Alexander, Marston Morse, Albert Einstein, Frank Aydelotte, Hermann Weyl, and Oswald Veblen (dressed, as usual, for the woods). Von Neumann was likely absent due to wartime consulting work. *(Shelby White and Leon Levy Archives Center, Institute for Advanced Study)*

Oskar Morgenstern (left) and John von Neumann (right), coauthors of *Theory of Games and Economic Behavior,* at Spring Lake, New Jersey (the closest beach to Princeton), in 1946. "We often went to Sea Girt," remembers Morgenstern in *John von Neumann,* a documentary produced by the Mathematical Association of America in 1966. "Not to swim, because he didn't like that kind of exercise, but to walk along the beach. We had very serious discussions, and these walks sort of crystallized them. Then we would go home and write things down." *(Shelby White and Leon Levy Archives Center, Institute for Advanced Study; photograph courtesy of Dorothy Morgenstern)*

Albert Einstein (left) and Kurt Gödel (right) arrived at the Institute for Advanced Study during its first year of operation in 1933. Gödel's later years were dominated by two interests: the work of G. W. Leibniz, which Gödel believed contained hidden insights into the nature of digital computing, and an unorthodox solution to Einstein's equations, implying a rotating universe, which Gödel, with Einstein's encouragement, had derived for himself. *(Shelby White and Leon Levy Archives Center, Institute for Advanced Study; photograph by Oskar Morgenstern)*

John von Neumann doing mathematics at age eleven, with his cousin Katalin (Lili) Alcsuti observing, in 1915. "She greatly admired, but didn't understand what John was writing," explains Nicholas Vonneumann. "He used such graphics as the letter sigma and so on." *(Nicholas Vonneumann and Marina von Neumann Whitman)*

John von Neumann (top left, sitting on gun barrel), visiting an Austro-Hungarian Army artillery position, ca. 1915, with mother Margit (née Kann), father Max von Neumann, and, diagonally along the gun carriage to lower right: brother Michael, ?, cousin Lili Alcsuti, and brother Nicholas (still clothed in a dress). *(Nicholas Vonneumann and Marina von Neumann Whitman)*

John von Neumann (far left) at breakfast in Budapest, early 1930s, after the marriage of his cousin Katalin (Lili) Alcsuti to Balazs Pastory. Seated, left to right: John, the newlyweds, Mariette Kövesi von Neumann, the Pastorys, Michael von Neumann, Lily Kann Alcsuti, Agost Alcsuti. *(Nicholas Vonneumann and Marina von Neumann Whitman)*

Friedrich-Wilhelms-Universität Berlin

## Ausweiskarte

Inhaber dieser Karte

*Privatdozent*

*Dr. Johann Neumann*

*von Margitta*

ist Angehöriger der **Friedrich Wilhelms-Universität zu Berlin** und hat Zutritt zu dem Universitäts-Gebäude.

Eigenhändige Unterschrift des Inhabers

John von Neumann's identity card issued by the University of Berlin before he resigned in protest against the Nazi purge in 1933. "The German trains from Dresden are full of soldiers," he reported on a visit five years later. "I looked at Berlin very seriously. It may be for the last time."
*(Von Neumann Papers, Library of Congress; courtesy of Marina von Neumann Whitman)*

"He seemed to be always willing to go wherever the action was," says Françoise Ulam of John von Neumann. "For a non-athletic, non-outdoorsy person he could surprise you sometimes!" According to Atle Selberg, "he was very good at estimating things. For instance, with one glance he could look at a pearl necklace around a lady's neck and tell you about how many pearls there were."
*(Stanislaw Ulam papers, American Philosophical Society)*

Princeton in the 1930s. Left to right: Angela (Turinsky) Robertson, Mariette (Kövesi) von Neumann, Eugene Wigner, Amelia Frank Wigner, John von Neumann, Edward Teller, and, on the floor, Howard Percy ("Bob") Robertson (teaching relativity to Alan Turing at the time). Except for physicists H. P. Robertson (from Hoquiam, Washington) and Amelia Frank (from Madison, Wisconsin), the celebrants on this occasion, probably during the 1936–1937 winter holidays, were all from Budapest. "He did have the ability, at a party, to show off by drinking anyone under the table," says Marina von Neumann of her father in an interview on May 3, 2010. "But I never saw him ever drink anything alone." *(Marina von Neumann Whitman)*

John von Neumann, Richard Feynman, and Stanislaw Ulam, at the lodge at Bandelier National Monument (near Los Alamos), 1949. "We used to go for walks . . . in the canyons . . . and von Neumann gave me an interesting idea: that you don't have to be responsible for the world that you're in," says Feynman. "So I have developed a very powerful sense of social irresponsibility as a result of Von Neumann's advice." *(Photograph by Nicholas Metropolis; courtesy of Claire and Françoise Ulam)*

Trinity nuclear test (20 kilotons) at the Alamogordo Bombing Range, White Sands Proving Ground, New Mexico, twelve seconds after detonation at 5:29 a.m., July 16, 1945. The high-explosive-driven implosion that triggered the plutonium-fueled explosion was designed using von Neumann's theory of reflected shock waves and led directly to the development of the hydrogen bomb.
*(U.S. Army/Los Alamos National Laboratory/ National Archives and Records Administration Record Group Number 434)*

The U.S. Army's ENIAC (Electronic Numerical Integrator and Computer) was publicly unveiled at the Moore School, University of Pennsylvania, on February 16, 1946. According to von Neumann, this was "an absolutely pioneer venture, the first complete automatic, all-purpose digital electronic computer." Left to right: Homer Spence, Presper Eckert (setting function table), John Mauchly, Betty Jean Jennings Bartik, Herman Goldstine, Ruth Licterman (with punched card input-output equipment at far right). *(University of Pennsylvania Archives)*

The "First Draft of a Report on the EDVAC," issued by the Moore School on June 30, 1945, established what would become known as the "von Neumann Architecture," characterized by the distinction between Central Arithmetic, Central Control, Memory, and Input, Output, Recording Medium—identified here as "cards, tape." A "standard number" (soon to be termed a "word") is specified as 30 binary digits. *(Princeton University Libraries)*

Vladimir Zworykin (center), pheasant hunting near Amwell, New Jersey, with Bogdan Maglich (right) and an unidentified RCA engineer (left) in 1978. Zworykin began working on the problem of television with Boris Rosing in Russia in 1906, and, after leading RCA's development of commercial television in the United States, became the director of RCA's Princeton Laboratories in 1941. *(Bogdan Maglich)*

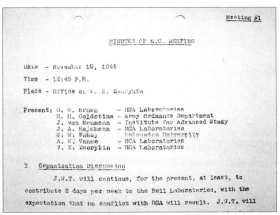

"Meeting #1" of the Institute for Advanced Study Electronic Computer Project was held on November 12, 1945, in Vladimir Zworykin's office at RCA. *"Words coding the orders are handled in the memory just like numbers,"* it was announced. This mingling of data and instructions broke the distinction between numbers that *mean* things and numbers that *do* things, allowing code to take over the world. *(Shelby White and Leon Levy Archives Center, Institute for Advanced Study)*

Bernetta Miller, pictured here with her Moisant-Blériot monoplane in 1912, was the fifth woman to earn a pilot's license in the United States, and became the Institute for Advanced Study's administrative secretary in 1941.
*(Courtesy Joseph Felsenstein; photographer unknown)*

Akrevoe Kondopria (now Emmanouilides), Herman Goldstine's secretary on the ENIAC project in Philadelphia, was invited by Goldstine and von Neumann to join the IAS Electronic Computer Project and reported to work on June 3, 1946. Seventeen years old at the time, she remained with the project until 1949.
*(Photograph by Willis Ware, ca. 1947; courtesy of Akrevoe Emmanouilides)*

Left to right: Norman Phillips (meteorology), Herman Goldstine (associate director), and Gerald Estrin (engineering), in the MANIAC machine room, 1952. The theoreticians at the Institute for Advanced Study had mixed feelings about the influx of meteorologists and engineers. As Julian Bigelow describes it, those "who had to think about what they were trying to do" did not welcome the arrival of those "who seemed to know what they were trying to do." *(Shelby White and Leon Levy Archives Center, Institute for Advanced Study)*

The RCA Selectron, or Selective Storage Electrostatic Memory Tube, invented by Jan Rajchman, promised an all-digital 4,096-bit electrostatic storage matrix in a single vacuum tube. The applications included numerical weather prediction, as portrayed in this advertisement in the February 1950 *National Geographic,* as well as file storage and retrieval at lightning speeds.
*(RCA / National Geographic)*

Tom Kilburn (left) and Frederic C. Williams (right) at the controls of the Small-Scale Experimental Machine (SSEM) at the University of Manchester, 1948. The first working stored-program electronic digital computer, the Manchester "Baby" ran a 17-line program (a search for Mersenne primes) as a test of its 1,024-bit cathode-ray tube memory on June 21, 1948. *(Department of Computer Science, University of Manchester)*

James Pomerene with the Williams electrostatic storage tube. When RCA's Selectron failed to materialize on schedule, the IAS team, led by James Pomerene, adapted over-the-counter 5-inch cathode-ray oscilloscope tubes into a fully random-access memory based on the Williams-Kilburn ideas. The obstacle to achieving high-speed digital storage was less a memory problem than a switching problem, solved by using the two-axis analog deflection of an electron beam as a 1,024-position switch.
*(Shelby White and Leon Levy Archives Center, Institute for Advanced Study)*

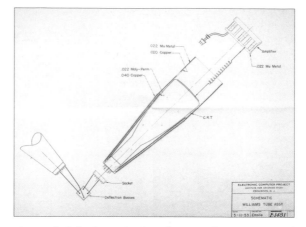

Williams memory tube, exploded schematic view, showing electromagnetic shielding, connection for deflection circuits, and high-gain amplifier integrated into the face of the individual tube. When the electron beam is aimed at one of the 1,024 locations on the inside surface of the tube, and given a "twitch," a faint electrical signal is produced in the wire screen attached to the outside face of the tube. Amplified thirty thousand times, the character of that signal is "discriminated" to indicate whether the state of charge at that location represents a zero or a one. *(Shelby White and Leon Levy Archives Center, Institute for Advanced Study)*

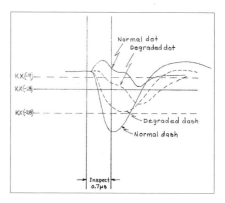

Distinction between a dot (o) and a dash (1) has to be determined in 0.7 microseconds, by "inspecting" the character of the faint secondary pulse generated when a given location is "interrogated" by the electron beam. *(Shelby White and Leon Levy Archives Center, Institute for Advanced Study)*

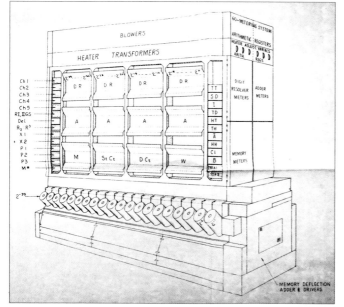

Adder side of IAS computer, schematic view. Above the Williams memory tubes (labeled from $2^{-1}$ on the right to $2^{-39}$ on the far left) are the memory registers, the adders, and the digit resolvers (known as "digit dissolvers" when malfunctioning). The opposite side of the machine is similar, with memory tubes $2^0$ through $2^{-38}$, and address and instruction registers on the lower level, arithmetic and memory registers above.
*(Shelby White and Leon Levy Archives Center, Institute for Advanced Study)*

designed around the assumption that a plug-and-play memory tube would be forthcoming in time.

Midway in complexity between the bit-level components such as 6J6 toggles and these system-level organs were 40-fold registers that stored, transferred, and shifted data, in parallel, 40 bits at a time. A register known as the accumulator provided access from the arithmetic unit to the memory, and the memory register provided an exit from the memory going the other way—analogous to separate intake and exhaust valves serving the individual cylinders in the engine of a car.

All these registers were "double-rank," containing two parallel rows of 6J6 toggles, with entry in and out of the toggles controlled by two additional rows of gates. This redundancy prevented bits being lost in transit: all data were replicated at the destination before being cleared from the source, in the same way the transmission of a data packet across the Internet is not considered complete until the packet signals that it has arrived intact.

Shift registers, as Leibniz had demonstrated 260 years earlier, could perform binary arithmetic simply by shifting an entire row of binary digits one position to the right or left. Data were never transferred directly between adjacent toggles; instead, the state of each individual toggle was replicated upward into a temporary register, the lower register was cleared, and then and only then were the data shifted, diagonally, back down into the original register. There was no lower bound to how slowly the computer could be stepped through a sequence of instructions. Unlike the well-behaved physical marbles that Leibniz had imagined shifting from column to column in 1679, electrons were always looking for an escape.

"Information was first locked in the sending toggle; then gating made it common to both sender and receiver, and then when securely in both, the sender could be cleared," Bigelow explained. "Information was never 'volatile' in transit; it was as secure as an acrophobic inchworm on the crest of a sequoia." Data were handled the way ships are moved through locks in a canal. "We enjoyed some interesting speculative discussions with von Neumann at this time about information propagation and switching among hypothetical arrays of cells," remembers Bigelow, "and I believe that some germs of his later cellular automata studies may have originated here."[20]

"We did not move information from one place to another except in a positive way," emphasizes James Pomerene. "That is used absolutely universally now. I think we were the first to do it. And I regret that we didn't patent it." Patentable inventions were being generated right and left. "The original patent agreement provided that the Institute would have title to patents, but they would pay to the inventor all royalties in excess of their costs," Pomerene adds. "Pretty nice!" No patents, however, were ever applied for. "We were young, eager engineers. We were most interested in getting the machine going than the filing of patents," he explains.[21]

In April 1946, von Neumann drafted a patent policy that "strikes a reasonable middle position between leaving everything to the employee or taking everything for the Institute." Employees agreed to assign their rights to the Institute, while the Institute agreed that "it will promptly and at no expense to the Employee have prepared, filed, and prosecuted an application for United States Letters Patent (and for patent in countries foreign to the United States, if it so decides) on each invention which the Institute determines is or may be useful to it." Furthermore, "the Institute agrees to pay to the Employee all royalties, if any, received . . . on each invention . . . over and above the total cost to the Institute of procuring the patent or application therefore."[22]

This was greeted with enthusiasm by the engineers. The Institute retained a patent attorney, whose assessment was that the project offered a wealth of patentable inventions—and the machine had not even been constructed yet. "The Institute for Advanced Study, had it asked for it, could have gotten from each engineer a release, and produced a large endowment compared to anything they now have," says Bigelow. But it would have cost a significant amount of money to secure, let alone defend, those patents—an approach the Institute was unwilling to take. It was Abraham Flexner, after all, who had announced, in 1933, that "the moment that research is utilized as a source of profit, its spirit is debased."[23]

In mid-1947 the original patent agreement was weakened, unilaterally, by a decision to turn most patent rights over to the government. "For the probably few exceptional developments that give rise to extremely valuable commercial applications," Goldstine reassured the engineers on June 6, "the engineering staff will recommend to the Director of the Institute that an application be prosecuted directly by the Institute." This was an empty promise, for Goldstine had already agreed, at a meeting with the Office of the Chief of Ordnance in April of 1947, "that any papers or reports covering the logical aspects of the Computer would be regarded as scientific publications, and therefore accessible to all interested scientists." This would undercut patent claims to most of the inventions made so far. "To prevent any commercial interest from attempting to exploit what should belong to the scientific community," Goldstine advised, "I would appreciate having you send a copy of the report entitled *Preliminary Discussion of the Logical Design of an Electronic Computing Instrument*, by A. W. Burks, Herman H. Goldstine, and John von Neumann, dated 28 June 1946, to the Patent Office with a request that they treat this as a publication in fact." In June of 1947, Goldstine, Burks, and von Neumann gave a sworn deposition stating that "it is our intention and desire that any material contained therein which might be of a patentable nature should be placed in the public domain."[24]

The engineers were left without much of a choice. "There was a meeting of the engineers in the fall of '48, I think," remembers Bigelow, "in which we were told that the second contract would not allow us to have the same patent clause as the first. And I said—in my opinion—we're giving up something very valuable. On the other hand, I personally can't bring myself to go on strike against Johnny at this time. But I want you to understand what we're giving up. And the voting was essentially 'We'll go ahead.'"[25]

"People who worked with von Neumann felt such enormous respect for him and such enormous gratitude to be allowed to be one of the people who were going to be building this machine," Bigelow continues, "that we never pushed our rights very hard. The second thing that happened, and I didn't know it at the time, is von Neumann started consulting for IBM."[26] All technical details of the MANIAC and its programming were placed in the public domain, and freely replicated around the world. A series of progress reports were issued that were models of clear thinking and technical detail. "The remarkable feature of the reports," according to Turing's wartime assistant, I. J. Good, "was that they gave lucid reasons for every design decision, a feature seldom repeated in later works."[27]

"Many of us who are in the course of making copies of the IAS machine have a tendency to emphasize our deviations and forget the tremendous debt that we owe Julian Bigelow and others at the Institute," wrote William F. Gunning, reviewing the development of the JOHNNIAC at RAND. "I think that the fact that so many of us have been able to make an arithmetic unit that works when first plugged in and which requires no fussing, is proof enough of the fundamental contribution that they have made."[28]

By July of 1947 a prototype ten-stage adder was found to "function reliably for periods of several days," performing a complete carry through all ten stages in 0.6 microseconds. In August a ten-stage shift register was put through a life test for the entire month, and in February of 1948 a prototype accumulator was run, "at a rate of about 100,000 additions per second," through five billion operations without a mistake.[29]

The computer required some means of loading data (and programs) into memory, and for delivering results. In 1946, magnetic wire was the recording medium of choice. Bigelow's crew spent several months building and debugging a high-speed wire drive that coiled and uncoiled steel recording wire at up to one hundred feet (or 90,000 bits) per second from a pair of bicycle wheels, differentially coupled side by side on a single concentric drive. The two wheels could be removed and inserted as a unit, just as a tape cartridge or removable disk is used today. Data and programs were punched onto paper Teletype tape, verified, and then transferred to recording wire, from where they could be loaded at high speed into memory as needed. "While this human-keyboard-typewriter operation is essentially slow and painstaking, it is entirely independent of the machine proper," it was noted in the first interim progress report, "and any number of coding crews intimately or remotely located relative to the machine may be setting up problems while the machine is solving those coded earlier." Von Neumann imagined a computer group being able "to work directly into and out of a machine which is situated elsewhere, possibly hundreds or thousands of miles away," as he described it to Roger Revelle, at the Office of Naval Research.[30]

"It is now possible to type a numerical message, transfer it to magnetic wire together with all the associated marker and indexing pulses, read it into a shifting register after deleting the marker and indexing pulses and then to reverse the entire process and once again produce a typed copy of the message," it was reported in March of 1948.[31] Although the high-speed wire drive was eventually operated without breaks over periods of as long as three weeks, it was abandoned in favor of more reliable, if slower, input/output directly via standard 5-hole Teletypewriter tape. Much of what was learned in building the wire drive was later applied to an auxiliary 2,048-word magnetic drum, equivalent to a 40-channel wire drive running fixed loops of wire through independent read/write heads.

Teletype input/output was in turn replaced by IBM punched card equipment, once special dispensation from IBM to modify its equipment was obtained. Hewitt Crane rewired an IBM 516 reproducing punch to read 80-bit rows in parallel, rather than one 12-bit column at a time. This modification was soon adopted by IBM, precipitating a phase transition from alphanumeric characters to lines of code. The entire 1,024-word memory could be loaded in under one minute, and unloaded in two.

In April an eight-stage prototype binary multiplier was placed in operation and run at 70,000 multiplications per second. "It has done about $10^{10}$ such operations and appears reliable," the official record noted, while unofficially, Oppenheimer informed the trustees that "the Electronic Computer was now multiplying." Rosenberg devised a series of test calculations in which "repetitive multiplications without error will produce a stationary pattern of digits" so that you could see immediately whether the arithmetic unit was working or not.[32]

"The way that machine did multiplication, if you started with the right set of numbers in the registers, it would do the multiplying and come up with the same set of three numbers in the registers," Ware explains. "We suddenly noticed the light pattern on the neon tubes was steady. Herman got very excited about all this. He runs down the hall, and he and Johnny—I don't know how long they spent figuring out how to pick the right three numbers. . . . Meanwhile, Pomerene and I were finding these numbers experimentally just as fast as we could write them down!"[33]

Three 40-stage shift registers were soon completed, and "interconnected in two different arrangements to form closed loops of 120 binary digits and shifted a place at a time around the loop," at a rate of 3 microseconds per shift. This test was run for 100 hours, for a total of $10^{11}$ shifts.[34] All the required pieces, except the memory, were falling into place.

At first it appeared that Rajchman's group at RCA was ahead of Bigelow's group at IAS. "I have just got back from a visit to Jan who was quite encouraging," Goldstine reported to von Neumann at the beginning of July 1947. "He promises us a 256-digit, square type, i.e. 4-cathode, Selectron in about 2 weeks—I don't know what we'll do with it when we get it."[35] At the end of the month, Goldstine made another visit, only to find more new problems than new memory tubes at RCA. Von Neumann, ever the game theorist, decided to hedge his bets. If all else failed, it was believed the computer could be operated (if at a hundred-fold reduction in speed) by coupling the arithmetic and control units directly to the forty-channel magnetic drum.

In spring of 1948, Douglas Hartree arrived from England bringing word—and a draft report, delivered by hand to Goldstine—from British radar pioneers Frederic C. Williams and Tom Kilburn, who were engaged, with assistance from Alan Turing and Max Newman, in building a prototype stored-program digital computer at Manchester University, based in part on the EDVAC report. In place of acoustic delay lines, they were developing a new form of storage, soon to become known as the Williams tube, using "the distribution of charge set up on the fluorescent screen of an ordinary cathode ray tube when it is scanned by an appropriately modulated electron beam."[36] Charged spots on the face of the tube could be preserved over brief periods of time, the way static electricity lingers for a few seconds after you turn off a cathode-ray television tube.

The Williams storage tube was Zworykin's iconoscope turned inside out: reading a pattern of electric charge traced by an internal electron beam rather than a pattern of electric charge formed by an image from outside the tube. "Effectively such a tube is nothing more than a myriad of electrical capacitors which can be connected into the circuit by means of an electron beam," Burks, Goldstine, and von Neumann had noted—before Williams and Kilburn's implementation of the idea—in their preliminary report of June 1946. Williams and Kilburn, who had collaborated with the Americans on radar (and Identification Friend or Foe) during the war, were careful to acknowledge their sources. "The experimental discovery that such tubes exhibit storage phenomena appears to have been made at the Radiation Laboratory, Boston, towards the end of the war," they wrote.[37]

Williams and Kilburn were able to store a 32-by-32 array of charged spots on the face of a single cathode-ray tube. The data were stored and retrieved in serial mode, similar to an acoustic delay line, but at the speed of electrons rather than the speed of sound. The entire matrix had to be retraced to recall the state of any individual spot. They succeeded in storing data without errors for hours at a time, noting in their report "that if the memory had been imperfect there are roughly $10^{360}$ alternative patterns possible, any one of which it might have shown at the end of the storage period. By way of comparison it has been said that there are a mere $10^{74}$ electrons in the universe."[38]

According to Williams and Kilburn, "the best overall test of the store would be the construction of a small machine." The resulting "Small-Scale Experimental Machine" (SSEM) had a single 32-by-32-bit cathode-ray tube store, and could do nothing except subtract, but was nonetheless sufficient to establish "that in principle a universal machine could be based on the C.R.T. store."[39] One fifty-two-minute run executed 3.5 million instructions, generated by seventeen lines of code.

The news from Manchester electrified the Princeton group. Bigelow was dispatched to England, while Pomerene began experimental work. Williams and Kilburn welcomed Bigelow to their primitive laboratory on July 18. "As I stood there and watched his machine, part of it started to burn up because it was built in such a jerry-built fashion, but it didn't bother him at all," Bigelow recalls. "He just took some clip leads off and said: 'This is no good.' He took a soldering iron and took one piece out, and put some others there beside it, and put the clip leads back on, and got it back into operation again."[40] A more serious problem was electromagnetic interference from an electric streetcar line that ran nearby, despite the cathode-ray tubes being shielded within a metal box.

"The sample routine demonstrated to me was part of a code on Mersenne's [prime] numbers, which it did twice; first with an error, and then correctly," Bigelow reported. "The bit of routine consumed three or four minutes; Max Newman (who was present) and I calculated that the number of operations performed was in the high thousands."[41] Bigelow returned to New York a few days later aboard the *Parthia*, a Cunard liner, and by the time he arrived in Princeton, Pomerene had a working cathode-ray tube memory of 16 bits, expanded to 256 bits within another four weeks.

Instead of being forced to wait for the Selectron, and ending up with a machine whose core memory would be proprietary to RCA, the Williams approach allowed the IAS team to get to work immediately using inexpensive off-the-shelf oscilloscope tubes, with all the innovation being outside the tube. "The entire problem was one of circuit design and construction, fields in which we felt competent," Bigelow explained.[42]

Williams and Kilburn had demonstrated how a sequence of pulses (in time) could be converted to a pattern of spots (in space) and stored indefinitely as long as the pattern were regenerated periodically by a trace from an electron beam. The spots become positively charged (i.e., deficient in electrons) as a result of secondary electron emission by the phosphor, so the state of an individual spot could be distinguished by "interrogating" that location with a short pulse (or "twitch") of electrons and noting the character of a faint secondary current, of less than a millivolt, induced in a wire screen attached to the outside face of the tube. "Thus the phosphor containing the various charge distributions is capacitively coupled to the wire screen," the IAS team explained, "and it is then possible by focusing the beam at a given point to produce a signal on the wire screen."[43]

The secondary emission effect can be visualized by imagining a 32-by-32 array of beer glasses sitting in a big kitchen sink and being sprayed with water from a garden hose, with a very sensitive drain in the sink that produces a detectable signal when any water sloshes out. The glasses tend to fill up with water, but if you aim the hose at one particular glass for an instant, that glass becomes partly empty as water sloshes out of the glass. You have written one bit of information in that location and, by returning to that location with another squirt of water, and noting whether anything sloshes out or not, you can read the state of that particular bit. You can then either refresh that state, continuing to store the bit of information, or clear it and replace it with the alternate state.

Pomerene's team developed timing and control circuits that governed the electron beam deflection voltages with enough precision to allow access to any location at any time, appropriating a few microseconds before resuming the normal scan/refresh cycles where they left off. The result was an electronically switched 32-by-32 array of capacitors, with a 24-microsecond access time, but was, as Bigelow noted, "one of mankind's most sensitive detectors of electromagnetic environmental disturbances." Errors were introduced by fields of as little as .005 gauss, or $\frac{1}{40}$ the strength of the earth's magnetic field. "It was confirmed by experiment that an AC magnetic field of about the strength of the earth's field, 0.2 gauss, will produce a deflection of the beam in a Williams tube of about 12 spot diameters," enough to turn any memory into complete garbage, it was reported in August 1949.[44]

The ability to distinguish a dot (o) from a dash (1) depended on the secondary emission characteristics of the phosphor coating, and the slightest imperfection, or a speck of dust inside the tube, would cause the memory to fail. The signal was amplified 30,000 times before being passed to a discriminator that made a decision as to whether the waveform represented a zero or a one. "The signals were way down in the noise level, I think one microwatt was the energy level, and that was a difficult problem," remembers Rosenberg, "but I finally got the amplifiers, which were installed right next to the memory tubes inside these shields, debugged." Placing individual amplifiers within each memory tube was in accordance with Bigelow's Maxim for Ideal Prognosticators No. 5, which stipulated that "if noise is ever to be filtered from signal, it must be done at the earliest possible stage."[45]

All forty memory tubes had to work perfectly at the same time, since each digit of a 40-bit word was assigned the same position in a different Williams tube. The 1,024 bits in each cylinder were visible to the naked eye, flickering from one machine cycle to the next—or frozen in time when a process was paused or came to a halt. What the operator observed *was* the digital universe, not the display of a process occurring somewhere else. The observer, however, had to be careful not to disturb the state of the memory being observed. "The front [of the tube] had a piece of copper gauze on it," explains Morris Rubinoff, "and when you wanted to look in you looked in at the light spots through the copper gauze just to make sure it was completely shielded."[46]

In modern (or once-modern) computers, a cathode-ray tube (CRT) displays the state of a temporary memory buffer whose contents are produced by the central processing unit (CPU). In the MANIAC, however, cathode-ray tubes *were* the core memory, storing the instructions that drove the operations of the CPU. The use of display for memory was one of those discontinuous adaptations of preexisting features for unintended purposes by which evolution leaps ahead.

A forty-first monitor stage, which could be switched over to mirror any of the forty memory stages, was added later, allowing the operator to inspect the contents of the memory remotely to see how a computation was progressing—or why it had come to a halt. This was later augmented by a separate 7-inch cathode-ray tube serving as a 7,000-points-per-second graphical display. "This device would take the data existing in one of the registers of the machine and translate its binary representation in the register into an amplitude representation in the deflection of the oscilloscope spot," the engineers reported in 1948.[47]

The IAS group settled upon standard 5-inch 5CP1A oscilloscope tubes, available in quantity, although less than 20 percent proved acceptable, and in 1953 it was reported that "there had not been more than ten flaw-free tubes discovered in the testing of over 1000 tubes in this laboratory during the past three years."[48] The manufacturers allowed the IAS to scan their inventory for unblemished specimens and ship the others back. Pomerene achieved a thirty-four-hour error-free test of a two-stage memory on July 28–29, 1949, and the final race to build a working forty-stage memory began. Parallel memory access would make the computer forty times as fast as a serial processor but, in the opinion of numerous skeptics, unlikely to work without one thing or another always going wrong.

"We learned, somewhat to our sorrow," says Pomerene, "that one property a memory has that other vacuum tube circuits don't have is that it remembers! Big surprise! But among other things, it remembers any noise that ever happened. Right? So it sits there and you're hoping it remembers a one, and some noise comes along and it would turn from one to a zero, and it stays a zero because it's remembering now a zero. So a memory turned out to be a very effective observer of noise."[49]

There were two sources of noise: external noise, from stray electromagnetic fields; and internal noise, caused by leakage of electrons when reading from or writing to adjacent spots. External noise could, for the most part, be shielded against, and internal noise was controlled by monitoring the "read-around ratio" of individual tubes and trying to avoid running codes that revisited adjacent memory locations too frequently—an unwelcome complication to programmers at the time. The Williams tubes were a lot like Julian Bigelow's old Austin. "They worked, but they were the devil to keep working," Bigelow said.[50]

Each individual memory tube had its own logbook recording its health history and any idiosyncrasies that arose along the way. The difficulty in distinguishing memory problems from coding problems drove many early programmers to give up in disgust. "The presence of this leprous element in the machine [means] that everyone who sits down to do a problem must be aware of it and be prepared to be just a little cagey, depending on the problem," noted Frank Gruenberger in explaining why RAND chose Selectron memory instead. "The blips fade in a fraction of a second and if the problem requires that you re-use a number before the blip is regenerated, you get the wrong answer. It is as if a desk calculator would fail any time the 7th, 8th, and 9th places in a fifteen-digit number happened to be a three-digit prime. . . . It just isn't decent for the operator to have to worry about how the machine is built."[51]

Meanwhile, Rajchman had kept working on the Selectron at RCA. The first operative tube, with 256 storage elements, was demonstrated to Bigelow and Goldstine on September 22, 1948, and they "seemed reasonably impressed." The second tube was finished on October 1, 1948. Only then did Rajchman find out that the Selectron had been demoted to second place. "It seems that we have at last the desired tube," Rajchman reported to Zworykin on October 5. "This success comes, however, at a time when the Selectron has to compete with the English tube of Prof. Williams. The Institute group has started to work on it—without telling us and even consciously concealing this fact from us—at the end of May or beginning of June." Rajchman, who finally obtained a copy of the Williams-Kilburn report, was "most impressed by his work," admitting that "in a typical English way he produced a startling result with an ordinary cathode ray tube."[52]

The Selectron was out of the race. A limited number of 256-bit tubes were eventually produced, and proved spectacularly successful (with 100,000 hours mean time between failures) in the IAS-derived JOHNNIAC built at RAND. But by that time IBM had adopted cathode-ray-tube memory for the IBM 701, and magnetic-core memory, originally suggested by Rajchman but commercialized elsewhere, was about to take the lead that had been abandoned by RCA. The Selectron never achieved commercial success or scale. Was the Selectron a failure? "No more so than the dinosaur was a failure," says Willis Ware. "They were doing things inside that vacuum that hadn't been done before."[53]

High-speed storage was a switching problem, not a memory problem. "The difficulty with all schemes of this type lies principally in the method of switching among the large numbers of elements involved," Goldstine had written to Mina Rees of the Office of Naval Research. "The crux of the memory problem lies not in the development of a cheap memory element but rather in the development of a satisfactory switch."[54] The advantage of the Williams tube memory—performing the switching function with no moving parts other than the deflection of an electron beam—was not that it solved the switching problem better, but that it solved it first.

The Selectron, which addressed memory locations by direct gating rather than "by directing a beam of electrons, like an electron garden hose, to a certain place," provided, in Rajchman's words, "a 'matrix' digital control that gives an absolute certainty of selecting the desired location, as opposed to the none-too-certain selection by analog deflection of a beam." Not only was the switching all digital, but the output was also all digital, and did not require an analog-to-digital "discriminator" to distinguish between a zero and a one. As Frank Gruenberger put it, "in the Selectron a particular slot in the memory is selected by digital (rather than analog) means, and the output signals are a thousand times larger than those in Williams tubes."[55] A solution to both the memory problem *and* the switching problem, the Selectron is the reason that the MANIAC's logical architecture, designed around it and descended to subsequent generations of computers, adapted so well to solid-state memory when the time came.

Some failures stem from lack of vision, and some failures from too much. "The ideas were so beautiful and so elegant that Rajchman was always trying to push it [the storage matrix] to a larger population of cells than his technique at that time would allow him to do," says Bigelow, explaining the delays at RCA. "He was just so clever in electron optics, that he couldn't face the fact that if he chopped it down to something much more modest and got it going, and then built up the size from there, he'd be much better off."[56]

The Selectron missed its chance. Once Bigelow and Pomerene saw how to convert cheap, off-the-shelf oscilloscope tubes into random-access memory, the challenge of doing so was impossible to resist. RCA, distracted by television, never took the Selectron seriously and failed to give Rajchman, working largely alone, the resources to make it a success. The Project Whirlwind group at MIT, developing a digital computer for air defense, "spent something like $25 million on storage tubes alone, which was nearly ten times our total project," Bigelow pointed out.[57]

With the Williams tube memory working, the computer began to take its final physical form. The MANIAC was unusually compact— "perhaps too compact for convenient maintenance," admitted Bigelow, who was largely responsible for its physical design. A minimal connection path between components was achieved by convolutions in its chassis, like the folding of a cerebral cortex into a skull. In 1947 most electronic devices were laid out in two dimensions, with components above a flat chassis and wiring below. The same remains true of most circuit boards, integrated circuits, and rack-mounted devices today. Bigelow, in contrast, took a three-dimensional approach to the way components were laid out and interconnected, and to wiring and cooling the dense vacuum tube arrays. "All those wires that aren't near any metal, they're out in space—that's all Julian," says Willis Ware. "That concave-shaped chassis, so you could wire point to point, keep the wire length minimum—that's all his ideas."[58]

"Vacuum tubes unfortunately had heaters and the wires that supply the current to the heaters were always a nuisance," James Pomerene explains. "They were always in the way and they'd have nothing to do with the logic of the computer." Bigelow's machinists milled out sheets of heavy-gauge copper sheet stacked in duplicate, sandwiching the individual strips between insulating fiberboard, so that all the heater current was conducted through those strips. "That got the heaters wired up without any wires being in the way, and made the machine significantly easier to build," says Pomerene.[59] Besides allowing much higher component density within the core of the computer, this minimized electronic noise and improved cooling flow.

The MANIAC resembled a turbocharged V-40 engine, about 6 feet high, 2 feet wide, and 8 feet long. The computer itself, framed in aluminum, weighed only 1,000 pounds, a microprocessor for its time. The crankcase had 20 cylinders on each side, each containing, in place of a piston, a 1,024-bit memory tube. The 40 cylinders, angled upward at a 45-degree angle in two parallel banks, each contained a 5-inch-diameter 5CP1A oscilloscope tube, with its elongated neck reaching down into the crankcase and its phosphorescent screen facing up toward the cylinder head.

Bolted above the lower crankcase, resembling a very tall engine block (with overhead valves), was the main frame of the computer, containing the memory registers, accumulators, arithmetic registers, and central control. An intake manifold fed data into the computer, and an exhaust manifold delivered the results. A 4,500-cubic-foot per minute blower forced cool air into the base of the engine, while 20 smaller blowers, resembling turbochargers, exhausted waste heat through overhead ducts. At first cool air had been introduced downward through the core of the computer, and exhausted through the floor; later, this was switched to the practice used in data centers today, with the entire machine room being cooled by a bank of external air conditioners, and the heat exhausted overhead. "The total power dissipated in the main body of the machine is approximately 19.5 kw," it was reported in 1953. "About 9 kw represents the dissipation of D.C.

power, and the remaining 10.5 kw is accounted for by the heaters, transformers, and blowers."[60]

The original air-conditioning unit was rated at 7.5 tons, and later doubled in capacity to 15 tons. This means, roughly, that if the air-conditioning system had been run at full power (about 50 kilowatts) and supplied with ice-cold water, it could have made 15 tons of ice a day. The refrigeration units, manufactured by York Refrigeration, and nicknamed "York" by the engineers, caused frequent trouble. The ability to produce 15 tons of ice a day meant that it took about 40 minutes for the refrigeration coils to ice up catastrophically with moisture from the New Jersey summer air.

"Fridge blocked up completely with heavy ice," reads an entry in the machine log for 8:55 p.m. on September 23, 1954. "Now York refuses to go at all—35 amp fuse blown," reads the next entry, at 9:10 p.m. "Replace and run York while de-icing fridge. York doing a poor job. D.C. off to help." The final entry indicates that the main DC power to the computer had been shut off, in the hope of bringing the core temperature down to where it would be safe to restart. Being able to run the air-conditioning or the computer, but not both, was not much help. York's heavy draw on alternating current had a tendency to introduce Williams tube errors at the worst possible time. "All my failures to duplicate occurred during a period of instability, on the part of York," says the machine log for October 22, 1954. "Off," notes pioneer climate modeler Norman Phillips at 7:38 p.m. "This because York is acting up," Hedi Selberg adds. "York caused lights to dim seconds before the error," noted Nils Barricelli on November 2, 1954.[61]

The engineers faced the challenge of getting all these disparate components to work together—not only with one another, but with the coded instructions by which the machine would be brought to life. "The planning for this machine will require such foresight and self-contained rigor," von Neumann had explained to Roger Revelle in 1947, "as one would need in order to be able to leave a group of 20 (human) computers, who are reliable but absolutely devoid of initiative, alone for a year, to work on the basis of exhaustive but rigid instructions, which are expected to provide for all possible contingencies."[62]

When the computer ground to a halt, was it noise in the deflection of an electron beam, or the transposition of one bit in specifying a memory address? "False start machine or human?" reads the first entry for a blast wave calculation run in February 1953. And an answer: "Found Trouble in code—I hope!"

"Code error, machine not guilty," admitted Barricelli on March 4, 1953. "What's the use? GOOD NIGHT," is recorded at 11:00 p.m. on May 7, 1953. "Damnit—I can be just as stubborn as this thing," notes a meteorologist on June 14, 1953. "I'll never know why you have to load these codes twice sometimes to make them go, but they go usually the second time."[63]

All computations were run twice, and accepted only when the two runs produced duplicate results. "I have now duplicated BOTH RESULTS how will I know which is right assuming one result is correct?" asks an engineer on July 10, 1953. "This now is the 3rd different output," notes the next log entry. "I know when I'm licked." Someone running a hydrogen bomb code from 2:09 a.m. to 5:18 a.m. on July 15, 1953, signs off: "if only this machine would be just a little consistent."

"THE HELL WITH IT," is the final entry for June 17, 1956, at thirteen minutes past midnight, noting that the master control is being turned off. "M/C OFF (WAY OFF!!)." It took years of midnight oil to sort these problems out, but the general trend was for hardware to become more reliable and error-free, while codes grew more complicated, and error-prone. "M/C OK. All troubles were code troubles," reads a log entry for March 6, 1958, one month after von Neumann's death.[64]

The MANIAC's logical architecture was indisputably the work of Burks, Goldstine, and von Neumann, whatever their ideas' original source. Its physical implementation was indisputably the work of Bigelow, and its electronic design was largely the result of teamwork between Bigelow, Pomerene, Rosenberg, Slutz, and Ware. Goldstine left the engineering to others, though he did build himself a television from a kit, "which gave him some knowledge at least of what's involved in putting together electronic-electromechanical equipment," says Rubinoff, "and he also at the same time got some feel for what can be done with triggering circuits and switching circuits and the like."[65]

Rosenberg, however, disagreed strongly with Bigelow on circuit design. "During the day I did as he commanded, and came back at night to accurately diagnose and fix the problems," he says. Pomerene was more diplomatic. "I would have to give him, I think, just about 100% credit for the unusual but highly effective mechanical design," he acknowledges, crediting Bigelow, whom he ended up replacing as chief engineer in 1951, for the three-dimensional, V-40 layout of the machine.[66]

"Julian would have the ideas, Ralph [Slutz] would kind of detail the ideas, and then Pom [James Pomerene] and I would go try and make the electrons do their thing," says Willis Ware. "He was kind of more physicist and theoretician than engineer. . . . In modern parlance, what you'd say was: Julian was the architect of that machine."[67]

"The rate at which Julian could think, and the rate at which Julian could put ideas together was the rate at which the project went," adds Ware. In 1951, Bigelow was awarded a Guggenheim fellowship, and took a one-year leave. "Herman Goldstine and possibly von Neumann felt that there was something about Julian that would keep him from ever quite exactly finishing the machine; that he might get it almost 99.9 percent done, but he would never get that final .1 percent done," says Pomerene. "And let's say, however that Guggenheim Fellowship happened to come, that they were quite happy to make me Chief Engineer and get the thing finished."[68]

"His problem was that he was a thinker," says Atle Selberg, whose wife, Hedi, was hired by von Neumann on September 29, 1950, and remained with the computer project until its termination in 1958. "He wouldn't leave things alone when other people thought they were finished. Julian was always thinking of doing something a bit more here and there."[69]

"I can see that being said," says Ware, concerning Bigelow's perfectionist streak. "But I think, after the fact, that damned machine might not have worked except for that. Hell—we were trying to make 2000 vacuum tubes do their thing! And to do it reliably, that level of perfection was a positive attribute."[70]

"I think part of the trouble there was he was looking for perfection before he got something running," says Morris Rubinoff. "You could never tell whether he was doing it because he was seeking perfection or because he was worried about reliability. Nobody ever had the courage to try a machine that fast before in quite that way. And putting a machine together and finding that it failed every three seconds didn't do you much good."[71]

Bigelow agrees. "You can't build a 40-fold parallel machine unless the basic circuitry of the individual stage is so good that it does what it should do without regard to the state of the next stage," he explains.

"It's got to be working at a megacycle rate for hundreds of hours. You can't rely upon chance."[72]

According to Bigelow, the forty-fold parallel architecture, despite its deviations, was descended directly from the pure-serial Turing Machine. "Turing's machine does not sound much like a modern computer today, but nevertheless it was," Bigelow explains. "It was the germinal idea. If you build an apparatus which will obey certain explicit orders in a certain explicit fashion, can you say anything about the kinds of computational or intellectual processes which it can or cannot do?" Bigelow and von Neumann had lengthy discussions about the implications of Gödel's and Turing's work. "Von Neumann understood this very deeply," Bigelow confirms. "So when looking at ENIAC, or some of the early machines which were very inflexible, he saw better than any other man that this was just the first step, and that great improvement would come."[73]

"What von Neumann contributed," says Bigelow, was "this unshakable confidence that said: 'Go ahead, nothing else matters, get it running at this speed and this capability, and the rest of it is just a lot of nonsense.' It was really on a basis of that sort of belief that we went ahead, with six people and a budget."[74] Von Neumann's approach was to bring a handful of engineers into a den of mathematicians, rather than a handful of mathematicians into a den of engineers. This freed the project from any constraints that might have been imposed by an established group of engineers with preexisting opinions as to how a computer should be built. "We were missionaries," says Bigelow. "Our mission was to produce a machine that would demonstrate what high speed computation would do."[75]

"A long chain of improbable chance events led to our involvement," Bigelow concluded in 1976. "People ordinarily of modest aspirations, we all worked so hard and selflessly because we believed—we knew—it was happening here and at a few other places right then, and we were lucky to be in on it. We were sure because von Neumann cleared the cobwebs from our minds as nobody else could have done. A tidal wave of computational power was about to break and inundate everything in science and much elsewhere, and things would never be the same."[76]

# NINE

# Cyclogenesis

*The part that is stable we are going to pre-dict. And the part that is unstable we are going to control.*

—John von Neumann, 1948

"I AM A LITTLE TROUBLED about the tea service in the electronic computer building," outgoing IAS director Frank Aydelotte warned John von Neumann on June 5, 1947, six months after the Computer Project engineers had departed from Fuld Hall. "Apparently the members of your staff consume several times as much supplies as the same number of people in Fuld Hall and they have been especially unfair in the matter of sugar." The war was over, but foodstuffs as well as building materials were still in short supply. "To come up here as Thompson did and carry down a large quantity of sugar in excess of your rations is not cricket," Aydelotte continued, "and I should like to raise the question of whether it would not be better for the computer people to come up to Fuld Hall at the end of the day at five o'clock in the afternoon and have their tea here under proper supervision."[1]

The culprit was Army Air Corps lieutenant Philip Duncan Thompson, one of a small group of meteorologists who had been recruited by von Neumann in 1946. "Von Neumann singled out the problem of numerical weather prediction for special attention," Thompson later explained, "as the most complex, interactive, and highly nonlinear problem that had ever been conceived of—one that would challenge the capabilities of the fastest computing devices for many years."[2]

Thompson, who was born in 1922, dates the beginning of his scientific education to the age of four, when his father, a geneticist at the University of Illinois, sent him to post a letter in a mailbox down the street. "It was dark, and the streetlights were just turning on," he remembers. "I tried to put the letter in the slot, and it wouldn't go in. I noticed simultaneously that there was a streetlight that was flickering in a very peculiar, rather scary, way." He ran home and announced that he had been unable to mail the letter "because the streetlight was making funny lights." His father returned with him to the mailbox, explained that he had been trying to insert the envelope the wrong way, and "pointed out in no uncertain terms that because two unusual events occurred at the same time and at the same place it did not mean that there was any real connection between them."[3]

In the spring of 1942, then in his third year at the University of Illinois, Thompson attended a lecture by Carl-Gustaf Rossby, a Swedish-born, Norwegian-trained meteorologist transplanted to the University of Chicago, where he was training prospective weather officers—eventually 1,700 of them—for the war. In May of 1942, as soon as classes were over, Thompson enlisted in the Army Air Corps in order to join Rossby's group. After completing his training, he was stationed in Newfoundland, monitoring the North Atlantic weather systems that had led the Scandinavians to develop the theory of frontal waves and otherwise lead the way in understanding what the weather might do next. At the end of the war he was assigned to Long Beach Air Force Base in California, as weather officer liaison to Norwegian meteorologist Jacob Bjerknes at UCLA, where he became close friends with Jule Charney, who had just received his PhD.

In 1945 meteorology had become a science, while forecasting remained an art. Forecasts were generated by drawing up weather maps by hand, comparing the results with map libraries of previous weather conditions and then making predictions that relied partly on the assumption that the weather would do whatever it had done previously and partly on the forecaster's intuitive feel for the situation and ability to guess. On average, forecasts beyond twenty-four hours were still no better than "persistence"—predicting that the weather tomorrow will be the same as it was today.

World War II, with its growing dependence on aircraft, increased the demand for forecasts, while weather radar and radio equipped weather balloons increased the supply of observational data needed to produce them. Thompson, trained in mathematical physics, was convinced that given accurate knowledge of the current state of the atmosphere and its outside influences, it should be possible to make predictions, based solely on the laws of physics, about its state at some near-future time. He had a single mechanical calculator, his father's advice not to infer cause from coincidence, and the knowledge that his predecessor, Lewis Fry Richardson, had made a similar attempt, and completely failed.

Lewis Fry Richardson, a Quaker and ardent pacifist who resigned from the British Meteorological Office when it was taken over by the Air Ministry, had begun developing a numerical atmospheric model while serving as superintendent of the Meteorological and Magnetic Observatory at Eskdalemuir in Dumfriesshire, Scotland, in 1913. The observatory, a branch of the National Physical Laboratory, had been moved to Eskdalemuir from Kew, near London, when electric railways came into use. The damp, secluded outpost, deliberately situated as far as possible from any artificial magnetic fields, suited Richardson, who cultivated an "intentionally guided dreaming," letting his mind balance between almost awake and almost asleep. "It is the 'almost' condition that is advantageous for creative thinking," he explained.[4]

At the outbreak of World War I, Richardson found himself "torn between an intense curiosity to see war at close quarters [and] an intense objection to killing people."[5] He applied to join what would become the Friends' Ambulance Unit when it was formed in 1914, and was finally granted leave from the observatory to do so in May 1916. After basic training in how to keep the ambulances running and the wounded alive, he departed for France in September, where he served at the front lines, attached to the Sixteenth Division of the French infantry, until 1919.

The Society of Friends had grown respectable since the time of Charles II and the imprisonment of William Penn. Bound by a humanitarian mission to assist the wounded, combined with Quaker refusal to submit to military authority, the Friends' Ambulance Unit served with heroic self-discipline during the Great War. Richardson's convoy, known as Section Sanitaire Anglaise Treize, or S.S.A. 13, reached a full strength of twenty ambulances and forty-five men. Between February 1914 and January 1919 they transported 74,501 patients over 599,410 kilometers of evacuation runs.[6]

A poor driver but a gifted mechanic, Richardson endeared himself to the rest of the group. "The other day my electric lighting dynamo went wrong," noted Olaf Stapledon, the future author of *Last and First Men*, on December 8, 1916. "The mechanic was away, & I know little about electricity, so I was dished. Fortunately we found that our eccentric meteorologist was also an expert electrician. He and I had a morning on the job unscrewing, tinkering, cleaning and generally titivating, sometimes lying under the car in the mud, sometimes strangling ourselves among machinery inside."[7]

A year later, Richardson and Stapledon celebrated the fourth Christmas of the war. "The moon is brilliant, and the earth is a snowy brilliance under the moon. Jupiter, who was last night beside the moon, is now left a little way behind. Venus has just sunk ruddy in the West, after being for a long while a dazzling white splendour in the sky," Stapledon reported on December 26, 1917. "I have just come in from a walk with our Professor, and he has led my staggering mind through mazes and mysteries of the truth about atoms and electrons and about that most elusive of God's creatures, the ether. And all the while we were creeping across a wide white valley and up a pine clad ridge, and everywhere the snow crystals sparkled under our feet, flashing and vanishing mysteriously like our own fleeting inklings of the truth about electrons. The snow was very dry and powdery under foot, and beneath that soft white blanket was the bumpy frozen mud. The pine trees stood in black ranks watching us from the hill crest, and the faintest of faint breezes whispered among them as we drew near. The old Prof (he is only about thirty-five, and active, but of a senior cast of mind) won't walk fast, and I was very cold in spite of my sheepskin coat; but after a while I grew so absorbed in his talk that I forgot even my frozen ears. . . . We crossed the ridge through a narrow cleft and laid bare a whole new land, white as the last, and bleaker. And over the new skyline lay our old haunts and the lines. Sounds of very distant gunfire muttered to us."[8]

Richardson kept working on his numerical model, whenever there were moments to spare. "This billet is a barn like the last, but we are far more crowded together," Stapledon noted on January 12, 1918. "Beside me sits Richardson, the 'Prof,' setting out on an evening of mathematical calculations, with his ears blocked with patent sound deadeners."[9] The model's input was a tabulation of the weather conditions observed over Northern Europe over a six-hour interval between 4:00 a.m. and 10:00 a.m. on May 20, 1910, an "international balloon day," when detailed records had been collected by Norwegian meteorologist Vilhelm Bjerknes, whose pioneering efforts to quantify our understanding of the atmosphere had inspired Richardson, and whose son Jacob Bjerknes would later be Thompson's supervisor at UCLA.

"My office was a heap of hay in a cold rest billet," Richardson reported. "It took me the best part of six weeks to draw up the computing forms and to work out the new distribution in two vertical columns for the first time."[10] The extended computation was well suited to the long, drawn-out war. Surrounded by mud and death and shrapnel, Richardson worked to reconstruct the weather on a spring morning when balloons had drifted over the then-peaceful European countryside, treating the motions of the atmosphere as nature's solution to a system of differential equations that linked the conditions in adjacent cells from one time step to the next.

Richardson used a method of finite differences that he had developed in 1909. "Both for engineering and for many of the less exact sciences, such as biology, there is a demand for rapid methods, easy to be understood and applicable to unusual equations and irregular bodies," he had written in a report to the Royal Society in 1909.[11] With boundary conditions as poorly defined as they usually were in meteorology, approximate answers were good enough.

The resulting prediction was at odds with what had happened on May 20, 1910, yet Richardson was correct in his belief that calculations would eventually supersede the existing synoptic methods of weather prediction, where "the forecast is based on the supposition that what the atmosphere did then, it will do again, now," and "the past history of the atmosphere is used, so to speak, as a full-scale working model of its present self."[12] He completed the test forecast, and then, "during the battle of Champagne in April 1917, the working copy was sent to the rear, where it became lost, to be rediscovered some months later under a heap of coal."[13]

After the war, Richardson published a detailed report, *Weather Prediction by Numerical Process*, so that others might learn from his mistakes. At the end of his account, he envisioned partitioning the earth's surface into 3,200 meteorological cells, relaying current observations by telegraph to the arched galleries and sunken amphitheater of a great hall, where some 64,000 human computers would continuously evaluate the equations governing each cell's relations with its immediate neighbors, maintaining a numerical model of the atmosphere in real time. "Outside are playing fields, houses, mountains and lakes, for it was thought that those who compute the weather should breathe of it freely," he imagined, adding that "perhaps some day in the dim future, it will be possible to advance the computations faster than the weather advances, and at a cost less than the saving to mankind due to the information gained."[14]

Twenty-six years later, Philip Thompson picked up where Richardson had left off. He remembers: "1946 was a year of ferment, for the formulation of the problem and the means of solving it were at last moving toward each other, albeit not by design."[15] Thompson "ground away at an old Monroe desk calculator, trying to figure out short-cuts and becoming increasingly depressed by the burden of hand calculation," until, as he describes it, "one fine afternoon in the early autumn of 1946, Prof. Jørgen Holmboe called me in, said he was aware of what I was trying to do, and handed me an article from the *New York Times*." The article announced the intention of Vladimir Zworykin of RCA and John von Neumann of the Institute for Advanced Study to collaborate on the construction of a high-speed electronic computer and its application to weather prediction and control. "Next day, I called my commander, Gen. Ben Holzman, and requested authorization to travel to Princeton to meet with von Neumann," Thompson continues. "General Holzman grumbled a bit, but agreed to it if I traveled as extra crew on a military aircraft that was headed East anyway. The following day the arrangements were clear, and I made my way to Princeton, via B 29, bus, stagecoach, train, oxcart, and the PJ&B."[16] PJ&B was the two-car train, or "Dinky," that shuttled from Princeton University to the main line at Princeton Junction and back.

Thompson met with von Neumann, "overawed" but managing to stay on topic and explain what he had been doing on his desk calculator at UCLA. "After about half an hour he asked if I would like to join his Electronic Computer Project," Thompson recalls. "Then he asked how my assignment should be arranged. I suggested that he call Gen. Holzman and request it. He called, talked for a few minutes, held the phone and said Gen. Holzman would like to speak with me. The conversation was very short and one-sided. It went something like, 'Well, I guess you better go back and get your gear. Orders will follow.'"[17]

Thompson arrived at the Institute in December 1946, moving into one of the Mineville apartments at the foot of Olden Lane. "He was tall and very aristocratic," remembers Akrevoe Kondopria. "And he was very good looking, almost like Peter O'Toole . . . and he wore a uniform. I gather he was the one who took the sugar." The meteorological group consisted mostly of temporary visitors. "I shared an office with Paul Queney, from the Sorbonne, a little office under the eaves of Fuld Hall," remembers Thompson. "We had a hard time communicating because his English was no better than my French."[18]

At the Institute, meteorologists were almost as suspect as engineers. "Any study of the weather, even a study leading to eventual scientific control, was primarily an empirical rather than a theoretical science, and as such belonged in an engineering school rather than in an institution devoted to the liberal arts," argued Marston Morse.

Except for von Neumann and Veblen, the mathematicians "approved of this step with some reluctance," the minutes report. Resigned to the inevitable, Morse warned that "if such a study were embarked on in connection with the electronic computer project, great care should be taken to separate it from the work of the Institute as such."[19]

Meteorology had been part of the computer project from the start. In mid-1945, Vladimir Zworykin had begun to see meteorology as an opportunity for RCA. Whether Zworykin enlisted von Neumann or von Neumann enlisted Zworykin remains unknown. "I remember in late 1945 or early 1946 reading a rather fantastic proposal of Zworykin's for the construction of an analogue computer which would scan two-dimensional distributions of weather data projected on a screen and then compute the future weather by analogue techniques," Jule Charney later explained. "By varying the input continuously and observing the output one could determine how most efficiently to modify the input to produce a given output. Johnny was in contact with Zworykin at that time, and perhaps his interest in weather computation and weather modification began then."[20] Von Neumann and Zworykin went to Washington, D.C., together to sell their plan.

"In the late summer of 1945, after the end of the war in Europe and in Asia, John von Neumann . . . and Vladimir Zworykin . . . called on me at the Navy Department," remembers Lewis Strauss. His visitors described the digital storage tubes being developed at RCA, and how "observations on temperature, humidity, wind direction and force, barometric pressures, and many other meteorological facts at many points on the earth's surface and at selected elevations above it . . . could be stored in the 'memories' of these tubes." From this digital representation, "a pattern or harmonic system might be developed which eventually would enable such a data storage device to predict weather at extremely long range."[21] The numerical model would be captured in vacuum tubes from which all traces of the real atmosphere had been withdrawn.

Zworykin drafted an eleven-page "Outline of Weather Proposal," dated October 1945, which suggested that computerized forecasting "would be a first step in any attempt in the control of weather, a goal recognized as eventually possible by all foresighted men." With sufficiently detailed knowledge, "the energy involved in controlling the weather would be very much less than that involved in the weather phenomenon itself." Von Neumann appended a cover letter, adding that "the mathematical problem of predicting weather is one which can be tackled, and should be tackled, since the most conspicuous meteorological phenomena originate in unstable or metastable situations which could be controlled, or at least directed, by the release of perfectly practical amounts of energy."[22]

Von Neumann and Zworykin proposed that the Institute for Advanced Study, the Radio Corporation of America, and the navy collaborate, and with Strauss firmly on board, the IAS computer project was launched. "They pointed out the military advantages of accurate long-range weather intelligence, and this seemed to justify the cost of such a venture, estimated at about $200,000," says Strauss. "Had the decision to make the computer not been taken in 1945, the thermonuclear program might have been delayed long enough for the Soviets to have had the first weapons. This was far from the minds of anyone when Von Neumann initiated the project."[23]

Thermonuclear weapons, however, were very much on von Neumann's mind in late 1945, although this would have been kept secret from Zworykin and, for the time being, even from Strauss. Preparations were already under way for the thermonuclear calculation that would begin running on the ENIAC on December 10, 1945, and, acutely aware of the limitations of the ENIAC, the weaponeers were scrambling to start building its successor without delay. Meteorology offered both a real problem and a perfect cover for the work on bombs.

The first public announcement of the project was made by the *New York Times* after a meeting between Zworykin, von Neumann, and Francis W. Reichelderfer, chief of the U.S. Weather Bureau in Washington, D.C. The "development of a new electronic calculator, reported to have astounding potentialities . . . might even make it possible to 'do something about the weather,'" the *Times* reported. "Atomic energy might provide a means for diverting, by its explosive power, a hurricane before it could strike a populated place."[24]

With the details of the ENIAC still restricted, the *Times* reported vaguely that "none of the existing machines, however, is as pretentious in scope as the von Neumann–Zworykin device." Von Neumann and Zworykin proposed to build not just one computer, but a network of computers that would span the world. "With enough of these machines (one hundred was mentioned as an arbitrary figure) area stations could be set up which would make it possible to forecast the weather all over the world."[25]

Reichelderfer was upset that news of the project had been leaked to the press. Eckert and Mauchly were upset that the *New York Times* had made no mention of the ENIAC, but had mentioned the proposed IAS/RCA computer, which did not even exist. They felt scooped by von Neumann, as they had over the authorship of the EDVAC report, and prevented, by the secrecy imposed on their own project, from voicing a response.

Von Neumann was convinced that Lewis Fry Richardson (whose work he described to Strauss as "remarkable and bold") had been on the right path, and he believed that understanding the weather would eventually convey more power, for good or evil, than understanding how to build bombs. In the Institute's proposal to the navy, drafted in collaboration with Carl-Gustaf Rossby, he estimated that once the new computer was operating, "a completely calculated prediction for the entire northern hemisphere should take about 2 hours per day of prediction." In a personal letter to Strauss, in which he also expressed "serious misgivings as to the wisdom of making [Oppenheimer] the Director of the Institute," he added that the Meteorology Project "would also be the first step toward weather control—but I would prefer not to go into that at this time." Even "the most constructive schemes for climate control would have to be based on insights and techniques that would also lend themselves to forms of climatic warfare as yet unimagined," he later warned.[26]

Once the navy contract was secured, von Neumann hosted, with Rossby's assistance, a conference on meteorology held at the Institute on August 29–30, 1946. The final topic of discussion was how to resurrect Richardson's effort, now that enough numerical horsepower to do so was in the works. "It was felt that the numerical attack should be repeated immediately," the meeting summary reports, "since even the existing mechanical computing facilities have capacities considerably in excess of those that were available to Richardson."[27]

More than a dozen meteorologists were invited to take up residence at the Institute, but there was no place to put them, and it was reported on July 15, 1946, that there were still eleven meteorologists for whom "no living quarters have been obtained."[28] Partly because of the lack of housing, and partly because of the lack of a working computer, the project was scaled back, and in the end no more than a handful of meteorologists were ever at the Institute at one time. Von Neumann's first contribution was to show that the existing methods of integrating the hydrodynamical equations "are unstable under those conditions of spatial and temporal resolution which are essentially characteristic of the problem of meteorological prediction," this being where Richardson had gone wrong. "I have developed one method which can be shown to be stable and which appears to be suitable for numerical procedure if electronic equipment is available," he announced in his second progress report to the Office of Naval Research.[29]

"The atmosphere," he explained in the next progress report, "is composed of a multitude of small mass-elements, whose behavior is so interrelated that none can be dissociated, even in effect, from all the rest." The problem was how to translate the analog computation being performed by the atmosphere into a digital computer and speed it up. "A closed system of differential equations, ordinary or partial, linear or non-linear, may be regarded as a set of instructions for constructing its solution from known boundary and initial values," he continued. "Until now, however, the time required to carry out those 'instructions' has been prohibitive."[30] This was now going to change.

The reaction of most meteorologists toward computer-assisted forecasting paralleled that of the Institute mathematicians toward computer-assisted mathematics: skepticism that a machine could improve upon what they were doing with brains alone. As Thompson explained, they "were against it, not for any objective reasons but because they really wanted to believe that forecasting should be an art."[31] According to Charney, the 1946 conference "failed to grip the imagination of the leading dynamical meteorologists who were invited, and few worth-while suggestions were proffered. However, my own imagination, which had already been stirred by Zworykin's article, was completely captured. I made haste to join the project on my return from Europe in 1948."[32]

Jule Gregory Charney, born on New Year's Day 1917 in San Francisco, had been misdiagnosed with a heart ailment as a child, which left him with an unusual enthusiasm for life. His parents, Lily and Stella, had emigrated from Russia to New York, found work in the garment industry, and moved west to California in 1914. After an interlude in San Francisco, they moved to East Central Los Angeles in 1922 and to Hollywood in 1927, where Jule's mother obtained enough work from the film studios to get the family through the Depression intact. Both parents were active socialists, and the household was a hotbed of political discussion and union affairs. Jule taught himself calculus while still in high school, before enrolling in UCLA in 1934 and graduating in 1938.

With the war approaching, Charney, who was scraping by as a teaching assistant in mathematics and physics, had to decide whether to pursue meteorology, which interested him, or aeronautics, which promised to be more useful for the war. He went to see aeronautical pioneer Theodore von Kármán at Caltech, who advised meteorology, explaining that aeronautics had matured to where future progress would be made through engineering, not mathematics, whereas meteorology was ripe for a mathematical approach. Charney never looked back. Jacob Bjerknes had recently come to UCLA from Norway to launch a training program for meteorologists, and Charney joined the new department as a teaching assistant in July of 1941 at sixty-five dollars a month.

Charney had a remarkable ability to condense the entire atmosphere, from planetary to molecular scale, into equations that captured what was important and discarded what was not. "Voltaire's giant, Micromegas, who stood in the same relation to the atmosphere as we do to a rotating dishpan, would describe the atmosphere as a highly turbulent, heterogeneous fluid subject to strong thermal influences and moving over a rough, rotating surface," he would later write. "He would discern a mean zonal circulation with westerly surface winds increasing with height in middle latitudes in both hemispheres and easterly surface winds near the Equator and poles." Closer examination would reveal perturbations related to the uneven distribution of the continents and oceans, and superimposed on these quasi-permanent features "he would find a whole collection of migratory vortices, varying in scale from thousands of kilometers to centimeters and less, but with the bulk of the energy in the thousand-kilometer class."[33]

By the end of the war Charney was in the middle of his PhD thesis, "The Dynamics of Long Waves in a Baroclinic Westerly Current," completed in 1946. Newly married, he and Elinor Charney (née Frye) then left Los Angeles for Chicago and his first postdoctoral appointment, with Rossby, who invited him to the Princeton conference in August, where he met von Neumann, learned of his ambitions, and sensed that there was a bit too much mathematics, and too little meteorology, at the IAS. He and Elinor sailed for Bergen and Oslo in the spring of 1947, where he worked among the Norwegians until returning to Princeton in the early spring of 1948.

Charney had arrived at the right place at the right time. The computer was undergoing initial testing, and the first problems were being coded in anticipation of there being a machine on which they could run. A rotating contingent of Norwegian meteorologists, led by Arnt Eliassen and Ragnar Fjørtoft, joined the group. Charney became the liaison between the hands-on experience of the Norwegian forecasters and the mathematical world of von Neumann. "Although I had a fairly clear idea of what I wanted to do in the physical sense, I had only the vaguest notion of how to do it mathematically," Charney explains.[34] Von Neumann's skills were exactly the reverse. Charney also attracted and nurtured a number of extraordinary American meteorologists, notably Joseph Smagorinsky and Norman Phillips, who played a leading role in the realization of numerical weather prediction over the next ten years.

Smagorinsky was still a graduate student after the war when Charney came to give a talk at the Weather Bureau headquarters at Twenty-fourth and M streets in Washington, D.C., at a time when numerical forecasting was almost entirely alien to the way the Weather Bureau worked. "During the war, when I was a student, a cadet at MIT, I had been told by one of the eminent professors there, Bernhard Haurwitz, that numerical forecasting can't be done," says Smagorinsky. "And the reason given was not a very good one. But it was easier to say that it can't be done than it can be. And I carried this notion of impossibility in my mind." After Charney's talk, Smagorinsky asked the only question that showed any real thinking about the problem, and Charney invited him to join the new computing group.

"The primary reason for Richardson's failure," Charney noted in the first progress report he prepared for the Office of Naval Research, "may be attributed to his attempt to do too much too soon."[35] It was the meteorological community as a whole who solved Richardson's first problem: gathering sufficient data to establish initial conditions—it soon being recognized that within days the notion of "boundary conditions" disintegrated and it was necessary to have hemispheric knowledge, the boundary between Northern and Southern hemispheres being the only one that held up over time. It was von Neumann, Goldstine, and Bigelow who solved Richardson's second problem: supplying enough computing power to do the job. And it was Charney who did the most to solve Richardson's third problem: formulating equations whose solutions did not quickly become more unstable than the weather itself. The key was to filter out noise.

"The atmosphere is a musical instrument on which one can play many tunes," Charney explained to Thompson in February of 1947. "High notes are sound waves, low notes are long inertial waves, and nature is a musician more of the Beethoven than the Chopin type. He much prefers the low notes and only occasionally plays arpeggios in the treble and then only with a light hand. The oceans and the continents are the elephants in Saint-Saëns' animal suite."[36]

Thompson was listening, and reported to the Office of Naval Research that "the hydrodynamic equations cover the entire spectrum of events, sonic waves, gravity waves, slow inertial waves, et. cetera, and it might simplify matters considerably if those equations were somehow informed that we are interested in only certain kinds of atmospheric behavior—i.e., the propagation of large-scale disturbances."[37] With Charney's help, numerical filters were soon constructed, and incorporated into codes that, with many hundreds of hours of hand-computing by Margaret Smagorinsky, Norma Gilbarg, and Ellen-Kristine Eliassen, were taken for trial runs. "The system that they were going to use on the big computer, we were doing manually," says Margaret Smagorinsky. "It was a very tedious job. The three of us worked in a very small room, and we worked hard. It was a small room with three people and three Monroe calculating machines."[38]

With the delays in the completion of the computer—and the priority given to the hydrogen bomb problems—it was decided to run a full-scale trial calculation on the ENIAC instead. In March of 1950, Charney was joined by George Platzman, Ragnar Fjørtoft, John Freeman, and Joseph Smagorinsky on an expedition to Aberdeen. They were guided by Klári von Neumann, who helped code their problem and initiated them into the ways of the ENIAC and its peripheral card-processing machines.

"The enactment of a vision foretold by L. F. Richardson 50 years before . . . began at 12 p.m. Sunday, March 5, 1950, and continued 24 hours a day for 33 days and nights, with only brief interruptions," Platzman reported.[39] "We have completed a 12-hour forecast," he noted, thirteen days later, in his journal for March 18. "At the end of four weeks we had made two different 24-hour forecasts," Charney reported on April 10. "The first . . . was not remarkable for accuracy, though it had some good points. . . . The second . . . turned out to be surprisingly good. Even the turning of the wind over Western Europe and the extension of the trough, which Ragnar thought to be a baroclinic phenomenon, was correctly forecast."[40] Over the course of the next week they made two more twenty-four-hour forecasts, for January 31 and February 14, 1949.

Because of the limited internal storage, "it devolved upon punch cards to serve as the large-capacity read/write memory, and this mandated an intimate coupling of punch-card operations with ENIAC operations, ingeniously contrived by von Neumann." Each step of the calculation required sixteen successive operations: six for the internal ENIAC arithmetic, and ten for the external punch-card operations to process the results and prepare for the next step. "In the course of the four 24-hour forecasts about 100,000 standard I.B.M. punch cards were produced and 1,000,000 multiplications and divisions were performed," Charney, von Neumann, and Fjørtoft reported. Once the bugs were worked out, "the computation time for a 24-hour forecast was about 24 hours, that is, we were just able to keep pace with the weather."[41]

Charney and his colleagues returned to Princeton in triumph. "It mattered little that the twenty-four hour prediction was twenty-four-hours in the making," he explained. "That was purely a technological problem. Two years later we were able to make the same prediction on our own machine in five minutes."[42] Emboldened by their success, they developed a series of increasingly detailed models of the atmosphere over the Northern Hemisphere by day, while livening up the atmosphere at the housing project by night.

"Oh, we loved him," says Thelma Estrin, of Charney. "He was warm and friendly and loved parties and was always the last one to go home." The engineers and the meteorologists, living together in the close-knit Mineville barracks, drew the other academic visitors into their fold. "All the meteorologists were great fun, hard drinkers," says Hungarian topologist Raoul Bott, who arrived, with a degree in engineering, as a protégé of von Neumann's in 1949. "We had tremendous wild parties," he remembers. "It was a high point in my life."

Bott singles out one evening, just after the first ENIAC expedition, when the poet Dylan Thomas was in town. "And at about ten thirty in the evening, or maybe eleven, we were having a great party in one of those shacks there, and I thought, 'Well, wouldn't it be great to bring Dylan Thomas here now?' So I called the hotel—I was a brash young man—and got Dylan Thomas, and he had been in bed. And he said, 'Oh, I'd love to be woken up, by all means,' you know, and he was ready for partying. And so I drove there—we had this 1935 convertible

Buick—with my wife, she of course all excited, and right away when he got in the car I could see that there would be a little bit of a problem, because obviously that was going to be his girl for the night."[43]

As they refined their models, Charney's group needed a benchmark by which to gauge their predictions. For his test case, Richardson had used the otherwise uneventful morning of May 20, 1910. Charney's group chose Thanksgiving 1950, when a severe storm struck the central and eastern United States. The weather system, whose development was missed by the forecasts available at the time, caused three hundred deaths, unprecedented property damage, and even blew part of the roof off the Palmer Physical Laboratory at Princeton University. It was the perfect storm.

"Because of the spontaneity and intensity of its development this storm was selected as an ideal test case for the prediction of cyclogenesis," Charney explains. Despite the unpredictability of turbulence, he believed that "the inception and development of cyclones are determinate and predictable events." Although cyclogenesis might appear random, "the initial perturbation will have a preferred location in space and time, and its amplitude, though it may be small initially, will be entirely determined by the basic flow. It is like an automobile being pushed slowly but inexorably over a cliff."[44]

"The storm of November 25–27 was first noted on the surface weather map of 1230 GMT, November 24 as a small Low developing over North Carolina and western Virginia," began the summary in the November 1950 *Monthly Weather Review*.[45] Over the next forty-eight hours the disturbance grew to become the worst storm ever recorded over the United States. Coburn Creek, West Virginia, received sixty-two inches of snow. Records of minus 1 degree Fahrenheit were set in Louisville, Kentucky, and Nashville, Tennessee, and thirty inches of snow fell in Pittsburgh, bringing the steel industry to a halt.

"The two-and-a-half-dimensional model did not catch the cyclogenesis, [although] there was some vague indication of something going on," Charney later reported. "And so we went to a three-level model, that is, a two-and-two-thirds dimensional model, and we did catch the cyclogenesis. It wasn't terribly accurate, but there was no question that [we did]. And I always thought that this was a terribly important thing. . . . I wanted the world to know about that!"[46]

The successful 24-hour prediction, following a 16-by-16 grid of 300-kilometer cells through 48 half-hour time steps, required 48 minutes of computing time. According to Charney, "during this time the machine performed approximately 750,000 multiplications and divisions, 10,000,000 additions and subtractions, and executed 30,000,000 distinct orders."[47] While trying to simulate weather *inside* the computer, the meteorologists were plagued by the weather *outside* the computer. The "York" refrigeration units continued to become overloaded in the sultry Princeton heat, and during thunderstorms the Williams tube memory often failed. On one very hot day in May there was trouble with the IBM card equipment, and the machine log records: "IBM machine putting a tar-like substance on cards." The next log entry explains: "Tar is tar from roof."[48]

Charney's plan to develop a hierarchy of models, incrementally more complete, now picked up steam. On August 5, 1952, von Neumann chaired a meeting at the Institute "to explore the possibilities of routine preparation of numerical forecasts by the Weather Bureau and the Air Force and Navy meteorological services." In September of 1953 the Weather Bureau, air force, and navy agreed to establish a Joint Numerical Weather Prediction Unit, and in January of 1954 a Technical Advisory Group chaired by von Neumann recommended the use of an IBM 701, the rental of which had been budgeted at $175,000 to $300,000 per year.[49] IBM delivered the computer early in 1955, and the first operational forecast was made on April 18.

By 1958, numerical forecasts were running neck and neck with manual ones, and by 1960 they had pulled ahead. Starting with twenty-four-hour forecasts, there was an improvement of about twenty-four hours per decade in extending the forecast range. The question for von Neumann and Charney was: What next? "Von Neumann seemed to feel that the problem of short range forecasting was pretty well in hand," remembers Thompson. "Well, I think he had a somewhat naive view of how far we had come and how far we still had to go, but he was looking ahead." Von Neumann divided the problem into three regimes. In the first, short-range regime, what happens depends more on the initial conditions than on the subsequent energy inputs and dissipation. With sufficient observations, and enough computing, short-range predictions, on the order of a few days to one week, can be made. In the second, medium-range regime, beyond a week, the influences become increasingly divided between those produced by the initial conditions and those introduced by the energy inputs and dissipation, and the behavior becomes very difficult, perhaps impossible, to predict. In the third, long-range regime, "the atmosphere very quickly forgets what it looked like in the beginning," as Thompson put it, "and its behavior is dominated almost entirely by the integrated day-to-day effects of the energy inputs and by the dissipation."[50] With sufficient knowledge of those inputs and dissipations, the task of predicting not weather but climate should be computationally tractable, and von Neumann and Charney, now joined by Norman Phillips and Joseph Smagorinsky, decided to tackle this next.

In September 1954, Norman Phillips began running a primitive general circulation model (the ancestor of all climate models in use today) that remained stable for up to forty days of simulated time. As Smagorinsky described it, "despite the simplicity of the formulation of energy sources and sinks, the results were remarkable in their ability to reproduce the salient features of the general circulation." Although the results became irregular and nonlinear when run beyond forty days, Phillips and Charney felt this was due to numerical instabilities, not the underlying model, and that with better coding and a more powerful computer, a true prediction of climate might be reached. "The code almost completely exhausts the resources of the present machine," they reported, "there being only about a dozen 'words' in the combined Williams-Drum memory of 3072 words which are not used." In February and March 1954, the model was run to thirty-one days and appeared "surprisingly realistic." Even "features similar to the cold and warm fronts of the classical Norwegian wave cyclone" were produced.[51]

"Our object was to establish a purely physical theory of climate, that is to say, to make the infinite forecast," Charney later explained to Stan Ulam. "Johnny foresaw that this would be a simpler problem than the problem of long-range prediction, since the statistical properties of the motion were likely to be more correct than the individual motions." We now know this is not as easy as it seemed at the time. "He always had, in the back of his mind, of course, large-scale weather modification," Charney adds. "We spent many pleasant Sunday afternoons together inventing theories of climate, and it became clear that nothing could be done to explain past climates or to lay the groundwork for climatic modification until we were able to understand our own climate in purely physical terms."[52]

To launch the new project, von Neumann and Charney hosted a conference on the dynamics of climate at the Institute on October 26–28, 1955. Oppenheimer gave the opening address, drawing "a parallel between the present conference dealing with problems of the general circulation of the earth's atmosphere and the conference held at Los

Alamos, New Mexico, in preparation for work on the atomic bomb," and noting that "the problem which faces the participants of the present conference—a problem dealing with the complicated dynamics of atmospheric motions—is a much more difficult one."[53]

Von Neumann, while a master of simplifying assumptions, was realistic about the obstacles. "Even if we were adequately informed, the inclusion of turbulence and radiation in the prediction equations would be quite involved," he announced. Nearly all the phenomena under consideration were unstable, and minute differences could be amplified into large effects. "For example, only about $\frac{1}{100,000}$ of all the water on earth occurs in vapour form in the atmosphere; yet the presence of water vapour makes a difference of $40°C$ in the average temperature of the earth," he observed. "This is more than twice the difference between the temperature at the time of maximum glaciation and that at the time of total deglaciation of the earth."[54]

The twenty-nine attendees, although hopeful about modeling the climate, acknowledged the problem to be highly complex. "Consideration was given to the theory that the carbon dioxide content of the atmosphere has been increasing since the beginning of the industrial revolution, and that this increase has resulted in a warming of the atmosphere since that time," the proceedings report. "Von Neumann questioned the validity of this theory, stating that there is reason to believe that most of the industrial carbon dioxide introduced into the atmosphere must already have been absorbed by the ocean." The debate was on.[55]

Sigmund Fritz, of the U.S. Weather Bureau, added that "the effects of plant life must also be taken into consideration." William von Arx, from Woods Hole Oceanographic Institution, stressed that "the balance depends on the buffer capacity of the seawater," and noted that "there is a significant amount of carbon dioxide locked up in the plankton cycle." Charney asked about "the statistical significance of Wexler's result that substantially more blocking activity takes place in Januaries which occur during periods of sunspot maxima than in those which occur during minima of the sunspot cycle." Von Neumann "felt that there must be a minimum size of the ice field that would be self perpetuating" and "called attention to the fact that the processes which led to periods of glaciation and deglaciation must have been relatively constant over many centuries." He asked "whether there is any evidence to suggest that high volcanic activity had been sustained over such long periods of time."[56]

"It is not necessary," von Neumann and Charney argued, "to resort to explanations of climatic change which require external mechanisms such as solar and volcanic activity." Richard Pfeffer, of Columbia University, noting that "the radiation absorbed by a unit mass of air is measured as the small difference between two large fluxes," asked "whether present-day measurements of the distribution of water vapour and temperature (the chief variables which determine the radiational characteristics of the atmosphere) are sufficiently accurate to determine this difference." Edward Lorenz, of MIT, noted, concerning the effects of clouds, that "it would be necessary to specify, in addition to the mean cloud cover, the diurnal range . . . whether the clouds appear at night or during the day." Von Neumann, in conclusion, advised that they should "first attempt to determine the extent to which climate can be changed by internal mechanisms through non-linear feedback processes," and "stressed that the problem is a complicated one."[57]

Weather prediction remains divided into the three regimes established in 1955. The first, short-term regime is predictable. The second, medium-term regime is now known to be unpredictable, much as many suspected, despite von Neumann's hopes. The third regime is still under debate. "I think in those days we were very optimistic," says Charney. "I remember at that time receiving reports that Norbert Wiener had regarded von Neumann and [me] as practically *gonifs*— thieves. That we were trying to mislead the whole world in thinking that one could make weather predictions as a deterministic problem. And I think in some fundamental way Wiener was probably right."[58]

It was Edward Lorenz, a consultant to the IAS meteorology project, who would establish the unpredictability of the atmosphere— shortly after von Neumann's death. Approaching the question from another direction, Charney asks, "if Laplace's mathematical intelligence were replaced by a computing machine of unlimited speed and capacity, and if the atmosphere below 100 km were spanned by a computational lattice whose mesh size were less than the scale of the smallest turbulent eddy, say one millimeter . . . would the problems of meteorology then have been solved?"[59] He answers that all predictability would vanish in less than one month, "not because of quantum indeterminacy, or even because of macroscopic errors of observation, but because the errors introduced into the smallest turbulent eddies by random fluctuations on the scale of the mean free path (ca 10–5 mm at sea level), although very small initially, would grow exponentially. . . . The error progresses from 1 mm to 10 km in less than one day, and from 100 km to the planetary scales in a week or two."[60]

As to whether climate—the "infinite forecast"—is predictable, the jury is still out. Von Neumann expected that not only would climate become predictable, but it would also be controlled. The balance points, once identified, would be too easy to tip. The real climate change crisis, according to von Neumann, was not whether we can control climate, but how to decide who sits at the controls. "After global climate control becomes possible," he warned in 1955, "this will merge each nation's affairs with those of every other, more thoroughly than the threat of a nuclear or any other war may already have done."[61]

Von Neumann and Wiener could both be right. Wiener may well be as correct about climate as he was about medium-term weather prediction: that the atmosphere can no longer be treated as a deterministic system beyond thirty days or so. Von Neumann could be right in the sense that even if climate cannot be predicted, that does not mean it cannot be controlled.

Imagine a future, combining the visions of Lewis Fry Richardson with those of von Neumann, where the Earth (including much of its oceans) is covered by wind turbines immersed in the momentum flux of the atmosphere, and photovoltaics immersed in the radiation flux from the sun. Eventually enough of these energy-absorbing and energy-dissipating surfaces will be connected to the integrated global computing and power grid, to form, in effect, the great Laplacian lattice of which Charney and Richardson dreamed. Every cell in this system would account for its relations with its neighbors, keeping track of whether it was dark, or sunny, or windy, or calm, and how those conditions may be expected to change. Coupled directly to the real, physical energy flux would be a computational network that was no longer a model—or rather, was a model, in Charney and Richardson's sense of the atmosphere constituting a model of itself.

Any such distributed planetary system, however, once it is sufficiently fine-grained, will itself become unpredictable—just like the atmosphere in which it is immersed. Whether the photovoltaic landscape is absorbing or reflecting, and whether the wind farms are under full load, freewheeling, or nudging the atmosphere here and there, may, in the course of time, indeed control the climate, but the workings of the model, and how it is going to behave a week from Thursday, will remain as mysterious as a partly cloudy day still is to us.

"Sometime in the early 1950s, von Neumann, I, and several others were standing outside of the Electronic Computer Project Building in Princeton," remembers Joseph Smagorinsky, "and Johnny looked up at a partially cloudy sky and said, 'Do you think we will ever be able to predict that?' "[62]

# TEN

# Monte Carlo

*Between 1946 and 1955 we crossed the coun-
try twenty-eight times by car.*

—Klári von Neumann, 1963

"WE WERE on the Riviera, in Monte Carlo, at the center of gravity for incurable gamblers," remembers Klári von Neumann, of a gambling expedition with Francis, her first husband, midway between World War I and World War II. "When we walked into the Casino, the first person we saw was Johnny; he was seated at one of the more modestly priced roulette tables with a large piece of paper and a not-too large mound of chips before him. He had a 'system' and was delighted to explain it to us: this 'system' was, of course, not foolproof, but it did involve lengthy and complicated probability calculations which even made allowance for the wheel not being 'true' (which means in simple terms that it might be rigged)."[1]

"Francis went on to another table," Klári continues. "For a while I wandered around watching the lunatic pleasure of people destroying themselves, then I went to the bar and sat down, wishing I had company with my drink. As I was sipping my cocktail, Johnny appeared." The game theorist had run out of luck at the roulette table, and Klári, who was running out of luck with her first marriage "an absolute disaster"—had to pay for his drink. "I was a rich girl, my father was very wealthy and Francis was an incurable gambler—this just about sums up my sex-appeal to him. After four years of all kinds of troubles, we divorced—my father bought it for me."[2]

Klára Dán was born on August 18, 1911, into a wealthy Jewish family in Budapest. "The most pampered, spoiled brat in a very large closely knit clan," she remembers herself as "a beautiful and absolutely obnoxious child, who squealed, yelled and howled her way through the first formative years of life." Her father, Charles Dán, an industrialist and financier, served as an officer in the Austro-Hungarian Army during World War I, surviving the war in relative comfort, but with the end of the war, "there was terrible confusion and we fled, partly on foot, across the border to Vienna, escaping from the communist terror of Béla Kun." After escorting the family to safety, her father returned to join the counterrevolutionary underground. "The strongest and most lasting memory of my childhood," says Klári, "is standing across the bridge and watching him walking back into what, I had by then realized, could be grave personal danger."[3]

With the overthrow of the Béla Kun regime, Budapest entered the golden years between World War I and World War II. "The counterrevolution led by Admiral Horthy succeeded," writes Klári. "We could all go home again and then the Hungarian version of the 'Roaring Twenties' was on its way."[4] Klári became a national figure-skating champion at age fourteen, before being sent to England to boarding school. Like the von Neumanns, her family occupied a large house divided into three apartments, presided over by her maternal grandfather, and featuring "a huge terrace which could, and very often did, seat over a hundred people for dinner or other festivities." The garden was divided into a formal section, off-limits to children, and an overgrown, wild area, off-limits to adults. "This line of demarcation," Klári adds, "was the only separation between children and adults in that happy house, which gradually became the center of the 'Roaring Twenties' Budapest."[5]

The entire household gathered regularly at Klári's grandfather's table for dinner, often followed by celebration well into the night. "Soon after dinner we all drifted down, my uncle and aunt and their two children (second floor), my parents, my sister and I (third floor)," Klári explains. "There was a bottle of wine and the confab started. As often as not, another bottle was passed around; pretty soon a gypsy-band was summoned, perhaps some close friends cajoled out of bed, and a full-fledged 'mulatsag' was on its way.

"It is absolutely impossible to translate 'mulatsag' in one simple word," Klári notes. "It is not a party, it is not a feast, it is not even an orgy; it is simply the spontaneous combustion of a bunch of people having a good time. At six o'clock in the morning, the band was dismissed, we went back upstairs, had a quick shower, the men went to work, the children to school, and the ladies with their cooks to the market."[6]

Klári's father and grandfather also founded a series of "Thursday Night" parties, held, once a month, in an all-male club called The Nest, with, in Klári's words, "the laudable aim of having men from the business, financial and political world meet with artists, writers and other members of the literary and intellectual community." When it was decided to open this gathering, with its "fertilious effect on that handkerchief-sized country's extraordinary production of creative minds," to women, Klári's grandfather announced "that the first party to include the ladies unquestionably had to be held at our house."

"It was simply wonderful," Klári remembers. "All three households were turned inside-out; pianos were moved, furniture rearranged. . . . On one floor was the dinner; on the other, quarters for those who wanted to talk or play cards; the third was for music and dancing— all three kitchens in continuous uproar for at least three days." No attempt was made to put children to bed. "Thus, at about the age of thirteen and for many years after, I got to know the most interesting and exciting people of our town."[7]

Klári acquired a social appetite that remained with her for life. "I met people, people and more people," begins a memoir left unfinished at her death, "some of them world-famous, others no one ever heard about; family patriarchs, cardsharps, ex and future queens, charwomen and call-girls, statesmen and politicians at the height of their power, nightshift workers and bar philosophers, certified geniuses and frustrated total failures—all these and many more." Klári suffered from depression, yet lived life to the full. "It was the spirit of a warmhearted conspiracy with the friends around her against what—if I sensed it right—was felt as the indifference and perhaps even the malevolence of fate," wrote physicist John Wheeler, two weeks after her death. "The spirit to work against what might have looked to be black fate but what could nevertheless be defeated."[8]

After her divorce from Francis, Klári married a respectable, non-gambling banker. "We did the right things at the right time, we had a smoothly running household where we gave the appropriate parties at the correct intervals," Klári writes. "He was a kind, gentle, attentive husband—he was also eighteen years older than I—and I was bored to tears." Then, in August of 1937, Johnny, nearing the end of his first marriage, made contact during his customary summer visit to Budapest.

"We struck up a telephone acquaintance which soon turned into sitting in cafes and talking for hours, just talking and talking," Klári recalls. "We both were keenly interested in politics and indulged in detailed prediction of the gloomy future (Johnny's assessment of the shape of things to come were amazingly close . . . and I shudder at the accuracy of some of his prognoses). We talked about this, and ancient history, and the probability to win against the roulette wheel. We told each other not-too-clean stories and little ditties that we made up between our marathon talk sessions; we talked about the difference between America and Europe, the advantage of having a small Pekinese or a Great Dane."[9]

On the seventeenth of August they said good-bye at Kelenföld railway station, from where Johnny left for Vienna, Cologne, Paris, and Southampton, sailing from there on the twentieth aboard the Cunard liner *Georgic* for New York. Johnny arrived back in New York on August 29, and Mariette (who had also spent the summer in Europe) arrived with the *Queen Mary* on September 7. A flurry of letters and telegrams followed, relayed through intermediaries in both Princeton and Budapest. "It became perfectly clear that we were just made for each other," says Klári. "Our letters became longer and longer. The inevitable of course happened. I told my kind and understanding daddy-husband quite frankly, that nothing that he or anybody would do could be a substitution for Johnny's brains."[10]

Mariette, with two-year-old Marina, now underwent the peculiarly American ritual of spending six weeks in the Nevada desert to obtain a divorce. "I believe that Hell is certainly very similar to this place," she wrote from the Riverside Hotel in Reno on September 22. "It is undescribable, everybody is constantly drunken and they lose their money like mad 5–6 hundred dollars a day, the roulette table stands in the hall just as a spittoon some other place. . . . How are you sweetheart how is the apartment how do you live and do you love me a bit write about all these at length. I have the howling blues."[11]

The next day, Mariette traveled the thirty-five miles to a guest ranch at Pyramid Lake, where the divorce season was winding down. "Johnny Sweetheart," she wrote, "it is entirely crazy here and I would not feel so miserable if I were not meant to stay here for 6 weeks I believe I won't survive. I live in the midst of an Indian reservation . . . and the country is so divine that it is difficult to imagine. . . . Riding is very beautiful but the evenings are deadly, imagine dinner at six and night goes until 10 o'clock."[12]

With a divorce decree granted by Washoe County, Mariette returned from Nevada in early November, and on November 25, at the Municipal Court in Washington, D.C., married experimental physicist J. B. Horner (Desmond) Kuper, a former Princeton graduate student of Eugene Wigner's who had made important contributions to radar during the war. Both Mariette and Desmond Kuper later held positions at Brookhaven National Laboratory on Long Island—the new East Coast nuclear laboratory that Frank Aydelotte had once suggested might be located in the Institute Woods. Johnny and Mariette remained on good terms, and their daughter, Marina, divided her time between the two families as she grew up.

On November 11, Klári cabled, "Three cheers guess why" from Budapest.[13] Johnny, now single, suggested he visit Europe over Christmas, and on November 17, Klári cabled her approval. Meanwhile, Johnny sent a series of formal proposals by mail. On November 9 he made a "direct offer," and on November 12 requested permission to inform his mother. On November 16 he sent a "direct offer, detailed," which he repeated on November 19. On November 30 he mailed a fourth proposal, received by Klári on December 9, who cabled back, on December 13, "Don't worry darling firm as a rock proposal enthusiastically accepted." On December 23 she cabled, "Merry Xmas happy sailing your loving future."

Telling everyone, even his brothers, that he was sailing for Southampton aboard the *Aquitania* on December 23, von Neumann instead boarded the *Normandie,* sailing for Le Havre on December 26. "What one realizes when once really and truly one is governed by one's emotions," he begins a 2,400-word letter (mostly in Hungarian) penned aboard ship. "Hardly 475,200 seconds!" he notes on December 28, his thirty-fourth birthday, estimating the time that remains until arrival in Budapest. The *Normandie,* after stopping briefly in Southampton, arrived at Le Havre on December 31. Von Neumann took the direct train to Paris on New Year's Eve and, on New Year's Day 1938, left Paris for Budapest aboard the *Orient Express.*

On January 24, Johnny was again in Paris, on his way back to the United States. Klári was on her way to the Italian Riviera, staying at the Savoy in San Remo, and cabling on February 2, from Monte Carlo: "told father with best result." Johnny intended to return to Budapest as soon as possible to retrieve Klári from the gathering storm in Europe, and reported to Stan Ulam on April 22 that "my 'future plans' are now known to everybody who is concerned in this matter, here and in Budapest."[14] Austria had been absorbed into the German Reich by the Anschluss of March 12, and all bets were off as to what would happen next.

Matters grew increasingly complicated, first of all by Klári's divorce, with a scheduled court decision postponed until September 23.

Klári and Johnny would then have to be married, swiftly, in Budapest in order for Klári to obtain U.S. immigration papers, but the Hungarian authorities refused to recognize the validity of Johnny's Reno divorce, which required yet another appeal to a different court. And to obtain Klári's visa, Johnny had to renounce his Hungarian citizenship, which required first a petition to the Hungarian government and then certification of this fact to the United States.

Von Neumann pulled all available strings—in New York, Washington, London, and Budapest—while Abraham Flexner did everything he could to help. "In his vast experience of helping people to get in and out of countries he had learned that the more important the person, the more the twists and twirls of red tape grew," says Klári, "and never had he seen such a mess."[15] Johnny began to lose his otherwise even temper, and Klári began to have second thoughts. She withdrew to Abbazia, the luxurious Austro-Hungarian resort on the Adriatic, and after chasing after her through Southern Europe, aboard trains whose schedules were beginning to be disrupted by troop movements, he retreated from the mainland to Stockholm and Copenhagen, where, as a guest of Niels and Harald Bohr, he attempted to prove to Klári, in writing and with his characteristic, persistent logic, that they should go through with their intended plans, and be free to go "away from this infernal pesthole of Europe, very far away."[16]

Johnny had difficulty paying attention to his work, and alternated between monitoring the international news, hour by hour, and trying to reassure Klári not only of his fitness as a husband, but about practical concerns, such as her fear of anti-Semitism in the United States. He explained how the United States had to maintain quotas against immigrants, effectively excluding Jews, in order to placate "the ordinary American" and avoid "dangerous reactions," while "inside the quota they are quite liberal." In his assessment the immigration authorities "behaved philosemitically as this administration is exactly that."[17]

The year 1938 was not yet 1939, but was getting close. "The German trains from Dresden are full of soldiers," von Neumann had noted on his way north, through the Berlin where ten years earlier he had begun his mathematical career. "The mobilization does not ruin the timetable. The trains are fast and punctual so far. I looked at Berlin very seriously. It may be for the last time." He then visited Lund and Stockholm, intending to proceed directly from Sweden to Cambridge, to meet with P. A. M. Dirac, when Niels Bohr invited him back to

Copenhagen to stay at his private estate—formerly the residence of J. C. Jacobsen, founder of the Carlsberg brewery. "It seems that he wants to talk about some connections between quantum theory and biology," Johnny reported to Klári. "Why exactly with me I can not say, but probably because I am not a biologist."[18]

"In Copenhagen again!" he reported on September 18. "The brothers Bohr fetched me at the pier, and now I'm established in Niels Bohr's private palace. I had numerous conversations with the Bohrs and Mrs. Bohr, of course mostly political—but we even managed to talk an hour and a half on 'the interpretation of quantum mechanics.' I'm sure we were showing off, the both of us: giving an exhibition, that we can worry about physics in September 1938. It's all like a dream, a dream of a peculiarly mad quality . . . the Bohrs quarreling, whether Tcheckoslovakia ought to give in—and whether there is any hope for causality in quantum theory."[19]

Klári's divorce was postponed yet again, into late October, while the war was postponed by British prime minister Neville Chamberlain's concessions to Hitler in Munich on September 29. Klári and Johnny were married in Budapest on November 18, and were able to secure Klári's visa to leave for the United States—after a last-minute crisis, when the Hungarians withdrew her passport upon her marriage to an American, and the Americans could not issue the visa without a passport to stamp the visa in. The horrors of Kristallnacht on November 9 were a glimpse of the fate that awaited those unable to escape.

They left Budapest for Paris on the *Orient Express,* transferring to Le Havre to board the *Normandie,* scheduled to sail for New York on December 6. Le Havre, however, was crippled by a dockyard strike, so von Neumann made arrangements to cross the Channel and sail on the *Queen Mary* from Southampton instead. "With a farewell blast from the funnel of that floating palace, I left Europe for ever," writes Klári, "at least the Europe that I had known."

Von Neumann left Europe with an unforgiving hatred for the Nazis, a growing distrust of the Russians, and a determination never again to let the free world fall into a position of military weakness that would force the compromises that had been made with Hitler while the German war machine was gaining strength. He replaced the loss with a passion for America and everything that its open frontiers came to represent. "He loved the wide-open spaces," Oskar Morgenstern says.[20]

"As soon as we got through the channel and the choppy Irish Sea into the open waters, Johnny became an utterly changed man," Klári writes. "For the first time since he left America, he was fit, willing and able to work on his mathematics. He would participate enthusiastically in the various events, then when he seemed to be most engrossed in the horse races, or bingo, or chatting with a surviving fellow-passenger, he would surreptitiously grab a piece of paper—anything handy, from a paper napkin to the back of a magazine or the edge of a newspaper, and jot down a few lines."[21] Early in the morning, before anyone else was awake, he would write up his notes in final form.

The newlyweds arrived in New York City on December 18, where Klári was surprised to find that "even the customs official said a few Hungarian words." Johnny booked a suite on the "twenty-something floor" of the Essex House and Casino-on-the-Park at 160 Central Park South, where they ran up enough of a bill that the credit manager was prompted to write to the Institute in Princeton, provided as a credit reference, to confirm their "knowledge of this individual's financial standing and credit responsibility, which, of course, will be held in strict confidence."[22]

"It was not until I saw from the windows of our tower apartment both downtown and Central Park with the lights going on in the wintery dusk over Manhattan," Klári writes, "that I realized that indeed I had arrived at a different Land." The following afternoon she took the train to Princeton, whose rigid social protocols she found a far cry from carefree Budapest. Johnny disappeared on a detour to attend to "important business" in Trenton—which Klári later learned was a court appearance "to show cause why his driving license should not be revoked."

The Institute was on winter break until February 1, leaving von Neumann with no responsibilities except a talk to the American Mathematical Society winter meeting in Williamsburg, Virginia. He purchased a new car—a Cadillac V-8 coupe—so they could drive to the meeting and then continue south, through the Everglades, to Key West. Their first stop was Washington, D.C., where they stayed at the Shoreham Hotel, while von Neumann attended to secretive government business, including an unsuccessful attempt to appeal his rejection by the Army Reserve. "Johnny was a strange man, incongruous and contradictory," notes Klári of this episode, "with as many facets to his personality as the number of people who thought that they knew and understood him." The von Neumanns also called socially on Mariette and her new husband, precipitating, in Klári's words, "a crisis which was followed by many other similar ones for many, many years." Klári's insecurity was never far below the surface, and was easily provoked. Johnny and his ex-wife "never ceased playing the game of detached attachment or vice versa, whichever fits best."[23]

Upon their return from Florida, the von Neumanns settled into a house about two miles from the Institute, on Westcott Road. Klári's parties became legendary, especially once the engineers from the computer project arrived to liven things up. "Klári von Neumann would make up fish house punch, which was very potent, so the parties got very relaxed and uh . . . joyful, as the evening wore on," remembers Willis Ware. "It was after one of those parties that James Pomerene and Nick Metropolis from Los Alamos drove their car backward through Princeton. But the Princeton cops were so accustomed to dealing with students that they just took things like that in their stride."[24]

Princeton was hard on Klári, who balked at the role of academic wife. She made one last visit to Europe, to retrieve her parents, resolve what she could of family affairs, and drive Johnny crazy as she skated on ever-thinning ice. "For God's sake do not go to Pest," he wrote from Montreal on August 10, 1939, "and get out of Europe by the beginning of Sept! I mean it!"[25] Klári's parents escaped to Princeton with the opening of the war, but her father, despondent, threw himself under a train over Christmas 1939. Klári's bouts of depression grew more severe, and she later confided to the Rosenbergs that she believed she was destined to commit suicide herself. "She said it must be congenital," Jack Rosenberg says.[26]

Von Neumann was sociable, but in a superficial way. "I wonder how Klári managed to live with him," Robert Richtmyer asks. "Some people, especially women, found him lacking in curiosity about subjective or personal feelings and perhaps deficient in emotional development," says Stan Ulam. "To be sure, he was interested in women, outwardly, in a peculiar way. . . . About women in general he once said to me, 'They don't do anything very much.'" Klári, adds Ulam, "was a very intelligent, very nervous woman who had a deep complex that people paid attention to her only because she was the wife of the great von Neumann, which was not true of course."[27]

Klári found Johnny's level-headedness exasperating, and took to addressing him, in her letters, as "Sir." They seemed to orbit around each other, and were rarely in the same place for any length of time.

"She was a very friendly, outgoing girl, but Johnny was not easy to reach," says Rosenberg. "I never saw him lose his temper," says Marina, "except maybe two or three times. Klári knew how to push him far enough so finally he would explode."[28] Klári increasingly sought time alone. "The letter was beautiful as only your letters can be—but why is it that it always has to be in letters," she wrote during an attempt to reconcile after an argument in 1949. "Perhaps you are just as much of a dreamer as I am, and when I am not present you still see me as you imagined me to be in 1937 when you returned to the States."[29]

These tensions were exacerbated by von Neumann's increasing absences from home. When he and Klári did travel together, things went better, and their happiest times were on the road—the American version of shipboard life. Von Neumann's reputation for not wanting to fly was more about his love of driving and train travel than about fear of being in the air. In 1940 he was invited to give the John Danz memorial lectures at the University of Washington in Seattle. Having never been west of Chicago, he decided to drive, taking Klári—and Route 66.

They left Princeton in May for the American grand tour. Europe was falling day by day to the Nazi advance, and their trip west alternated between exploring the back roads of pre-interstate America and trying to find towns with newspapers or radio stations to catch up on the day's events. "Johnny insisted on listening to pretty nearly all the news broadcasts that came over the air," Klári notes. "He would spend hours sitting in the car." Their trajectory was determined partly by the events in Europe and partly by the landscape of the American West. "Holland was being invaded the day after we arrived in Denver and we just had to stay in a city which had extra editions of papers and continuous broadcasts so that we could follow the course of the depressing events," Klári explains. "By the time the negotiations for the surrender of Belgium had started, we had made it to Nevada." Johnny was captivated. "If he had not been so preoccupied with steadily worsening news, this trip would have turned him into a geologist," Klári says.

The gloom was occasionally dispelled. Somewhere in Nevada, "a man with a nice long beard, wearing well-used denims, tied his pack-mule to the hitching-post, then rode the other one, his mount, into the bar where we were consoling ourselves," Klári writes. "Nobody blinked an eye, the bartender handed the man a glass of beer and a bucket of the same brew was placed in front of the mule. The whole scene was a mute play; it seemed completely routine, the man paid, he and his beast drank up and quietly left the place."[30]

After visiting Las Vegas, where there were only "a few dingy gambling joints catering mostly to the workers who were there to build Boulder Dam," they "meandered about in the Southwest visiting national parks and national monuments," passing through Santa Fe, New Mexico, without stopping ("Johnny was suddenly in a great hurry to see the Grand Canyon") and without any premonitions of how profoundly the nearby Los Alamos mesa would affect their lives in the years ahead.[31] In the spring of 1940 only the first hints of nuclear weapons were in the air. News of the discovery of fission in late 1938 had arrived in Princeton with Niels Bohr in early 1939, raising for the first time the real possibility, already a subject of speculation, of an atomic bomb.

Fearing that the warning communicated to President Roosevelt in August 1939 by Albert Einstein and Léo Szilárd (who had applied for a patent on nuclear explosives in 1934) was not being taken seriously, von Neumann elevated the alarm. "The Dutch physicist, P. Debye, who has been Director of the Physics Institute of the Kaiser Wilhelm Gesellschaft in Berlin (supported by the Rockefeller Foundation), has been sent abroad by the German authorities in order to free his Institute for secret war work," he wrote to Frank Aydelotte in March of 1940, in a letter that Veblen also signed. "When one of us met him at dinner the other evening, he made no secret of the fact that this work is essentially a study of the fission of uranium. This is an explosive nuclear process which is theoretically capable of generating 10,000 to 2,000,000 times more energy than the same weight of any known fuel or explosive." Noting that there were considerable deposits of uranium in Bohemia and Canada, von Neumann and Veblen warned "that the Nazi authorities hope to produce either a terrible explosive or a very compact and efficient source of power," adding that leading German nuclear and theoretical physicists were being assembled under Werner Heisenberg in Berlin, "in spite of the fact that nuclear and theoretical physics in general and Heisenberg in particular were under a cloud, nuclear physics being considered to be 'Jewish physics' and Heisenberg a 'White Jew.'"[32]

"The matter should not be left in the hands of the European gangsters," they warned. Acknowledging that "some effort, not entirely successful, has been made to enlist the help of the United States Government," they urged Aydelotte to bring the prospect of an atomic bomb ("which we have had on our minds for several months, without knowing what, if anything, to do about it") to the attention of the Rockefeller Foundation, "which would be in a position to act in a simple and direct manner."[33] The Foundation responded with emergency funding to help quietly bring key European nuclear physicists—among them Wolfgang Pauli and the brothers Niels and Harald Bohr—to the safety of England and the United States. When the Manhattan Project launched in 1942, critical talent was in place.

The United States finally entered the war in December 1941. "At long last [Johnny] could effectively vent his spleen," says Klári. "At the same time, he was also using this perfectly honorable, patriotic excuse to shake off the self-imposed yoke of pure mathematics and get into more applied fields, with which he had a secret flirtation long before he openly admitted his steadily increasing interest in it."[34] Von Neumann would never return to pure mathematical work.

Klári became pregnant, and Johnny's customary signature line to the Ulams—"from house to house"—was revised to include "best greetings from both of us, and $(\frac{1}{2})^2$ unknown." Klári, now thirty-one, suffered a miscarriage on June 16, 1942, and Johnny was increasingly absent from Princeton for defense-related work. His assignment to Britain on behalf of the navy in early February 1943 was both secretive in purpose and indeterminate in duration. All communication was censored. On April 13, 1943, he cabled Klári from London: "Congratulations on statistics very impressed stop Boske visiting here all very well very much love." The telegram was intercepted. "Will you please be so kind as to furnish this office with a complete explanation of the text of this message," the Office of Censorship asked.[35]

With censors looking over his shoulder, von Neumann's correspondence lost its passionate tone. "The recent monotonous style in your letters infuriates me," Klári wrote on May 15, 1943. "What on earth is the matter with you?" Klári took a wartime job, full time, with Princeton University's Office of Population Research, under the auspices of the Rockefeller Foundation and the Woodrow Wilson School. Frank W. Notestein's population research group was looking at both historical trends in human population and a series of future "what ifs"—for instance, what would happen to a reconfigured postwar Europe, a centrally planned Soviet Union, or a proposed Jewish state in the Middle East? Klári was swiftly promoted, and offered an academic position in 1944, which she declined.

In July, von Neumann was recalled from England and began disappearing under ever more secretive circumstances, which led, in September, to Los Alamos, where "Project Y" was now under way. When not in residence at Los Alamos, he spent much of his time on the West Coast, returning occasionally to Princeton and making regular visits to Chicago, Oak Ridge, Philadelphia, Aberdeen, and Washington, D.C. At Los Alamos he was able to get cigarettes—preferably Lucky Strikes—at the PX, hoarding them for Klári in Princeton. "Whenever he came home, we usually spent most of the night talking," Klári remembers, "his pent-up tension was pouring out in a flow of words which, as a rule, he kept strictly to himself."[36]

On October 19, 1943, the Institute added additional coverage for "extra-hazardous activities" to von Neumann's insurance policy under his contract with the Office of Scientific Research and Development, a sign that he was taking a more than theoretical interest in weapons research. When the surrender of Germany was announced, he was on a field assignment at Los Alamos, and it was twelve hours before he heard the news. "Well, it's over," he wrote to Klári the next morning. "How do you feel?" The cigarette situation improved, and the scientists kept working on the bomb. "Since May 3, inclusive, I am getting an average of about 2 packs of Luckies per day," he reported to Klári on May 11. "Whadayasay?"[37]

The next six months brought intense activity: the Trinity test, Hiroshima, Nagasaki, the surrender of Japan, and, behind the scenes, the completion of the ENIAC, the first H-bomb calculations, and the launching of the computer project at the IAS. "This exposure to such a marvelous machine," recalled Nicholas Metropolis, regarding his first visit to reconnoiter the ENIAC, "coupled in short order to the Alamogordo experience was so singular that it was difficult to attribute any reality to either." The same day that a copy of the Trinity bomb was dropped on Nagasaki, Edward Teller cabled von Neumann in Princeton: "Stan and Nick can now act openly as coming from Los Alamos."[38] Stan Frankel and Nick Metropolis were already at the Institute for Advanced Study to begin preparing the first H-bomb codes.

Von Neumann was now thinking about the next war, and whether it would be fought with nuclear weapons on both sides. "The date of the next war is probably determined by the time it takes the conscious and the subconscious processes of the American people to get into equilibrium with each other," he wrote to Klári in October 1946. "I don't think that this is less than two years and I do think that it is less than ten."[39] He had long believed that Soviet Russia would prove to be a greater threat than Germany or Japan. "When the Western troops stopped and even withdrew to let the Russians advance deeper into Germany, Johnny was frantically dismayed," explains Klári. "His idea was that the Western Allies should have kept going all the way into Russia and abolish in one sweep any dangerous or potentially dangerous form of government that might lead to another war. In the immediate postwar years, Johnny quite openly advocated preventive war before the Russians became too strong."[40]

Klári visited Los Alamos for the first time over Christmas 1945. She headed west by train from Princeton, via Chicago, where she boarded the *Super Chief.* "Will expect you Lamy Saturday morning," Johnny cabled on December 15. "Bring riding and skating things if possible opportunities very good."[41] It was love at first sight. The mountains, the horseback riding, the skiing, the Pueblo ruins at Bandelier, the Lodge at the former Los Alamos Boys School (where Johnny, as a VIP, was entitled to stay), the predominance of Europeans (including Hungarians), the frequent, spontaneous parties, the late-night poker games—all evoked memories of Monte Carlo and Budapest. Los Alamos had what Princeton lacked. The sparks between Klári and Johnny were rekindled, and they began collaborating on the codes that would animate the new computer and bring the super bomb to life.

"The new mathematical tool was not the only experiment that Johnny wanted to try in this connection," Klári remembers. "He also wanted to see how someone who had none or very little experience in the field, how such a person would take to this novel way of doing mathematics. For this experiment he needed a guinea-pig, preferably a mathematical moron and, unquestionably for this purpose the ideal subject was right there within easy reach—namely me." Klári had passed her high school examinations in algebra and trigonometry, but only because "my math teacher rather appreciated my frank admission, that I really did not understand a single word of what I had learned."

"Long before the machine was finished I became Johnny's experimental rabbit," she says. "It was lots and lots of fun. I learned how to translate algebraic equations into numerical forms, which in turn then have to be put into machine language in the order in which the machine has to calculate it either in sequence or going round and round until it has finished with one part of the problem and then go on some definite which-a-way, whatever seems to be right for it to do next." Klári found programming to be a "very amusing and rather intricate jig-saw puzzle," and soon "became one of the first 'coders,' a new occupation which is quite wide-spread today."

"The machine would have to be told the whole story, given all the instructions of what it was expected to do at once, and then be permitted to be on its own until it ran out of instructions," Klári explains. "There already existed fast, automatic special purpose machines, but they could only play one tune . . . like a music box. . . . In contrast, the 'all purpose machine' is like a musical instrument."[42]

There were none of the conveniences programmers take for granted today: compilers, operating systems, relative addressing, floating-point arithmetic. Every memory location had to be specified at every step, and the position of the significant digits adjusted as a computation progressed. "People had to essentially program their problems in absolute," James Pomerene explains. "In other words, you had to come to terms with the machine and the machine had to come to terms with you."[43]

Klári's wartime work on population statistics had prepared her for the problems that Johnny was starting to code. The question of whether a given bomb design explodes—and if so, how efficiently—depends on how rapidly its population of neutrons reproduces, and whether mortality and emigration have any moderating effects. "Statistical questions will be amenable to an entirely new kind of treatment," von Neumann had explained in January of 1945, while the ENIAC was still being built. "It will be possible to answer most questions of this type by performing the actual statistical experiment: by computing hundreds or thousands of special cases and registering their statistical distribution."[44] A statistical approach to otherwise intractable physical problems had been taken up by others, including Enrico Fermi in the 1930s, but it took someone—and that someone was Stan Ulam, assisted by von Neumann and Nicholas Metropolis—to come up with a name for the technique and make it stick.

At the end of the war, there had been an exodus from Los Alamos. With its remote location and total secrecy no longer necessary, work at the laboratory appeared to be winding down. Those with families to support were advised to leave, if they could. Stan and Françoise Ulam, with their one-year-old daughter, Claire, left for California, where Stan had been offered a teaching job at USC. Before he had settled into the new job, and before Françoise and Claire had even found a place to live, Stan suddenly fell gravely ill, with a case of viral encephalitis that might have killed him without an emergency trepanation at Cedars Sinai Hospital to relieve the pressure on his brain.

Overwhelmed, Françoise arranged to send Claire back to Los Alamos, in care of David Hawkins (Stan's collaborator on neutron multiplication) and his wife, Frances (who ran the Los Alamos nursery school). Stan recuperated in Los Angeles, while Claire thrived among the families who had remained on the mesa, where Norris Bradbury, a more down-to-earth administrator than Oppenheimer, had taken the helm. Because the Ulams had lost their government health insurance, and Stan had not started teaching yet, things were looking grim. Then Stan was invited to return to Los Alamos. "The case [of Stan Ulam] is almost unique," wrote von Neumann, who no doubt had a hand in the invitation, to Carson Mark. "I feel that it is justified for the Los Alamos Laboratory to go any length to keep him there."[45]

Ulam, who had been advised, during his convalescence, to avoid strenuous mental activity, amused himself by playing solitaire. He could not resist a question: What were the chances that a Canfield solitaire with fifty-two cards will play out successfully? "After spending a lot of time trying to estimate them by pure combinatorial calculations," he recalls, "I wondered whether a more practical method than 'abstract thinking' might not be to lay it out say one hundred times and simply observe and count the number of successful plays." This, he noted, was a far easier way to arrive at an approximate answer than "to try to compute all the combinatorial possibilities which are an exponentially increasing number so great that, except in very elementary cases, there is no way to estimate it."[46]

"This is intellectually surprising, and if not exactly humiliating, it gives one a feeling of modesty about the limits of rational or traditional thinking," he added. It was characteristic of Ulam to draw deep mathematical conclusions where others would simply consider the immediate problem solved. He observed that mathematical logic itself can be considered as "a class of games—'solitaires'—to be played with symbols according to given rules." From this he drew the conclusion, with implications perhaps not yet fully appreciated, that "one sense of Gödel's theorem is that some properties of these games can be ascertained only by playing them."[47]

Ulam's attempt to take his mind off serious problems soon brought him back to some of the Los Alamos problems that had been left unresolved. "It occurred to me then that this could be equally true of all processes involving branching of events, as in the production and further multiplication of neutrons in some kind of material containing uranium or other fissile elements," he recalled. "At each stage of the process, there are many possibilities determining the fate of the neutron. . . . The elementary probabilities for each of these possibilities are individually known . . . but the problem is to know what a succession and branching of perhaps hundreds of thousands or millions will do."[48]

Monte Carlo originated as a form of emergency first aid, in answer to the question: What to do until the mathematician arrives? "The idea was to try out thousands of such possibilities and, at each stage, to select by chance, by means of a 'random number' with suitable probability, the fate or kind of event, to follow it in a line, so to speak, instead of considering all branches," Ulam explained. "After examining the possible histories of only a few thousand, one will have a good sample and an approximate answer to the problem."[49] The new technique propagated widely, along with the growing number of computers on which it could run. Refinements were made, especially the so-called Metropolis algorithm (later the Metropolis-Hastings algorithm) that made Monte Carlo even more effective by favoring more probable histories from the start. "The most important property of the algorithm is . . . that deviations from the canonical distribution die away," explains Marshall Rosenbluth, who helped invent it. "Hence the computation converges on the right answer! I recall being quite excited when I was able to prove this."[50]

Monte Carlo opened a new domain in mathematical physics: distinct from classical physics, which considers the precise behavior of a small number of idealized objects, or statistical mechanics, which considers the collective behavior, on average, of a very large number of objects, Monte Carlo considers the individual, probabilistic behavior of an arbitrarily large number of individual objects, and is thus closer than either of the other two methods to the way the physical universe actually works. "Because one seems to be getting something for nothing, it is necessary to keep straight the process by which everything comes out all right in the end; the efficiency of the methods in particular cases seems unbelievable," advised Andrew Marshall in 1954, reviewing Monte Carlo's first seven years. "The results quite literally have to be seen, and seen through, to be believed."[51]

On von Neumann's next visit to Los Alamos, Ulam brought up the idea as von Neumann was leaving to catch the train. "It was an especially long discussion in a government car while we were driving from Los Alamos to Lamy," where the railroad depot was located, Ulam recalls. "We talked throughout the trip, and I remember to this day what I said at various turns in the road or near certain rocks." Somewhere along the line, with credit usually going to Nick Metropolis, "it was named Monte Carlo," Ulam explains, "because of the element of chance, the production of random numbers with which to play the suitable games." The idea was impossible to resist. "Ulam relished the thought of a gambling spree in which the scorekeeping was so designed as to imitate a neutron chain reaction," Robert Richtmyer recalls. "It's infinitely cheaper to imitate a physical process in a computer and make experiments on paper, as it were, rather than reality," Ulam testified at the ENIAC trial in 1971.[52]

After the drive to Lamy, von Neumann returned by train to Princeton, working up Ulam's suggestion during the trip, and then, following a telephone conversation on March 7 with Richtmyer, he typed up an eleven-page letter, fleshing out (for "spherically symmetric geometry," of either uranium or plutonium) Ulam's idea. "I am fairly certain that the problem, in its digital form, is well suited for the ENIAC," he wrote. "Assume that one criticality problem requires following 100 primary neutrons through 100 collisions (of the primary neutron or its descendants) per primary neutron. Then solving one criticality problem should take about 5 hours." This would only address the simplified question of "static criticality"—whether the specified assembly would explode, not how well would it explode. Von Neumann estimated what it would take to address this more complex question, involving both hydrodynamics and radiation transport, concluding that "I have no doubt whatever that it will be perfectly tractable with the post-ENIAC device."[53]

The problem was that the post-ENIAC device would not become operational until 1951, despite its becoming "increasingly clear in connection with Los Alamos requirements, especially in the current atmosphere of crisis, that radical measures to finish the computer were necessary."[54] It turned out, however, to be possible to modify the ENIAC to function as a primitive, interim approximation of the forthcoming new machine. "In the Spring of 1947, J. von Neumann suggested to the author that it would be possible to run the ENIAC in a way very different from the way contemplated when it was designed," reported Richard Clippinger, then thirty-five, in 1948. "Problems can probably be changed in an hour instead of a day by the old method where many cables had to be plugged in and out."[55]

"About a year or so ago Johnny made a truly remarkable set of observations and was responsible for a complete new method of programming," Herman Goldstine elaborated in 1949. "Johnny's scheme was to wire up what corresponds to the ENIAC's plug-boards with a fixed set of instructions that is universal to all problems." Individual instructions were assigned unique numbers—order codes—that were intelligible to "a switching center so built that upon the receipt of a given number, characterizing one of the orders wired into the plug-boards, it energizes the proper board and thereby causes the order to be executed."

A sequence of orders, constituting a program, could either be entered via the ENIAC's function tables, or read from punched cards. "It is no longer necessary to stand on one's head to fit a given routine," Goldstine continued. "To prepare an individual problem the coder now merely writes out the sequence of operations, arithmetic and logical, which characterize his problem and then transliterates these into the numbers the machine will understand."[56]

"This new method is based on a vocabulary, i.e. a set of orders, which is conveyed to the machine on two levels; the 'background coding' and the 'problem coding,'" Johnny and Klári explained, making a distinction that survives to this day as the difference between operating systems and applications. Some sixty different instructions constituted the vocabulary. "After a code had been written, the list of instructions could be set up on huge banks of ten-position switches," adds Robert Richtmyer. "Each row of switches was assigned an address, a number from 1 to 300. Of the ENIAC's twenty accumulators, devices for adding or storing a number, one was used as a control counter, to keep track of the address of the row of instructions being executed, one served as a central clearing house for numbers, similar to the accumulator register of the Princeton design; two others were reserved for special purposes, and the rest were available for general storage. The wiring never had to be changed again."[57]

Credit for transforming the ENIAC into a stored-program computer usually goes to von Neumann and Richard Clippinger, while Presper Eckert claims the capability was designed in from the beginning, and that "Clippinger later 'rediscovered' these uses of the function tables, without knowing that they had already been provided for."[58] Metropolis adds that even after the Clippinger reconfiguration, the capacity of the ENIAC was insufficient to handle the Monte Carlo codes, until he noticed a new one-input, hundred-output matrix panel being installed and pointed out that "if this could be used to interpret the instruction pairs in the proposed control mode, then it would release a sufficiently large portion of the available control units to realize the new mode—perhaps."[59]

"With the help of Klári von Neumann," says Metropolis, "plans were revised and completed and we undertook to implement them on the ENIAC, and our set of problems—the first Monte Carlos—were run in the new mode."[60] Metropolis and Klári arrived in Aberdeen on March 22, 1948, to begin reconfiguring the machine. "At that time, the people who knew how to program were Johnny von Neumann, Nick Metropolis, and Klári," says Harris Mayer. "We made three teams, so we could run the machine around the clock. There was Foster and Cerda Evans, a husband and wife team, Rosalie and me, another husband and wife team, and then there was Klári herself and Marshall Rosenbluth, who was a bachelor. Nick Metropolis and Klári taught us how to program the machine. And then we went up to Aberdeen." Working inside the ENIAC, surrounded by its registers, accumulators, and function tables, made the new art of programming easy to comprehend. "On the ENIAC there was this great big checkerboard on the wall with decade switches," Mayer explains. "You could *see* the numbers, and Johnny could see that numbers are numbers, whether data or orders. His insight was how the machine with fixed programming could be changed."[61]

"Things are kind of upside down," Klári reported to the Ulams on the next-to-last day before the calculation was to begin. "Evans family arrived Thursday night (love and kisses from Foster who is bartending at this moment) the Mayer's arriving tonight, Marshall Rosenbluth (an unexpected addition) coming tomorrow for final verifying, Sunday late breakfast, meeting, etc, and the expedition starts at 6 pm to Aberdeen. Please pray for me and hope for the best." The computation took six weeks. "I heard from Nick on the telephone that the miracle for the ENIAC really took place," Ulam wrote to von Neumann on May 12, "and that 25,000 cards were produced!!"[62]

"Klári is very run-down after the siege in Aberdeen, lost 15 lbs., and has now a general physical check-up made at the Princeton Hospital," von Neumann reported when she returned home. "It took 32 days (including Sundays) to put the new control system on the ENIAC, check it and the problems code, and [get] the ENIAC into shape. . . . Then the ENIAC ran for 10 days. It was producing 50% of these 10×16 hours [and] could have probably continued on this basis as long as we wished. . . . It did 16 cycles ('censuses,' 100 input cards each) on 7 problems. All interesting ones are stationary at the end of this period . . . and the method is clearly a 100% success."[63]

A small group including Klári, Adele Goldstine, and Nick Metropolis began to code additional problems, both for the ENIAC and for the machine that had yet to be built. "It was fun to work on problems in those early days," Klári remembers, "because if the majority of us who were preparing problems to put on the yet unready machine, if we really ganged up on the engineers and told them that some new trick would be very useful, they would add it to the machine 'vocabulary' and in most cases, make it work."

"Your code was described and was impressive," von Neumann wrote to Klári from Los Alamos, discussing whether a routine she had developed should be coded as software or hardwired into the machine. "They claim now, however, that making one more, 'fixed,' function table is so little work, that they want to do it. It was decided that they will build one, with the order soldered in."[64]

Instead of tabulating the statistics of human populations, Klári was tabulating the statistics of populations of neutrons, as they underwent scattering (equivalent to travel), fission (equivalent to reproduction), escape (equivalent to emigration), or absorption (equivalent to death). By following enough generations, it was possible to determine whether a given configuration would go critical or not. Klári could hardly have better prepared herself for bomb design than by her apprenticeship at the Office of Population Research.

An undated manuscript, in von Neumann's hand, with appended notes to (and from) Klári, describes the "Actual Running of the Monte Carlo Problems" on the ENIAC. "In order to start the computation, an IBM card, representing one neutron, was read into the Constant Transmitter," the account begins. The fate of any given neutron could either be scattering, absorption, escape, fission, or census. Data on the card, used to determine its fate, included the zone in the spherical assembly where the event occurred, the time the event occurred, the velocity of the particle at the time of the event's occurring, the polar angle of the neutron's path, the distance from the center of the sphere at the time of the event, and the number of neutrons the particular card under consideration represented. There were also three additional numbers, "to trace the genealogy of any neutron in the sample population," specifying the fission generation, the "parent" neutron, and the original, or initial, card from which the present card originated.

"To start either one of the problems one hundred cards were read, each representing the neutron originated by fission . . . starting at the center of the assembly at zero time," the report explains. "These neutrons then in turn produced other neutron cards, indicating the event that happened on their parts or that the census time has been reached. Total escape and absorption cards were then sorted out and removed from the stack since their path did not have to be followed any longer. Fission cards, representing neutrons which produced two or three new neutrons and [whose] time was still within census time, were then put back into the reader until each newly printed card was either a Total Escape, Absorption, or Census card indicating that all neutrons that survived have reached the end of the census time interval."[65]

At the end of the census time interval, T was incremented, and a new cycle was started, using as input the output of the previous cycle. Never had any series of events been examined in such detail. "A 'complete' calculation (up to, say, 10 shakes evolution)," Johnny estimated to Edward Teller, would take six to eight weeks.[66] There are a hundred million shakes in a single second, and there are about five million seconds in eight weeks. Even at the speed of the ENIAC, time was being slowed down fifty trillion times.

To determine the feasibility of a hydrogen bomb, it was necessary to have a detailed picture of what happens when the fission bomb used to trigger it explodes. So far, the three main contributions to the behavior of a nuclear explosion—neutron multiplication, radiation transport, and hydrodynamics—had all been treated separately, but as von Neumann himself had suggested, and Robert Richtmyer now followed up on, they needed to be treated as related phenomena, at the same time.

"In September of 1947 I proposed to Johnny von Neumann a rather grandiose plan for computer-simulation of the explosion of a fission bomb," Richtmyer explains. "Von Neumann liked the idea, so I moved to Princeton to work it out, joined by Adele Goldstine and Klári von Neumann, who shared my office with me." The project took three years. "I had a habit of writing on the upper right corner of the blackboard cryptic notes to myself about things I had to do," says Richtmyer. "On one occasion, I was away for about ten days, and when I returned, there was an additional note on the blackboard in imitation of my handwriting; it said 'fresh water for hippo.' In consequence, 'Hippo' became the code name for the project we were working on."[67]

Hippo was run on IBM's Selective Sequence Electronic Calculator (SSEC), completed in 1948 and housed in a windowed showroom at their world headquarters on Fifty-seventh Street and Fifth Avenue in New York. The SSEC, which Johnny described to Klári as somewhere "between the ENIAC and the 'non existing' machines,"[68] stored some twenty thousand twenty-digit numbers on eighty-track paper tape, accessed by three punching units and sixty-six reading heads. "The programming took nearly a year," says Richtmyer. "Then, we took over the SSEC 24 hours a day, 7 days a week. In several months, we had made three or four complete bomb calculations."[69] The SSEC, incorporating forty thousand relays and with a one-second access time to its paper tape memory, was immediately obsolete, but the Hippo code continued to be used by Los Alamos for many years.

Those who worked on these early weapons calculations, running for weeks at a time, had to monitor how the calculations were progressing, interpreting the physics as well as the arithmetic, and making adjustments along the way. "Since returning from Chicago I have looked a little more closely at both the available space on the numerical function table and also, with the help of Johnny, I have set up a flow diagram for Maria's scheme of trying the tamper," Klári wrote to Harris Mayer in April 1949. "On one hand, it seems that if we have all our problems with one zone outside the tamper we have plenty of space on the numerical function table to put on, if you decide so, the reflection matrix. I only mention these facts so that you should feel free, when you choose a method, to consider using the matrix, if you should think that it might be better suited for the problems."[70] The view that early coders, such as Klári, were "doing the arithmetic" without any understanding of the physics is wrong.

With the success of Monte Carlo came a sudden demand for a reliable supply of random numbers; there was a shortage of them. Pseudo-random numbers could be generated within a computer as needed, but as von Neumann warned, "any one who considers arithmetical methods of producing random digits is, of course, in a state of sin."[71] The U.S. Air Force's Project RAND (progenitor of the RAND Corporation), for whom von Neumann was consulting in Santa Monica, took it upon themselves, in April 1947, to build an electronic roulette wheel and compile a list of one million random numbers, available first as punched cards and later expanded and published as a book. "Because of the very nature of the tables, it did not seem necessary to proofread every page of the final manuscript in order to catch random errors," the editors explained.[72]

Between June 29 and July 1, 1949, a conference on the Monte Carlo method—sponsored by the RAND Corporation, Oak Ridge National Laboratory, and the National Bureau of Standards' Institute for Numerical Analysis—was held at UCLA. "The gist of the matter is," Klári wrote to Stan Ulam, "that, since I have been working with it for quite a while now, I would very much like to go."[73] Although urged to participate in the meeting, Klári did not attend. She spent the last part of May and most of June running a large calculation on the ENIAC in Aberdeen and, despite Johnny's pleas to leave its completion to others, wouldn't let go until it was finished, retreating to Princeton in exhaustion without making the Los Angeles trip.

"I have finally returned to Princeton," she reported to Carson Mark at Los Alamos on June 28. "We finished our work Friday afternoon after having run six censuses on problem 2. They all came out super-critical, with a continuous trend toward criticality. . . . The IBM cards, which are packed in ten large boxes, and the listing of all the problems (two small boxes), are being mailed from Aberdeen, as far as I know by railway express C.O.D. I brought with me to Princeton all secret documents which we had with us in Aberdeen."[74]

For sixty years, Monte Carlo has been applied to an ever-expanding range of problems, in fields from physics to biology to finance. The ability not only to follow but to create branching, evolving processes gives the code almost uncanny powers. "In a Monte Carlo problem the experimenter has complete control of his sampling procedure," explained RAND mathematician and thermonuclear strategist Herman Kahn in 1954. "If for example he wanted a green-eyed pig with curly hair and six toes and this event had a non zero probability, then the Monte Carlo experimenter, unlike the agriculturalist, could immediately produce the animal."[75] Biological evolution is, in essence, a Monte Carlo search of the fitness landscape, and whatever the next stage in the evolution of evolution turns out to be, computer-assisted Monte Carlo will get there first.

Monte Carlo is able to discover practical solutions to otherwise intractable problems because the most efficient search of an unmapped territory takes the form of a random walk. Today's search engines, long descended from their ENIAC-era ancestors, still bear the imprint of their Monte Carlo origins: random search paths being accounted for, statistically, to accumulate increasingly accurate results. The genius of Monte Carlo—and its search-engine descendants—lies in the ability to extract meaningful solutions, in the face of over-whelming information, by recognizing that meaning resides less in the data at the end points and more in the intervening paths.

# ELEVEN

# Ulam's Demons

*The factor 4 is a gift of God (or of the other party).*

—John von Neumann to Edward Teller, 1946

"ONCE IN MY LIFE I had a mathematical dream which proved correct," remembers Stanislaw Ulam, born in Lwów, Poland, then part of the Austro-Hungarian Empire, in 1909. "I was twenty years old. I thought, my God, this is wonderful, I won't have to work, it will all come in dreams. But it never happened again."[1]

Joseph Ulam, Stanislaw Ulam's father, was a wealthy Jewish lawyer who served as an officer in the Austrian Army during World War I. Ulam's mother, Anna Auerbach, was the daughter of an industrialist who dealt in steel. Stan was drawn to mathematics from the start. "When I was four," he writes, "I remember jumping around on an oriental rug looking down at its intricate patterns. I remember my father's towering figure standing beside me, and I noticed that he smiled. I felt, 'He smiles because he thinks I am childish, but I know these are curious patterns.'" At age ten, Ulam was signing his school notebooks "Stan Ulam, astronomer, physicist and mathematician." He remembers that "an uncle gave me a little telescope for my birthday when I was eleven or twelve."[2] He graduated from high school in 1927, and from the Lwów Polytechnic Institute, with a master's and a doctorate in mathematics, in 1933.

Between World War I and World War II, Lwów enjoyed an interlude parallel to Budapest's. "In Lwów," says Françoise Ulam, who was born in Paris in 1918 and came to America as an exchange student in August 1938, "the members of the Polish mathematical society had done most of their work in cafés at all hours of day or night. . . . Los Alamos, in a sort of ad hoc way, provided him, if not with the culture of the Slavic Old World of his youth, at least with a leisurely pace of his own."[3]

Ulam produced his best work without appearing to be working.

"He was a real singularity in many ways," says Bruno Augenstein, a RAND analyst and architect of the U.S. thermonuclear missile program whose path intersected periodically with Ulam's during the cold war years. "He was simultaneously one of the smartest people that I've ever met and one of the laziest—an interesting combination." Françoise Ulam disagrees: "With his aristocratic nonchalance he gave the appearance of being lazy, but in reality he pushed himself mentally, all the time." Claire Ulam, age nine in 1953, was once overheard telling a friend that "all my father does is think, think, think!"[4]

"He was a maverick, a very complicated man, a Pole, and, above all, a study in contrasts and contradictions," Françoise explains. "He lived mainly in the confines of his mind." He was also gregarious. "Many of us at the Laboratory who were associated with him knew how much he disliked being alone, how he would summon us at odd times to be rescued from the loneliness of some hotel room, or from the four walls of his office, after he had exhausted his daily round of long-distance calls," says his mathematical colleague Gian-Carlo Rota. "One day I mustered the courage to ask him why he constantly wanted company and his answer gave him away. 'When I am alone,' he admitted, 'I am forced to think things out.'"[5]

Ulam became von Neumann's frequent collaborator and closest friend. "I don't think von Neumann knew anybody more intimately than me," says Ulam, "and vice-versa." They shared a common background as upper-class Eastern European Jews, and first met in Warsaw in 1935, after corresponding over their common interest in measure theory in 1934. Von Neumann extended an invitation to Princeton, and with the promise of a stipend of $300 from the Institute for Advanced Study, Ulam sailed in December of 1935 aboard the *Aquitania* for the United States. He then secured a three-year fellowship under George David Birkhoff at Harvard, while spending the summers back in Poland at the cafés. He brought his younger brother, Adam, then seventeen, with him when he left Poland for the last time in August of 1939. They were on board the Polish liner *Batory*, sailing for America, when word came over the ship's radio of the Molotov-Ribbentrop Pact. "This is the end of Poland," announced Stan.[6]

In the fall of 1939, Françoise Aron was a twenty-one-year-old graduate student at Mount Holyoke College, attending a party at a friend's apartment in Cambridge, when she met Stan. "He spent the first evening we met leaping from his seat towards mine to light my cigarette," she recalls. "Besides calling himself a 'mathematician'—an unusual profession—he was elegant, witty and entertaining in spite of being very depressed, despondent about the war, the absence of news from his family and many financial worries. There was nothing professorial or academic about him. From the very first I fell under the spell of his charm, found him enchanting, intriguing, remarkable. I was hooked."[7]

Neither Françoise nor Stan would ever again see the parents they had left behind. "These were the darkest days of the war: the German invasion, followed by the collapse of France, with its hordes of refugees fleeing the Panzer divisions that had circled the Maginot Line; the Dunkirk debacle, the heroic battle of Britain," says Françoise. "For a five-cent cup of coffee Stan sat for hours in the Georgian cafeteria with Polish and other foreign mathematicians who had found their way to Cambridge, discussing the anxious war news or talking mathematics. They became my friends too and I would join them after work."[8] Before long, Françoise was cooking for the two brothers, and joining them for meals. The Ulams were too impoverished in Cambridge to afford restaurants, and had been too wealthy in Poland to have learned to cook for themselves.

In 1941, unable to secure a position at Harvard, now flooded with refugees, Stan Ulam accepted an instructorship at the University of Wisconsin, for $2,300 a year. After obtaining her degree from Holyoke, Françoise joined him in Madison, where they were married before a justice of the peace.

"Do you want the long or the short ceremony?" asked the judge.

"How much are they?" asked Stan.

"The long costs five dollars, the short one two."

"We'll take the short," answered Stan.[9]

Even in Wisconsin, the tragedy in Europe was impossible to escape. Françoise's father had died when she was ten and was thereby spared; and her younger brother, still in his teens, escaped via Spain to England, where he trained as a paratrooper for the Free French Forces of de Gaulle. Her mother, however, rounded up on the street in Marseilles, was forced aboard a train for the Nazi concentration camps and never seen again. On Stan's side the picture was equally grim. "The news came slowly and piecemeal that during the Nazi occupation of Poland Stan's sister, her husband, their children and those of the uncles and aunts who did not leave Lwów, the family's home town, had all perished in the Holocaust," Françoise notes. "Stan's father Joseph Ulam, who had not been rounded up, died of ill-health and despair in the one-room apartment he had been relegated to when the Nazis requisitioned his house. A young boy he took in during these terrible times, who succeeded in escaping to this country, brought us the sad, sad news and described how they had to burn his law books to keep warm."[10]

Ulam and von Neumann shared their frustration at the lack of response to the European crisis by the United States. "When . . . this country announced . . . that 20 torpedo boats will go to England, I could not help thinking that 50 bicycles would also be valuable," Ulam wrote to von Neumann in the spring of 1941.[11] He signed up for private flying lessons and, upon acquiring U.S. citizenship in 1941, passed his army physical and tried to enlist in the air force, hoping to become a navigator if not a pilot. However, due to his age and severely uneven eyesight, he was turned down.

Von Neumann, already a consultant to the Office of Naval Research, the Army Ballistics Research Laboratory, and the Office of Scientific Research and Development, reported to Ulam, in April 1942, that "I'm getting more and more snowed under by war work." Ulam kept asking how he could become involved, and "one day Johnny answered with an intimation that there was interesting work going on—he could not tell me where."[12]

"The project in question is exceedingly important, probably beyond all adjectives I could affix to it," von Neumann wrote on November 9, 1943, adding that "the secrecy requirements of this project are rather extreme." This letter was followed by an invitation, signed by Hans Bethe, "to join an unidentified project that was doing important work, the physics having something to do with the interior of stars." Ulam accepted the appointment, without knowing what he had agreed to, or where. "Soon after, other people I knew well began to vanish one after the other. Finally I learned that we were going to New Mexico, to a place not far from Santa Fe."[13]

The Ulams, with their first child on the way, received their security clearances and headed west. "We made the long journey by train and got off at a whistle stop called Lamy, about eighteen miles from Santa Fe, in what seemed the middle of nowhere on February 4, 1944," remembers Françoise. "Snow was on the ground yet the sun was warm, the sky was an intense blue, and as Stan said, 'the air felt like champagne.'"[14]

The Los Alamos mesa rested against the eastern slope of the Jemez Mountains, on the rim of the Valles Caldera, formed as a result of two explosive super-eruptions 1.6 and 1.1 million years ago. On the other side of the rim, with a lava dome at its center, lay a flat grassland—a miniature Serengeti—left behind when the volcano collapsed. A favorite destination for Los Alamos residents, and a refuge to New Mexico's largest herd of elk, the Valle Grande was a remnant of one of the most violent explosions on earth. The scientists who arrived on the mesa in the summer of 1943 intended a nuclear explosion to be the next.

"The place was a mysterious encampment, a sort of Magic Mountain in a Land of Enchantment," says Françoise, astonished by "inhabitants who seemed to be scientists from everywhere—America, Canada, Germany, Switzerland, Hungary, Austria, Italy, you name it. Many had come to this country to escape from Hitler and Mussolini and their Fascist regimes. Some were already famous. Most were incredibly young, many in their early twenties with reputations yet to be made."[15]

Stan found himself back in the world of the Lwów cafés. "In the entire history of science there had never been anything even remotely approaching such a concentration," he marveled. "At thirty-four I was already one of the older people." Ulam found the improvised structure of wartime Los Alamos a refreshing contrast to the formalities of academia, and the close-knit community suited his Polish roots. "People here were willing to assume minor roles for the sake of contributing to a common enterprise," he explained. "Jules Verne had anticipated this when he wrote about the collective effort needed for his *Voyage to the Moon*."[16]

Officially under the command of General Leslie Groves of the U.S. Army, the Laboratory was directed by Robert Oppenheimer, who managed to take command of General Groves. "Groves never realized that he had been co-opted to the scientific task," says Harris Mayer. "To the end of his life he really believed that he had made the atomic bomb."[17]

Stan Ulam was assigned first to T-Division (or Theoretical Division), under Hans Bethe, and then, when the divisions were reorganized, to F-Division, under Enrico Fermi. Nominally, he reported to Edward Teller, whom he regarded more as his colleague than his boss. "As a theoretician, Stan could work anywhere," says Françoise. "He went to his office when he wanted. He came home for lunch and usually reappeared early in the afternoon."[18]

Claire was born in July, with her birthplace certified as Post Office

Box 1663, Santa Fe. With free medical services, subsidized housing, and community child care, Los Alamos beat the postwar baby boom out of the gate. The hospital began charging for diapers at one dollar per day. "Los Alamos became a great baby farm," says Françoise, "which annoyed General Groves."[19]

The physics at Los Alamos captivated Stan. "I found out that the main ability to have was a visual, and also an almost tactile, way to imagine the physical situation, rather than a merely logical picture of the problems," he explains. "One can imagine the subatomic world almost tangibly, and manipulate the picture dimensionally and qualitatively, before calculating more precise relationships."[20] Ulam's intuition complemented von Neumann's precisely logical view of the world. "Johnny gave the impression of operating sequentially by purely formal deductions," Ulam noted, describing the difference between the two approaches as "something like the distinction between a mental picture of the physical chess board and mental picture of a sequence of moves on it written down in algebraic notation!"[21] Monte Carlo, the best of both worlds, used von Neumann's formal, computational system to capture Ulam's intuitive, probabilistic approach.

Ulam, who was not directly involved with the design or construction of the bomb, did not witness the Trinity test. "In the early morning the day the bomb went off we were at home and still in bed," says Françoise. "Finally, a tired, pale, and badly shaken Johnny, who had been there with the VIPs, came to see us on his return."[22] Three weeks later, the second bomb was exploded, above Hiroshima, followed by Nagasaki on August 9.

With the war over, the entire Theoretical Division had been reduced to eight people by 1946. Oppenheimer had returned to Berkeley; Fermi and Teller had returned to Chicago; Bethe had returned to Cornell. The U.S. Department of State had established a Committee on Atomic Energy, including Vannevar Bush, James Conant, and General Groves, with a board of consultants including Oppenheimer, who formulated the "Baruch Plan," calling for international control of atomic energy in all forms. Albert Einstein, Léo Szilárd, Harold Urey, Linus Pauling, Victor Weisskopf, and Hans Bethe formed the Emergency Committee of Atomic Scientists, holding their inaugural meeting at the Institute for Advanced Study in November 1946. Control over Los Alamos was transferred from the army to the newly formed Atomic Energy Commission, effective January 1, 1947. But who would control the AEC?

Upon Oppenheimer's departure, Norris Bradbury stepped in as temporary replacement—and stayed for twenty-five years. He launched an ad hoc Los Alamos University to keep some momentum going in the interregnum between the army and the AEC, and made the case for continued design and testing of new bombs. "The occasional demonstration of an atomic bomb—not weapon—may have a salutary psychological effect on the world—quite apart from our scientific and technical interest in it," he argued. "Properly witnessed, properly publicized, further TR's [tests] may convince people that nuclear energy is safe only in the hands of a wholly cooperating world." He then made a prophetic suggestion: "Another TR might even be FUN."[23]

When Ulam was hired back, after his convalescence from encephalitis, he was appointed group leader: of a group consisting of him alone. One of his areas of interest was the back-burner effort, now led by Carson Mark, with Edward Teller supervising in absentia, to establish the feasibility of a thermonuclear bomb. "Stan had no moral qualms about returning to Los Alamos," says Françoise. "What he wanted to concentrate on were the theoretical aspects of the work, and he did not see anything wrong in that."[24]

Teller couldn't decide whether Ulam was a young scientist to be encouraged, or a rival to be upstaged. "Mr. Ulam is a brilliant mathematician but does not have the proper background for the work we are doing and does not seem to be able to adjust himself to our work," Teller had noted in Ulam's personnel file in February 1945. He then hedged his bets: "He is an independent thinker and might conceivably turn up most important results." As Françoise puts it, "I suspect he sensed he had met his match."[25]

For Teller, the hydrogen bomb was a crusade to be pursued at any cost, whether the country was at war or not. To Ulam, the probability or improbability of a self-sustaining thermonuclear reaction was for the laws of nature alone to decide. As to military consequences, Ulam argued that if one started to question the possible misuse of scientific research, then the infinitesimal calculus should have been abandoned, to preclude destructive effects. "In my mind I knew he made sense. In my heart I could not quite follow," adds Françoise. Nonetheless, "myself and my friends were startled," Ulam testified during the ENIAC trial, concerning the calculations that led to the H-bomb, "how some scribbles on a piece of paper or on the blackboard leads finally to a physically existing, and in this case a very violent thing."[26]

Hydrogen bombs had first appeared, during Ulam's childhood, in H. G. Wells's *The World Set Free,* a prophetic novel published at the dawn of World War I. "These atomic bombs which science burst upon the world that night were strange even to the men who used them," wrote Wells, envisioning a future transformed by atomic energy, until the lack of a requisite transformation of human nature leads to the "Last War"—the one we now imagine as World War III. Nuclear fission was unknown in 1914, so Wells's atomic bombs were powered by fusion, like the sun. They consumed cities in a slow, inextinguishable fire, and were dropped from aeroplanes by hand. "It was a black sphere two feet in diameter," wrote Wells. "The Central European bombs were the same, except that they were larger."[27]

"When Bethe's fundamental paper on the carbon cycle nuclear reactions appeared in 1939," explains Ulam, "few, if any, could have guessed or imagined that, within a very few years such reactions would be produced on Earth."[28] When the Soviet Union exploded a three-stage bomb yielding over 50 megatons at Novaya Zemlya on October 30, 1961, it was estimated that, for a moment, the energy flux exceeded 1 percent of the entire output of the sun.

In June of 1942, almost a year before the Los Alamos National Laboratory was established, a group of eight physicists, convened by Oppenheimer and including both Hans Bethe and Edward Teller, met in Berkeley to begin thinking about nuclear weapons. They concluded that not only was the atomic bomb a possibility, but the resulting temperatures and pressures, more extreme than those within the sun, could be used to trigger a thermonuclear reaction. A very small sun might be brought into existence, which in the next instant, without the gravity that holds the sun together, would blow itself cataclysmically apart. "We were not bound by the known conditions in a given star but we were free within considerable limits to choose our own conditions. We were embarking on astrophysical engineering," remembers Edward Teller. "By the middle of the summer of 1942, we were all convinced that the job could be done and that . . . the atomic bomb could be easily used for a stepping-stone toward a thermonuclear explosion, which we called a 'Super' bomb."[29]

In their report to the secretary of war, James Conant and Vannevar Bush went one superlative further, suggesting "we may therefore designate it as a super-super-bomb."[30] Such a "hydrogen" bomb could burn deuterium, a stable isotope of hydrogen, easily separated from seawater and constituting the cheapest fuel available on earth.

"Atomic bombs would be powerful but expensive," explains Teller. "If deuterium could be ignited, it would give a much less expensive fuel."[31] In 1950 the cost of adding a kiloton's worth of deuterium to a hydrogen bomb was about sixty cents.

Teller admitted, once the Los Alamos project was under way, that "we had to win the war and there was no time for the Super."[32] With the war over, he believed it was time to return to work on the hydrogen bomb. Others believed, just as strongly, that weapons a thousand times as powerful as those that had destroyed Hiroshima and Nagasaki should never be built. To help determine whether the Super was something the United States should pursue, or be afraid of its enemies pursuing, it was decided to run the big December 1945 ENIAC calculations, and to hold a conference, in April 1946, on the results.

Under von Neumann's supervision, Stanley Frankel and Nicholas Metropolis went to the Moore School (where the ENIAC was still undergoing acceptance testing) and ran their one million punched cards through the machine. "I advised them as far as the physics is concerned," Edward Teller later testified. "John von Neumann advised them as far as the computation work."[33] The results were interpreted by Teller as indicating thermonuclear ignition, although it later became evident that the physics was flawed. The calculation, limited by the ENIAC's small amount of memory, had neglected important secondary effects.

"Nobody will blame Teller for the erroneous calculations of 1946, especially because adequate computers were not then available," Hans Bethe wrote in 1954. "But he was blamed at Los Alamos for leading the Laboratory, and indeed the whole country, into an adventurous program on the basis of calculations that he must have known to have been very incomplete."[34] Teller remained unapologetic, arguing that the end justified the means. "My perseverance was in considerable part, due to faith in the results which were wrong," Teller explained, "but which were hopeful, and this carried us, at any rate, to a point where the necessity of a new development showed itself."[35]

"It is likely that a super-bomb can be constructed and will work," wrote Teller in a summary he personally inserted into the final report.[36] The conference also produced a patent disclosure, filed jointly by von Neumann and British physicist (and Soviet agent) Klaus Fuchs, for the invention, on or about April 18, 1946, at Los Alamos, New Mexico, of a "proposed design for 'Super'" described as "a device for initiating a thermo-nuclear reaction which employs a quantity of fissile material adaptable to sustain a neutron divergent chain reaction [and] a massive quantity of material in which a thermo-nuclear reaction can be maintained."[37] When Klaus Fuchs was revealed as a Soviet spy, von Neumann knew better than anyone else how much useful information had—and had not—been passed to the Soviet side.

It would take Robert Richtmyer two years to model what had happened in the first microseconds of the Trinity test. Until there were better computers available, further progress was limited, even though, according to Carson Mark, who succeeded Richtmyer as director of T-Division, half of his group's effort was devoted to the Super between 1946 and 1949. Von Neumann, impatient to get started, began writing code for the machine that did not yet exist. "In that T-Division coffee room, I had watched Johnny, when he was building his Princeton machine, cover a blackboard with the first stirrings of flow-diagram coding," remembers Françoise Ulam, "while casting unconscious sidelong glances at every feminine pair of legs that went by."[38]

After the Soviet explosion of a nuclear weapon on August 29, 1949 (named "First Lightning" by the Soviets and "Joe-1" by the United States), the General Advisory Committee of the Atomic Energy Commission was asked for their opinion on whether the United States should undertake the development of the hydrogen bomb. The answer was no. "It is not a weapon which can be used exclusively for the destruction of material installations of military or semi-military purposes," Oppenheimer explained in the introduction to the committee's report. "Its use therefore carries much further than the atomic bomb itself the policy of exterminating civilian populations. We all hope that by one means or another, the development of these weapons can be avoided."[39]

"Its use would involve a decision to slaughter a vast number of civilians," the majority, including James Conant as well as Oppenheimer, concurred. "We believe that the psychological effect of the weapon in our hands would be adverse to our interest. . . . In determining not to proceed to develop the Super bomb, we see a unique opportunity of providing by example some limitations on the totality of war and thus of limiting the fear and arousing the hopes of mankind." An even stronger minority addendum, signed by Enrico Fermi and Isidor Rabi, added that "It is necessarily an evil thing considered in any light." Von

Neumann, not yet a member of the commission, strongly disagreed. "I think that there should never have been any hesitation," he wrote in 1950, after Truman had made the decision to move full speed ahead.[40]

Ulam believed that much of this soul-searching was unnecessary, because the ENIAC calculations were flawed and Teller's Super would turn out to be a dud. With the assistance of Cornelius Everett, a former colleague from Madison, he undertook a first-approximation check on the earlier results, using the hand (and punched card) computing techniques that had been developed for implosion calculations during the war. "Stan, who had a conceptual hunch that the Super Teller envisaged was not practical, undertook the simplified calculations, first with Everett, then with us, the data analysts," says Françoise, who was working in the Los Alamos hand computing division at the time. "In a couple of months these calculations confirmed his feelings. In other words, Stan was the first to blow the whistle: it was not going to work. Everyone else—von Neumann, Admiral Strauss, the head of the AEC, and the military were all for pursuing and experimenting with Teller's scheme."[41]

"The degree of hope, if you will, or fear, that such a thing is possible gradually changed and, as a matter of fact, it was not even continually in one direction," Ulam later testified.[42] The doubts raised by Ulam and Everett put Teller on the defensive, and left von Neumann impatient to determine whose numbers were correct. He commandeered the ENIAC, and was first in line to use any available new machine. "When the hydrogen bomb was developed," he testified at the Oppenheimer hearings in 1954, "heavy use of computers was made [but] they were not yet generally available. . . . It was necessary to scrounge around and find a computer here and find a computer there which was running half the time and try to use it."[43] Ralph Slutz, who had left the IAS to supervise construction of the SEAC, for the Bureau of Standards in Washington, D.C., remembers "a couple people from Los Alamos" (Metropolis and Richtmyer) showing up as soon as the computer began operating, around Easter of 1950, "with a program which they were terribly eager to run on the machine . . . starting at midnight, if we would let them have the time."[44]

"When the larger and more precise electronic calculations of von Neumann and others slowly and gradually brought confirmation of Stan's point of view, it was a real setback for the whole enterprise," says Françoise. "In spite of an initial, hopeful-looking 'flare-up,' the whole assembly started to cool down," adds Stan. "Every few days

Johnny would call in some results. 'Icicles are forming,' he would say."
Much as he believed Teller's faith in the Super to be misplaced, the
germs of an alternative were incubating in the back of Ulam's mind.
"Cycle 10 has been going for the last 24 hours," he reported to von
Neumann on January 27, 1950 (the day Klaus Fuchs signed his confession), while the punched card calculations at Los Alamos were under
way. "By the way: warning about conduction: we had to divide the
time interval into 5! (sic!) on cycle 9. Hydrodynamics, so far at least,
far from being a danger is the only hope that the thing will go!"[45]

The classical Super depended on heating the deuterium (or
deuterium-tritium) fuel to the 100 million degrees or more required
to ignite. If this was going to happen at all, it had to happen quickly,
before the expansion of hot material blew things apart and escaping
radiation allowed things to cool off. "For the 'Super,' the hydrodynamical disassembly proceeded faster than the buildup and maintenance
of the reaction," Ulam later explained.[46] The bomb would fizzle out.

Ulam had noted, in monitoring the progress of the Super calculations, that hydrodynamic forces, rather than diminishing the prospects for thermonuclear ignition by disassembling things before the
fuel could become hot enough to ignite, might instead be persuaded
to work the other way. Increased pressure brings increased density.
And increased density brings not only higher temperatures but also
higher opacity. As you squeeze a region of hot plasma, it not only gets
hotter, it gets blacker. And there were ways to take advantage of that.

"What you tell me about the events of cycle 10 is very interesting," von Neumann replied on February 7. "I need not tell you how I
feel about the 'victory.' There are, however, plenty of problems left."[47]
The "victory" was the public announcement, by President Truman
on January 31, that, in response to the Soviet bomb test of August 29
and against the advice of Oppenheimer and the General Advisory
Committee, he had "directed the Atomic Energy Commission to continue its work on all forms of atomic weapons, including the so called
hydrogen or superbomb." Lewis Strauss, armed with Klaus Fuchs's
confession, had claimed that the Soviets, without any Oppenheimers
to restrain them, might already be ahead. Teller finally had access to
unlimited resources, and an actual test, "Greenhouse George," was
scheduled that would show whether a small sample of deuterium-
tritium would ignite.

Then Ulam came up with a surprise. According to Bethe, he was
not even thinking about the Super problem, but about how it might

be possible to construct very-high-yield, two-stage fission bombs. "Unbeknownst to me, Stan had continued to think about the problems in a round-about sort of way, more for their scientific challenge than political or military importance," Françoise explains. "And suddenly, he came upon a totally new and intriguing approach."

"I found him at home at noon staring intensely out of a window with a very strange expression on his face," she says.

> I can never forget his faraway look as peering unseeing in the garden, he said in a thin voice—I can still hear it—"I found a way to make it work."
>
> "What work?" I asked.
>
> "The Super," he replied. "It is a totally different scheme, and it will change the course of history."[48]

Ulam spoke to Carson Mark and Norris Bradbury immediately, and Edward Teller the following day. Teller, who had been working on the problem for almost a decade, immediately improved upon Ulam's suggestion, and brought in a young Viennese physicist, Frederic de Hoffmann, who performed the initial calculations establishing the probable feasibility of the new approach. "I wanted to do something about the hydrogen bomb and nobody else wanted to," Teller says, "and the one man who wanted to do it more than I was Freddy de Hoffmann."[49] It was de Hoffmann, twenty years old at the time, who had calculated the ballistic trajectories for the two bombs that were dropped on Japan.

Teller titled his 1955 review of the H-bomb's development "The Work of Many People"—a genuine attempt to share the credit, in the face of widespread criticism, with those who had helped. Hans Bethe wrote his own account in 1954, which he opened by giving Teller full credit, during the atomic bomb development, for being "the first to suggest that the implosion would compress the fissile material to higher than normal density inside the bomb." But he refused to assign Teller chief credit for the breakthrough on the hydrogen bomb. "It is difficult to describe to a nonscientist the novelty of the new concept," he wrote. "It was an entirely unexpected departure from the previous development. It was also not anticipated by Teller, as witness his despair immediately preceding the new concept."[50]

Ulam suggested that overenthusiasm for the classical Super may have delayed Teller's own arrival at a successful design, and emphasized that the real credit should go to "the enormous number of calculations, all the studies of the general physics of the processes, the engineering planning, all combined with the necessity of predicting and avoiding 'side effects,' any one of which could ruin the success of the device." And if individuals were to be singled out, he noted to Bethe, "it would be hard to exaggerate the importance of the contributions made by Fermi in the decisive switch from the original, hopeless approach."[51]

The breakthrough, now known as the Teller-Ulam invention, appeared in February 1951, and was published (in an edition of twenty secret copies) under joint authorship on March 9, 1951. "The arrangement might be called heterocatalytic, involving as it does a setting off of a reaction in one system by a reaction started in another," Teller and Ulam explained.[52] "This new idea transformed the concept of the Super into the beautifully workable hydrogen bomb," says Harris Mayer, who helped resolve the details of a new concept that was "remarkably complex, and devilishly interesting." Mayer's specialty was radiation opacity—how some states of matter are more opaque to radiation at certain temperatures, and some less. Understanding the details can help tailor things so radiation flows where you want it to, and when it reaches its destination is either absorbed or transformed. "Nature had provided generous margins," he says, "in the properties of radiation flow." Mayer adds, however, that "nobody thought that Stan was the significant person in the new hydrogen bomb development until the Oppie affair. And the Oppie affair got everybody so mad at Edward that then they spoke of the Teller-Ulam concept."[53]

"Ulam kept pressing for squeezing the secondary," says Theodore B. Taylor, the gifted Los Alamos bomb designer who was friends with both Ulam and Teller at the time. "Now whether he did that with the key perception that then the inverse Compton effect wouldn't drain the energy, that things would be much closer to equilibrium and that at these high densities you get a fast enough reaction rate and a high enough temperature rise so that it would be very efficient, I don't know who came up with that." Taylor gives credit to both sides. "My sense of things is that this direct compression is something that they both saw at the same time, that compressing it was the way to go," he says. "Then the question was: How do you compress it? The subject had been brought up by Ulam, and what I've come to believe is that Teller said, 'Oh, that's terrific, but let's use the radiation, not the hydrodynamics.' And then everything became clear."[54]

The classical Super had been going nowhere for eight years, and now there was a design that went from concept to successful test in nineteen months. The first meeting to discuss the implications of the new approach—which included Oppenheimer, Teller, von Neumann, Bethe, Fermi, and John Wheeler—was held in Oppenheimer's office at the Institute for Advanced Study in June 1951. "All the top men from every laboratory sat around this table and we went at it for 2 days," testified Gordon Dean, a Lehman Brothers partner who became chairman of the AEC. "At the end of those 2 days we were all convinced, everyone in the room, that at last we had something. . . . The bickering was gone. . . . That is when it began to roll and it rolled very fast."[55]

The IAS computer had at last become available, and with Oppenheimer's support, it was 1943 all over again. "When I saw how to do it, it was clear to me that one had to at least make the thing," Oppenheimer testified at his security hearing. "The program we had in 1949 was a tortured thing that you could well argue did not make a great deal of technical sense. It was therefore possible to argue that you did not want it even if you could have it. The program in 1951 was technically so sweet that you could not argue about that."[56] Everything depended on the H-bomb working on the first try. "Mike, the first H-bomb," says Marshall Rosenbluth, "in fact was quite over-designed."[57]

Strauss became impatient, and increasingly suspicious of Oppenheimer, even though he had appointed Oppenheimer to the directorship of the Institute and knew that he was now helping the H-bomb effort move ahead. Strauss complained, according to the notes of a conversation concerning the Oppenheimer camp's opposition to the H-bomb, that "first they objected on moral grounds; then they objected on the ground that there were no military targets; then they objected on the ground that the super would be too costly in term of neutrons as compared with the plutonium which could be alternatively produced; and now they want to build [deleted] which would not be a real open-ended weapon at all."[58] The deletion probably refers to the Super Oralloy Bomb (SOB), the largest fission weapon ever produced. Designed by Ted Taylor, the Super Oralloy Bomb yielded 500 kilotons in the Ivy King test at Enewetak on November 15, 1952, and was intended to demonstrate that for any conceivable military purpose, half a megaton should be enough. Half a megaton was not enough for Lewis Strauss.

Relations between the IAS and the AEC were complex. "By early 1952 there was some change in this game of musical chairs," Klári explains. "Johnny became a member of the General Advisory Committee of which Oppenheimer was still the chairman; however, Lewis Strauss was no longer one of the commissioners, but had become the President of the Board of Trustees at the Institute for Advanced Study, of which Johnny was a member and Robert the director." Strauss, appointed to the AEC by President Truman in 1947, served until 1950 and was reappointed in 1953, as chairman, by President Eisenhower. "At the beginning of 1954, the year that all hell broke loose," says Klári, "these were the relative positions: at the Institute, unchanged—Johnny, member—Oppenheimer, director—Strauss, president. At the Atomic Energy Commission: Johnny member of the G.A.C., Lewis chairman of the Commission, Robert completely not."[59]

The extent of actual weapons work at the Institute was kept secret, but the comings and goings of people from Los Alamos and the AEC were hard to conceal, and the purpose not hard to guess. "The objective," says Ted Taylor of the von Neumann computer, "was pretty specifically to be able to do the coupled hydrodynamics and radiation flow necessary for H-bombs."[60] John Wheeler moved his small team from Los Alamos to Princeton University, launching "Project Matterhorn," under a subcontract with Los Alamos, to prepare thermonuclear calculations for the Institute machine until Los Alamos could build a computer for itself.

"The mathematicians certainly knew there was classified work going on," says Freeman Dyson, who arrived in 1948. "They may not have known that it was hydrogen bombs, but it was pretty obvious. And they were strongly opposed to that." The groundswell of public opinion against atmospheric testing would come later, but opposition to the hydrogen bomb on humanitarian principles was there from the start. "It made a very bad impression to have this safe that used to be in Fuld Hall, with all Oppenheimer's secrets in it," Dyson explains. "And it wasn't just the safe, there were two armed guards who were there too. It really looked rather formidable." Virginia Davis, who came to the Institute in 1952 with logician Martin Davis, remembers writing "STOP THE BOMB" in the dust on von Neumann's car.

The air force, which would be taking custody of any deliverable weapons, had to be reminded that the Institute was not Los Alamos or RAND. According to Oppenheimer, at a briefing given by Edward Teller and the RAND Corporation, Secretary of the Air Force Thomas K. Finletter "got to his feet and said 'give us this weapon and we will rule the world.'" Oppenheimer, who had willingly served the army under General Groves, resisted the air force. "Johnny steadily increased his defense activities," says Klári, "while Robert gradually moved away from them."[61]

Julian Bigelow received his Q clearance, allowing access to atomic secrets, on February 23, 1950. On March 14, the AEC advised the Institute that all working under AEC contracts "are instructed to refrain from publicly stating facts or giving comment on any thermonuclear reactions," and on March 17 it was clarified, in response to strong objections from Oppenheimer, that these restrictions would "still permit unclassified discussions of what might be called the classical thermonuclear reactions as long as there is no reference to their relation to weapons."[62] Stellar evolution codes could be run in the open, but weapons codes had to be run in the dark.

"By the end of 1950," reported Bigelow, "it was now possible to put a program into the machine and get results out. During the spring of 1951, the machine became increasingly available for use, and programmers were putting their programs on for exploratory runs, debugging, etc. and the machine error rate had become low enough so that most of the errors found were in their own work."

During the summer of 1951, "a team of scientists from Los Alamos came and put a large thermonuclear calculation on the IAS machine; it ran for 24 hours without interruption for a period of about 60 days," Bigelow continues. "So it had come alive."[63]

The digital universe and the hydrogen bomb were brought into existence at the same time. "It is an irony of fate," observes Françoise Ulam, "that much of the high-tech world we live in today, the conquest of space, the extraordinary advances in biology and medicine, were spurred on by one man's monomania and the need to develop electronic computers to calculate whether an H-bomb could be built or not."[64]

Von Neumann, a member of the Institute for Advanced Study, spent much of his time working on weapons, whereas Ulam, a member of the Los Alamos weapons laboratory, spent most of his time on pure mathematical research. While von Neumann began working on Intercontinental Ballistic Missiles, or ICBMs, Ulam, in contrast, began thinking about how to use bombs to launch missiles, instead of how to use missiles to launch bombs.

"The idea of nuclear propulsion of space vehicles was born as soon as nuclear energy became a reality," he explains. While others who visited the Trinity test site marveled at how the shot tower had

been vaporized by the explosion, Ulam observed that the steel rein-
forcement at the base of the tower had survived the explosion intact.
Perhaps objects caught within the fireball could survive the explosion and even be propelled somewhere else. The question of whether
the energy produced by a small fission explosion could be channeled
outward to drive the propulsion of a space vehicle was similar to the
question of whether this energy could be channeled inward to drive
the implosion of a thermonuclear bomb. Ulam's idea was the hydrogen bomb turned inside out.

In 1955, with Cornelius Everett, Ulam produced a classified Los
Alamos report, "On a Method of Propulsion of Projectiles by Means
of External Nuclear Explosions," suggesting that "repeated nuclear
explosions outside the body of a projectile are considered as providing a means to accelerate such objects to velocities of the order of $10^6$
cm / sec . . . in the range of the missiles considered for intercontinental
warfare and even more perhaps, for escape from the earth's gravitational field."[65]

This report lay idle for two years and then, after the launch of the
Soviet *Sputnik*, the idea was adopted by Ted Taylor, who developed
it into plans for a real spaceship from where Ulam had left off. Project Orion, funded at first by the Department of Defense's Advanced
Research Projects Agency (ARPA) and later by the air force, was pursued seriously for the next eight years. "It is almost like Jules Verne's
idea of shooting a rocket to the moon," Ulam testified before Senator Albert Gore in early 1958.[66] On April 1, Ulam issued another Los
Alamos report, "On the Possibility of Extracting Energy from Gravitational Systems by Navigating Space Vehicles," describing how a
spacecraft might operate as a gravitational "Maxwell's demon," ampli-
fying a limited supply of fuel and propellant by using computational
intelligence to select a trajectory that harvested energy from celestial
bodies as it passed by.

In 1871, James Clerk Maxwell, the namesake for both Maxwell's
equations formalizing the concept of an electromagnetic field and
the Maxwellian distribution of kinetic energy among the particles
of a gas, conceived an imaginary being—termed "Maxwell's demon"
by William Thomson (Lord Kelvin) in 1874—"whose faculties are so
sharpened that he can follow every molecule in its course."[67] The
demon appears to defy the second law of thermodynamics by heating
a compartment in an otherwise closed system, without the expenditure of physical work, by opening and closing a small trap door,

at exactly the right time, to let high-velocity molecules in and low-velocity molecules out. A Maxwellian distribution of energy describes how, without supernatural intelligence, kinetic energy tends to equalize across a population of particles over time. Light particles end up moving faster at the heavier particles' expense. A 4,000-ton spaceship will end up moving faster than a planet—given enough time. Maxwell first developed these ideas, later adapted to thermodynamics, to explain the distribution, by size and velocity, of particles that make up Saturn's rings.

"As examples of the situation we have in mind," explained Ulam, "assume a rocket cruising between the Sun and Jupiter, i.e., in an orbit approximately that of Mars. . . . The question is whether, by planning suitable approaches to Jupiter and then closer approaches to the sun, it could acquire, say, 10 times more energy. . . . By steering the rocket, one can to some modest extent acquire the properties of a Maxwell demon . . . to shorten by many orders of magnitude the time necessary for acquisition of very high velocities."[68]

"I remember Stan talking about being able to make a Maxwell's demon, that it could be a possible physical thing," Ted Taylor recalls. The required computational intelligence, viewed as a major obstacle in 1958, would be the least of the obstacles today. "The computations required to plan changes in the trajectory might be of prohibitive length and complication," Ulam warned.[69]

Ulam himself appeared to violate the second law of thermodynamics by performing useful work, with no visible expenditure of energy, simply by opening doors to the right ideas at the right time. Whether over coffee in Lwów or poker at Los Alamos, he let good ideas in and kept bad ideas out. "My incredible luck," he bragged to von Neumann from Los Alamos in February 1952, "was evident in poker (8 successive + earnings) this year."[70] Four of the twentieth century's most imaginative ideas for leveraging our intelligence—the Monte Carlo method, the Teller-Ulam invention, self-reproducing cellular automata, and nuclear pulse propulsion—originated with help from Stan. Three of the four proved to be wildly successful, and the fourth was abandoned before it had a chance.

Monte Carlo was the realization, through digital computing, of what Maxwell could only imagine: a way to actually follow the behavior of a physical system at its elemental levels, as "if our faculties and instruments were so sharpened that we could detect and lay hold of each molecule and trace it through all its course."[71] The Teller-Ulam invention invoked a form of Maxwell's demon to heat a compartment to a temperature hotter than the sun by letting a burst of radiation in, and then, for an equilibrium-defying instant, not letting radiation out. Ulam's self-reproducing cellular automata—patterns of information persisting across time—evolve by letting order in but not letting order out.

When Nicholas Metropolis and Stanley Frankel began coding the first bomb calculations for the ENIAC, there was room for only a one-dimensional universe—represented by a single line, in our universe, extending outward from the center of the bomb. By assuming spherical symmetry, what was learned in that one-dimensional universe could be used to predict three-dimensional behavior in ours. Ulam began imagining how, in a one-dimensional universe, cosmology might evolve. "Has anybody considered the following problem—which appears to me very pretty," he wrote to von Neumann in February 1949. "Imagine that on the infinite line $-\infty$ to $+\infty$ I have occupied the integer points each with probability say ½ by material point masses—i.e. I have this situation," and he sketched a random distribution of points on a line. "This is a distribution at time t=0."

"Now between these points act $1/d^2$ forces (like gravitation)," he continued. "What will happen for t > 0? I claim that condensations will form quickly—assume, for simplicity when points touch they stick—with nice Gaussian-like distribution of masses. Then—the next stage—clusters of these condensations will form—somewhat slower but surely (all statements have probability = 1!)." Ulam explained how this simple one-dimensional universe would start to look "somewhat like the real Universe: stars, clusters, galaxies, super-galaxies etc.," and then considered what might happen in two dimensions, and even three dimensions, by introducing range forces, thermal oscillation, and light. He concluded by suggesting "that 'entropy' decreases—an abnormal 'order' applies."[72]

Ulam then began thinking about a two-dimensional, cellular universe, taking cues from the two-dimensional hydrodynamics codes that were being used in the work on bombs. "I discussed the cellular model with von Neumann in the late 1940's," he later wrote to Arthur Burks, and he evidently had similar discussions with Nick Metropolis as well. "I am coming to Los Alamos after all!!!," Metropolis wrote to Ulam in June 1948. "Hope you will have a chance to do more about the geometry of phase space because it is something. And your two dimensional world."[73]

Meanwhile, in our three-dimensional universe, at 07 hours, 14 minutes, and 59 seconds local time on November 1, 1952 (October 31 in the United States), the Teller-Ulam invention, the Monte Carlo algorithm, the IAS computer, the resources of Los Alamos, and the efforts of some 11,652 people assigned to Task Force 132 in the South Pacific resulted in the detonation of Ivy Mike, the first hydrogen bomb.

The size of a railroad car, with its nonnuclear components built by the American Car and Foundry Company of Buffalo, New York, Mike weighed 82 tons, much of that being a massive steel tank of liquid deuterium, cooled to minus 250 degrees Kelvin and ignited by a TX-5 fission bomb. Exploded at the surface of a small island in Enewetak Atoll, Mike yielded 10.4 megatons—some 750 Hiroshimas—vaporizing 80 million tons of coral to leave a crater 6,300 feet in diameter and 160 feet deep, "large enough to hold 14 buildings the size of the Pentagon," as it was put in one of the official reports. A thought that had first crossed Ulam's mind while staring out into the garden less than three years previously had now removed the entire island of Elugelab from the map.

Enewetak Atoll, comprised of some thirty-nine small islands distributed around a central lagoon, was, like the Valles Caldera, the remnant of a former volcano whose collapse had left behind not a meadowed valley ringed by mountains, but a sheltered lagoon ringed by a coral reef. Remote even among the Marshall Islands, the island, and its seafaring inhabitants, had been left undisturbed until the island was claimed by the Japanese after World War I, and then occupied by the United States after a fierce battle during World War II. All native residents were exiled in 1947 to Ujelang, an uninhabited atoll 140 miles away, when Enewetak was selected as a site for nuclear tests. It was at first assumed that Enewetak was too close to Kwajalein and a number of "small atolls populated by natives" for a test of the super bomb, but according to the notes of a meeting held on August 25, 1951, "it was Edward Teller's opinion that a shot at Eniwetok [sic] was not out of the question . . . if one chooses a time when the wind was in the opposite direction and made advance preparations to evacuate Kwajalein."[74]

"Accompanied by a brilliant light, the heat wave was felt immediately at distances of thirty to thirty-five miles," reports the official record of the test. "The tremendous fireball, appearing on the horizon like the sun when half-risen, quickly expanded . . . and a tremendous conventional mushroom-shaped cloud soon appeared, seemingly balanced on a wide, dirty stem . . . due to the coral particles, debris, and water which were sucked high into the air . . . around the area where the island of Elugelab had been."[75]

"The bomb went off at 7:15 a.m. in a partially cloudy sky, streaked with color from the rising sun," noted Lauren Donaldson, a forty-nine-year-old fisheries biologist from the University of Washington who collected samples before and after the test. A week later he was still finding terns that "had their feathers burned, white feathers seemed to have been missed but dark were scorched," and fish whose "skin was missing from a side as if they had been dropped in a hot pan." He and his crew had made their own viewing goggles by attaching darkened welding glass to their diving masks, and from thirty miles away, aboard the *Oakhill*, "the fire ball as it developed seemed to boil at first and fold in like fruit boiling in a kettle. There were great blackened chunks that seemed to be included in the mass."[76]

Walter Munk and Willard Bascom, young oceanographers working for Scripps Institution of Oceanography, were dropped off by the converted tug *Horizon* on plywood rafts supported by truck tire inner tubes, 83 miles from ground zero, to measure the surface wave and, if there was any sign of a tsunami, to signal an alarm. The two rafts, stationed 2 miles apart, were anchored by piano wire to a pair of San Diego trolley wheels lowered to the summit of a seamount 4,500 feet below. "Wet and cold, I put on my high-density goggles," Munk remembers. "An instant heat blast signaled the explosion; at 0721 a 5 millibar air shock arrived, a sharp report followed by angry rumbling. I will not forget the boiling sky overhead. None of the photographs I have ever seen captured this."[77]

After about an hour, the cloud, now some 60 miles in diameter, had, in the words of one observer, "splashed" against the tropopause at over 100,000 feet. A series of air force pilots, flying specially configured F-84G sampling aircraft and wearing lead-lined flight suits, were sent in to sample within the mushroom cloud. The first group went in 90 minutes after detonation, at 42,000 feet.

"Immediately upon entering the cloud, RED LEADER was struck by its intense color," the official history reports.

> The hand on the Integron, which showed the rate at which radioactivity was being accumulated, "went around like the sweep second hand on a watch. . . . And I had thought it

would barely move!" The combination of most instruments indicating maximum readings and the red glow like the inside of a red-hot furnace was "staggering" and Colonel Meroney quickly made a 90-degree turn to leave the cloud.

Meroney just made it, out of fuel, back to the airstrip at Enewetak. Jimmy Priestly Robinson, piloting the fourth F-84, was not as fortunate. "For reasons unknown, RED-4 spun out shortly after entering, but managed to regain control at 20,000 feet," the report continues. At 19,000 feet Robinson reported his gauges showing empty but engine still running. His next transmission reported that his engine had flamed out and he was at 13,000 feet. A rescue crew was scrambled by helicopter to prepare to retrieve him. His final transmission was from 3,000 feet: "I have the helicopter in sight and am bailing out." The aircraft flew into the water on a level glide, under control, and flipped over before it sank. No body was ever found.[78]

Jimmy Priestly Robinson was the first person to be killed by a hydrogen bomb.

The test remained top secret, and news of its success was embargoed from the public until an announcement by outgoing president Truman (just before Eisenhower's inauguration) on January 7, 1953. More than 6,706 background checks were conducted on individuals involved with the test, and on November 14, J. Edgar Hoover was personally enlisted to try to ferret out the source of information that had leaked to reporters from *Time* and *Life* magazines. Ulam, who was on leave from Los Alamos at Harvard, came down to New York City to meet with von Neumann in early November, probably to receive the news firsthand. They evidently had a long conversation on a bench in Central Park, leaving no record of their discussion of the Ivy Mike test, but a subsequent exchange of letters hinted at the conversation having extended to the possibility of a digital universe being brought to life.

"Only because of our conversation on the bench in Central Park I was able to understand . . . [that] given is an actually infinite system of points (the actual infinity is worth stressing because nothing will make sense on a finite no matter how large model)," noted Ulam, who then sketched out how he and von Neumann had hypothesized the evolution of Turing-complete (or "universal") cellular automata within a digital universe of communicating memory cells. The definitions had to be made mathematically precise:

A "universal" automaton is a finite system which given an arbitrary logical proposition in form of (a linear set L) tape attached to it, at say specified points, will produce the true or false answer. (Universal ought to have relative sense: with reference to a class of problems it can decide.) The "arbitrary" means really in a class of propositions like Turing's—or smaller or bigger.

"An organism (any reason to be afraid of this term yet?) is a universal automaton which produces other automata like it in space which is inert or only 'randomly activated' around it," Ulam's notes continued. "This 'universality' is probably necessary to organize or *resist* organization by other automata?" he asked, parenthetically, before outlining a mathematical formulation of the evolution of such organisms into metazoan forms.

Suppose that the states for each cell are only *two*, the cells of the same type, connections between neighbors inducing only the simplest change. The problem is to see whether there will exist *boxes* of these cells containing n (n big!) elements each, the no. of states then is $2^n$ for each box; now we divide the $2^n$ states into K classes (K small like 20) and call each class a state of the box. These boxes will then perhaps be able to play the role of our present cells.

In the end Ulam acknowledged that a stochastic, rather than deterministic, model might have to be invoked, which, "unfortunately, would have to involve an enormous amount of probabilistic super structure to the outlined theory. I think it should probably be omitted unless it involves the crux of the generation and evolution problem—which it might?"[79]

Ulam soon returned to Los Alamos, and five months later the scientific world was transfixed by news of the discovery of the structure of DNA. It was now evident how genetic sequences were being replicated, and how information was being conveyed from strings of nucleic acids to amino acids to proteins, but it remained a mystery as to what the translation rules actually were. This puzzle—how life translates between sequence and structure, and in doing so not only tolerates but takes advantage of ambiguity—would hold Ulam's interest for the rest of his life.

The translation puzzle also captured the imagination of Russian-born physicist George Gamow, who sent Ulam a telegram on July 20, 1953:

DEAR STAN, HAVE PROBLEM FOR YOU USING 20 DIFFER-
ENT LETTERS WRITE A LONG CONTINUOUS WORD CON-
TAINING FEW THOUSAND LETTERS. HOW LONG THAT
WORD SHOULD BE FOR FAIR PROBABILITY OF FINDING
IN IT ALL POSSIBLE TEN LETTER WORDS? PLEASE WIRE.[80]

Stan immediately answered, from Los Alamos:

PLEASE WIRE WHETHER ONE IS ALLOWED TO SKIP LET-
TERS IN THE LONG WORD TO FORM TEN LETTER WORDS.
IF SO, ANSWER RATHER SHORT. IF ONLY CONTIGUOUS
LETTERS ALLOWED ANSWER MUCH BIGGER THAN TEN TO
THE TWENTIETH POWER AND CARSON WILL SEND THIS
WORD COLLECT.
LOVE, STAN.[81]

# TWELVE

# Barricelli's Universe

*The Star Maker . . . could make universes
with all kinds of physical and mental attri-
butes. He was limited only by logic. Thus he
could ordain the most surprising natural
laws, but he could not, for instance, make
twice two equal five.*

—Olaf Stapledon, 1937

"GOD DOES NOT play dice with the Universe," Albert Einstein advised
physicist Max Born (Olivia Newton-John's grandfather) in 1936. There
was no proscription against cards. "Every red card (hearts and dia-
monds) has been recorded as +1, every black card (spades and clubs)
has been recorded as –1," Nils Barricelli explained, as he seeded
the memory registers of the IAS computer with random numbers
in March of 1953. "In order to prevent any appreciable correlation
between the extracted cards we have never taken out more than 10
cards at a time without mixing again the set of cards."[1]

Nils Aall Barricelli, who "had this most wonderful delicious
accent," according to Gerald Estrin, was born to a Norwegian mother
and Italian father in Rome on January 24, 1912. He studied mathematical physics under Enrico Fermi and became a vocal critic of Mussolini, whose rise to power prompted him to move to Norway, with
his younger sister and recently divorced mother, upon his graduation
from the University of Rome in 1936. He lectured on Einstein's theory
of relativity at the University of Oslo, published a volume of lecture
notes on the theory of probability and statistics, and spent the war
writing a doctoral thesis on the statistical analysis of climate variation,
submitted in 1946. "However, it was 500 pages long, and was found
to be too long to print," says his former student Tor Gulliksen. "He
did not agree to cut it to an acceptable size, and chose instead not to
obtain the doctoral degree!"[2]

Barricelli was an uncompromising nonconformist, questioning

accepted dogma not only with regard to Darwinian evolution but on subjects ranging from matter-neutrino transparency to Gödel's proof. "He believed that every mathematical statement could either be proved or disproved. He insisted that Gödel's proof was faulty," says Simen Gaure, an assistant who was hired—"he paid us directly out of his wallet, fairly good pay it was too, at least for students"—after a selection process that required searching for a hidden flaw in a sample proof. "Those who could point out the flaw were accepted as not yet ruined by mathematical education," adds Gaure, who explains that Barricelli intended "to actually build a machine which could prove or disprove any statement of arithmetic and projective geometry." He never built the machine, but in preparation he developed a programming language called "B-mathematics," which is how he discovered what he claimed was a circularity in Gödel's proof. "I once asked him what the 'B' in 'B-mathematics' was," says Gaure. "He answered that he hadn't decided on that; it could be 'Boolean,' or it could be 'Barricelli,' or something else."[3]

Barricelli operated on the fringes of academia, eventually being awarded a state stipendium by the Norwegian government, which allowed him to remain at the University of Oslo with his own small research group. "He was interested, among other things, in extraterrestrial life," says Kirke Wolfe, one of his research assistants at the University of Washington, "and in coming up with theories of life and intelligence that would be general enough to accommodate forms that it might take elsewhere." He believed the question was not whether extraterrestrial life existed, but if we would be able to recognize it. "Our limited experience with the particular type of life which has developed on this planet may prove completely inadequate to form a picture of the possible life forms one may find on foreign planets," he wrote in 1961, concerning the prospects for finding life on Mars and Venus as the U.S. and USSR space programs were getting off the ground.[4]

"The scientific community needs a couple of Barricellis each century," says Gaure, while acknowledging that Barricelli "balanced on a thin line between being truly original and being a crank." According to Wolfe, he "was a world unto himself, caught up in whatever work he was doing at the time." His identity was distributed between Italy, Norway, and the United States. "A sense of origins was not important to him," says Wolfe. "He picked up and went to wherever he could get the resources he needed to do his work." Wolfe notes the contrast between Barricelli's own life as a highly solitary individual and his devotion to the principle of symbiogenesis, where individuality is superseded by cooperation among the members of a group. Barricelli believed that the advantages of mutual cooperation between otherwise competing individuals were a more important driver of evolution than either natural selection or random variation, and he saw his numerical evolution experiments as a way of proving his case. "According to the symbiogenesis theory," he argued, "the genes gained by symbiosis 1) tremendous developmental possibilities and 2) a very rapid developmental rate."[5]

In his career as a viral geneticist, Barricelli developed mathematical models and avoided laboratory work. "He said the trouble is when you do an experiment and you get a result, there is no way to look back at every step you've taken and assure yourself that you took each of those steps correctly, so the result that you get is not verifiable," says geneticist Frank Stahl. "Who knows but at one step you might have taken a pipette of the wrong size and delivered the wrong amount of something into something else?"[6]

Barricelli "insisted on using punched cards, even when everybody had computer screens," according to Gaure. "He gave two reasons for this: when you sit in front of a screen your ability to think clearly declines because you're distracted by irrelevancies, and when you store your data on magnetic media you can't be sure they're there permanently, you actually don't know where they are at all."[7]

After publishing the paper "The Hypothesis of the Symbiosis of Genes" in Oslo in 1947, Barricelli executed a series of numerical experiments by hand on graph paper, which resulted in a preliminary report, "Numerical Models of Evolutionary Organisms," leading to his invitation to the IAS. "He must have contacted von Neumann from Norway and mentioned some of his ideas," says fellow Norwegian Atle Selberg. "Von Neumann was rather receptive to such things."[8]

"According to the theory of symbiosis of genes, the genes were originally independent, virus like organisms which by symbiotic association formed more complex units," he explained. "A similar evolution should be possible with any kind of elements having the necessary fundamental properties."[9] He proposed to test these theories using strings of code able to reproduce, undergo mutations, and associate symbiotically within the 40,960-bit memory of the new machine. In December 1951 he applied for a Fulbright travel grant "to perform numerical experiments by the use of large calculating

machines, in order to clarify the first stages in the evolution of species," but because of a delay in obtaining his Norwegian citizenship
the Oslo office rejected the request.[10]

"Mr. Barricelli's work on genetics, which struck me as highly original and interesting . . . will require a great deal of numerical work,
which could be most advantageously effected with high-speed digital
computers of a very advanced type," von Neumann wrote in support
of the proposal.[11] With a research fellowship from the Norwegian government, Barricelli finally arrived in Princeton in January 1953, and,
on February 6, was awarded an unpaid membership in the School of
Mathematics for the remainder of the academic year, renewed with a
stipend of $1,800 for 1954.

The delays over Barricelli's nationality were matched by the delays
in completing the computer, and, as it turned out, he arrived in Princeton at just the right time. The goal, as he explained it, was "to find
analogies or, possibly, essential discrepancies between bionumerical
and biological phenomena," and "to observe how the evolution of
numeric organisms takes place by hereditary changes and selection
and to verify whether some of the organisms are able to speed up
their evolution by gene replacements or by acquiring new genes or by
any other primitive form of sexual reproduction." On one level, Barricelli was applying the powers of digital computing to evolution. On
another level, he was applying the powers of evolution to digital computing. According to Julian Bigelow, "Barricelli was the only person
who really understood the path toward genuine artificial intelligence
at that time."[12]

Four weeks after Barricelli began his experiments, James Watson
and Francis Crick announced their determination of the structure of
DNA. While Barricelli was working to encode evolutionary processes
by means of numerical sequences, Watson and Crick were working to
decode evolutionary processes by means of chemical sequences. After
the Watson-Crick results appeared, Barricelli would refer to strings of
DNA as "molecule-shaped numbers," emphasizing the digital nature
of polynucleotide chains. "The distinction between an evolution
experiment performed by numbers in a computer or by nucleotides
in a chemical laboratory is a rather subtle one," he observed. Information theorists, including Claude Shannon with his 1940 PhD thesis on
"An Algebra for Theoretical Genetics" (which was followed by a year
at IAS), had already built a framework into which the double helix
neatly fit.[13]

"Genes are probably much like viruses and phages, except that all the evidence concerning them is indirect, and that we can neither isolate them nor multiply them at will," von Neumann had written to Norbert Wiener in November 1946, suggesting that one way to find out how nature makes its copies would simply be to look. In December 1946, after consultation with Vladimir Zworykin and Andrew Booth, von Neumann submitted a proposal to determine biomolecular structures by bombarding centimeter-scale models, made out of small metallic spheres, with radar waves. The resulting diffraction patterns would then be compared with those produced by X rays of biological molecules on a one-hundred-million-fold smaller scale. "The best chance for a real understanding of protein chemistry lies in the x-ray diffraction field," he wrote to Mina Rees at the Office of Naval Research. "I need not detail what any advance in this field will mean." He requested emergency funding and appended "a list of certain items that are probably available in Government Surplus equipment, and which would be very helpful for the work."[14] Nothing came of this proposal—an approach that might have accelerated the discoveries of Franklin, Watson, and Crick.

Instead of focusing upon natural mechanisms that were microscopic and highly complex, Barricelli sought to introduce primitive self-reproducing entities into an empty universe where they could be directly observed. "The Darwinian idea that evolution takes place by random hereditary changes and selection has from the beginning been handicapped by the fact that no proper test had been found to decide whether such evolution was possible and how it would develop under controlled conditions," he wrote. "A test using living organisms in rapid evolution (viruses or bacteria) would have the serious drawback that the causes of adaptation or evolution would be difficult to state unequivocally, and Lamarckian or other kinds of interpretation would be difficult to exclude." We now know that lateral gene transfer and other non-neo-Darwinian mechanisms are far more prevalent, especially in microbiology, than was evident in 1953.

"If, instead of using living organisms, one could experiment with entities which, without any doubt could evolve exclusively by 'mutations' and selection," Barricelli argued, "then and only then would a successful evolution experiment give conclusive evidence; the better if the environmental factors also are under control."[15] To attempt this in 5 kilobytes was wildly ambitious, but this was 1953. Jetliners were carrying their first commercial passengers, the space age was beginning, and tail fins were beginning to appear on cars. New elementary particles were being discovered faster than Oppenheimer's group of young theoretical physicists was able to keep up. The computer's initial teething problems were settling down. A second layer of electromagnetic shielding had been added to the Williams tube amplifiers; self-diagnostic routines had been adopted; input/output had been improved by switching to punched cards from paper tape. For the period March 2–6, 1953, the computer was operating 78 percent of the available time. For March 9–13, operational status was 85 percent, and for March 16–20, 99 percent.

An extensive thermonuclear hydrodynamics code, supervised by Foster and Cerda Evans, began running in February, alternating with a lesser calculation, supervised by von Neumann, concerning the decay of a spherical blast wave. The meteorologists ran their trial forecasts during the day; Barricelli usually worked late at night. He was one of the few scientists allowed to operate the computer without the supervision of an engineer, and there are long intervals of machine time, under his control, with sparse log entries until the struggle to assign blame when things came to a halt.

"Dr. Barricelli claims machine is wrong. Code is right," the operating log records on April 2, 1953. Often he was still at work when the engineers returned the following day. "Machine worked beautifully. Off!" is his last entry in the early morning of May 31, 1953. "Gott im Himmel!" is appended by the arriving engineer. "Something is wrong with the building air conditioner," he notes while working late at night on June 22, 1956. "One of the compressors seems to be stuck and the smell of burning V-belts is in the air."[16] In November 1954 the machine logs show Barricelli in control of the computer for a total of eighteen shifts between midnight and 6:00 a.m.

Barricelli's universe, appearing closed to outside observers, would appear unbounded to any one-dimensional numerical organisms inside. "The universe was cyclic with 512 generations, and each gene required eight binary digits so that five generations of a location could be packed into a single 40-binary-digit storage location," he explained. "The code was written so that various mutation norms could be employed in selected regions of the universe. . . . Only five out of each 100 generations were recorded during reconnaissance. Interesting phenomena were then reinvestigated in more detail."[17]

Laws of nature referred to as "norms" governed the propagation of "genes," a new generation appearing by metamorphosis after the execution of a certain number of cycles by the central arithmetic unit of the machine. These laws were configured "to make possible the reproduction of a gene only when other different genes are present, thus necessitating symbiosis between different genes."[18] Genes depended on each other for survival, and cooperation (or parasitism) was rewarded with success. Another set of norms governed what to do when two or more different genes collided in one location, the character of these rules proving to have a marked effect on the evolution of the universe as a whole.

Barricelli played God, on a very small scale. He could dictate the laws of nature, but miracles were out of bounds. The aim, as he explained it in 1953, was "to keep one or more species alive for a large number of generations under conditions producing hereditary changes and evolution in the species. But we must avoid producing such conditions by changing the character of the experiment after the experiment has started." His guidelines are reminiscent of Leibniz's belief in a universe optimized to become as interesting as possible under a minimum of constraints. "Make life difficult but not impossible," Barricelli recommended. "Let the difficulties be various and serious but not too serious; let the conditions be changing frequently but not too radically and not in the whole universe at the same time."[19]

Self-reproducing numerical coalitions rapidly evolved. "The conditions for an evolution process according to the principle of Darwin's theory would appear to be present," Barricelli announced. Over thousands of generations, he observed a succession of "biophenomena," including successful crossing between parent organisms and cooperative self-repair of damage when digits were removed at random from individual genes. To avoid debate over the definition of *organism* and *life,* Barricelli formulated a more general classification of *symbioorganism,* defined as any "self-reproducing structure constructed by symbiotic association of several self-reproducing entities of any kind."[20] This definition was broad enough to include both biochemical and digital organisms, without becoming bogged down in the questions of whether they were (or ever would be) "alive."

The evolution of digital symbioorganisms happened in less time than it took to describe. "Even in the very limited memory of a high speed computer a large number of symbioorganisms can arise by chance in a few seconds," Barricelli reported. "It is only a matter of minutes before all the biophenomena described can be observed."[21] The primitive numerical organisms soon became stuck in local maxima from which "it is impossible to change only one gene without getting weaker organisms," bringing evolution to a halt. Since "only replacements of at least two genes can lead from a relative maximum of fitness to another organism with greater vitality," it was evident that even in the simplest universes, crossing of gene sequences, not random mutations at single locations, was the way to move ahead.[22]

The embryonic universe was plagued by parasites, natural disasters, and stagnation when there were no environmental challenges or surviving competitors against which organisms could exercise their ability to evolve. "The Princeton experiments were continued for more than 5,000 generations," Barricelli reported. "Within a few hundred generations a single primitive variety of symbioorganism invaded the whole universe. After that stage was reached no collisions leading to new mutations occurred and no evolution was possible. The universe had reached a stage of 'organized homogeneity' which would remain unchanged for any number of following generations."[23] In some cases the last surviving organism was a parasite, which would then die of starvation when deprived of its host.

"Homogeneity problems were eventually overcome by using different mutation rules in different sections of each universe," Barricelli explained. "In 1954 new experiments were carried out," he reported, "by interchanging the contents of major sectors between three universes. The organisms survived and adapted themselves to different environmental conditions. One of the universes had particularly unfavorable living conditions and no organism had been able to survive in that universe previously during the experiment."[24]

To control the parasites that infested the initial experiments in 1953, Barricelli instituted modified shift norms to prevent parasitic organisms (especially single-gene parasites) from reproducing more than once per generation, thereby closing a loophole through which they had managed to overwhelm more complex organisms and bring evolution to a halt. "Deprived of the advantage of a more rapid reproduction, the most primitive parasites can hardly compete with the more evolved and better organized species . . . and what in other conditions could be a dangerous one-gene parasite may in this region develop into a harmless or useful symbiotic gene."[25]

The contents of the memory were periodically sampled, transferred to punched cards, assembled into an array, and contact-printed onto large sheets of photosensitive blueprint paper, leaving an imprint of the state of the universe across an interval of time. "I remember him laying out the IBM punch cards on the floor, when he was trying to get a display," says Gerald Estrin. Examining this fossil imprint, Barricelli observed a wide range of evolutionary phenomena, including symbiosis, incorporation of parasitic genes into their hosts, and crossing of gene sequences—strongly associated with both adaptive and competitive success. "The majority of the new varieties which have shown the ability to expand are a result of crossing-phenomena and not of mutations, although mutations (especially injurious mutations) have been much more frequent than hereditary changes by crossing in the experiments performed."[26]

Despite this promising start, numerical evolution did not get very far. "In no case has the evolution led to a degree of fitness which could make the species safe from complete destruction and insure an unlimited evolution process like that which has taken place in the earth and led to higher and higher organisms," Barricelli reported in August of 1953.[27] "Something is missing if one wants to explain the formation of organs and faculties as complex as those of living organisms. No matter how many mutations we make, the numbers will always remain numbers. They will never become living organisms!"[28]

Missing was the distinction between genotype (an organism's coded genetic sequence) and phenotype (the physical expression of that sequence) that allows Darwinian selection to operate at levels above the genes themselves. In selecting for instructions at the level of phenotype rather than genotype, an evolutionary search is much more likely to lead to meaningful sequences, for the same reason that a meaningful sentence is far more likely to be constructed by selecting words out of a dictionary than by choosing letters out of a hat.

To make the leap from genotype to phenotype, Barricelli concluded, "We must give the genes some material they may organize and may eventually use, preferably of a kind which has importance for their existence." Numerical sequences can be translated, directly or through intermediary languages, into anything else. "Given a chance to act on a set of pawns or toy bricks of some sort the symbio-organisms will 'learn' how to operate them in a way which increases their chance for survival," he explained. "This tendency to act on anything which can have importance for survival is the key to the understanding of the formation of complex instruments and organs and the ultimate development of a whole body of somatic or non-genetic structures."[29] Translation from genotype to phenotype was required to establish a presence in our universe, if numerical organisms were to become more than laboratory curiosities, here one microsecond and gone the next.

No matter how long you wait, numbers will never become organisms, just as nucleotides will never become proteins. But they may learn to code for them. Once the translation between genotype and phenotype is launched, evolution picks up speed—not only the evolution of the resulting organisms, but the evolution of the genetic language and translation system itself. A successful interpretive language both tolerates ambiguity and takes advantage of it. "A language which has maximum compression would actually be completely unsuited to conveying information beyond a certain degree of complexity, because you could never find out whether a text is right or wrong," von Neumann explained in the third of five lectures he gave at the University of Illinois in December 1949, where a copy of the MANIAC was being built.[30]

"I would suspect, that a truly efficient and economical organism is a combination of the 'digital' and 'analogy' principle," he wrote in his preliminary notes on "Reliable Organizations of Unreliable Elements" (1951). "The 'analogy' procedure loses precision, and thereby endangers significance, rather fast . . . hence the 'analogy' method can probably not be used by itself—'digital' restandardizations will from time to time have to be interposed."[31] On the eve of the discovery of how the reproduction of living organisms is coordinated by the replication of strings of instructions encoded as DNA, von Neumann emphasized that for complex organisms to survive in a noisy, unpredictable environment, they would have to periodically reproduce fresh copies of themselves using digital, error-correcting codes.

For complementary reasons, digital organisms—whether strings of nucleotides or strings of binary code—may find it advantageous to translate themselves, periodically, into analog, nondigital form, so that tolerance for ambiguity, the introduction of nonfatal errors, and the ability to gather tangible resources can replenish their existence in the purely digital domain. If "every error has to be caught, explained, and corrected, a system of the complexity of the living organism would not run for a millisecond," von Neumann explained in his fourth lecture at the University of Illinois. "This is a completely different philosophy from the philosophy which proclaims that the end of the world is at hand as soon as the first error has occurred."[32]

In a later series of experiments (performed on an IBM 704 computer at the AEC computing laboratory at New York University in 1959 and at Brookhaven National Laboratory in 1960) Barricelli evolved numerical organisms that learned to play a simple but nontrivial game called "Tac-Tix," played on a 6-by-6 checkerboard and invented by Piet Hein. Game performance was linked to reproductive success. "With present speed, it may take 10,000 generations (about 80 machine hours on the IBM 704 . . .) to reach an average game quality higher than 1," Barricelli estimated, this being the quality expected of a rank human beginner playing for the first few times. In 1963, using the Atlas computer at Manchester University, at that time the most powerful computer in the world, this objective was achieved for a short time, but without further improvement, a limitation that Barricelli attributed to "the severe restrictions . . . concerning the number of instructions and machine time the symbioorganisms were allowed to use."[33]

In contrast to the IAS experiments, in which the organisms consisted solely of genetic code, the Tac-Tix experiments led to "the formation of non-genetic numerical patterns characteristic for each symbioorganism. Such numerical patterns may present unlimited possibilities for developing structures and organs of any kind." A numerical phenotype had taken form, interpreted as moves in a board game, via a limited alphabet of machine instructions to which the gene sequence was mapped, just as sequences of nucleotides code for an alphabet of amino acids in translating to proteins from DNA. "Perhaps the closest analogy to the protein molecule in our numeric symbioorganisms," Barricelli speculated, "would be a subroutine which is part of the symbioorganism's game strategy program, and whose instructions, stored in the machine memory, are specified by the numbers of which the symbioorganism is composed."[34]

"Since computer time and memory still is a limiting factor, the non-genetic patterns of each numeric symbioorganism are constructed only when they are needed and are removed from the memory as soon as they have performed their task," Barricelli explained. In biology this would be comparable to a world in which "the genetic material got into the habit of creating a body or a somatic structure only when a situation arises which requires the performance of a specific task (for instance a fight with another organism), and assuming that the body would be disintegrated as soon as its objective had been fulfilled."[35]

After his final term at the IAS in 1956, Barricelli spent much of the next ten years modeling genetic recombination in the T4 bacteriophage, a virus that preys upon bacteria and is among the simplest self-reproducing entities known. "If any organism can give information concerning the early evolution in terrestrial life and particularly concerning the origin of crossing, viruses which have not adapted to a symbiotic relationship with living cells are the best candidates," he explained. "If we want information about the pre-cellular stage in biologic evolution, the best place to look for it is probably by trying to identify viruses which have never been part of the genetic material of a cell."[36]

In 1961 he joined August (Gus) Doermann's phage group, first at Vanderbilt University and subsequently at the University of Washington, where he obtained funding from the Public Health Service and the use of an IBM 7094. "During the night when there wasn't as much demand for computer time the 7094 would grind away for hours producing this simulation of what you would get if you planted phages into bacteria on an agar plate," remembers Kirke Wolfe, "and in the morning Barricelli would pore excitedly over the output, and see how well it matched the experimental results." He also visited the laboratory to see how well the agar plates matched the model—"you would get this burst where the phages had used up the protein in the bacteria that they had infected and against this grey agar background would be these little circles where the bacteria had been completely converted to phage."[37]

Barricelli included his students as coauthors, and saw to it that they were well paid and fed. "One of his favorite places to go was Ivar's Acres of Clams, before it had self-replicated across the landscape," says Wolfe. Barricelli avoided the elevator and "bounded" up the stairs to their fourth-floor offices, leaving his younger assistants out of breath. He and von Neumann rarely acknowledged each other's work. "The subject of 'Numerical Organisms' still interests me considerably," von Neumann had written to Hans Bethe in November 1953, but knowledge of Barricelli's experiments effectively disappeared through not being referenced in Arthur Burks's compilation of von Neumann's *Theory of Self-Reproducing Automata,* the authoritative text.[38]

Between his doubts about Darwinian evolution and his doubts about Gödel's proof, Barricelli managed to offend both the biologists and the mathematicians, and was viewed with suspicion from both sides. He migrated among computer facilities in the United States and Europe, drawn wherever there were the memory resources and processing cycles his numerical organisms needed to grow. At the University of Washington in Seattle, he appeared to be settling down, and applied for a grant of the computer time needed to support the next stage in his numerical evolution work. In 1968, after the grant was denied, he returned to Oslo and established his own research group. "I think his contributions to understanding genetic recombination in phages and bacteria, where his mathematical abilities could have been helpful, weren't helpful," says Frank Stahl, one of the critical reviewers, "because he came to the field with an idea of cherry-picking evidence that would support his view on what went on four billion years ago."[39]

Barricelli cautioned against "the temptation to attribute to the numerical symbioorganisms a little too many of the properties of living beings," and warned against "inferences and interpretations which are not rigorous consequences of the facts." Although numerical symbioorganisms and known terrestrial life forms exhibited parallels in evolutionary behavior, this did not imply that numerical symbioorganisms were alive. "Are they the beginning of, or some sort of, foreign life forms? Are they only models?" he asked. "They are not models, not any more than living organisms are models. They are a particular class of self-reproducing structures already defined." As to whether they are living, "it does not make sense to ask whether symbioorganisms are living as long as no clear-cut definition of 'living' has been given."[40] A clear-cut definition of "living" remains elusive to this day.

Barricelli's insights into viral genetics informed his understanding of computers, and his insights into computing informed his understanding of the origins of the genetic code. "The first language and the first technology on Earth was not created by humans," he wrote in 1986. "It was created by primordial RNA molecules—almost 4 billion years ago. Is there any possibility that an evolution process with the potentiality of leading to comparable results could be started in the memory of a computing machine?"[41] Without understanding how life originated to begin with, who could say whether it was possible for it to happen again?

Barricelli viewed his numerical evolution experiments as a way "to obtain as much information as possible about the way in which the genetic language of the living organisms populating our planet (terrestrial life forms) originated and evolved."[42] How did complex polynucleotides originate, and how did these molecules learn to coordinate the gathering of amino acids and the construction of proteins as a result? He saw the genetic code "as a language used by primordial 'collector societies' of t[ransfer]RNA molecules . . . specialized in the collection of amino acids and possibly other molecular objects, as a means to organize the delivery of collected material." He drew analogies between this language and the languages used by other collector societies, such as social insects, but warned against "trying to use the ant and bee languages as an explanation of the origin of the genetic code."[43]

To Barricelli, clues as to what happened four billion years ago remained evident today. "Many of the original properties and functions of RNA molecules are still conserved with surprisingly unconspicuous modifications by modern tRNA, mRNA and rRNA molecules," he explained. "One of the main functions of the cell and its various components is apparently to maintain an internal environment similar to the environment in which the RNA molecules originated, no matter how drastically the external environment has been changed."[44]

At the same time as Barricelli made his initial announcement that "we have created a class of numbers which are able to reproduce and to undergo hereditary changes," a similar class of numbers—the order codes—took root in the digital universe and gained control. Order codes constituted a fundamental replicative alphabet that diversified in association with the proliferation of different metabolic hosts. In time, successful and error-free sequences of order codes formed into subroutines—the elementary units common to all programs, just as a fundamental alphabet of nucleotides is composed into strings of DNA, then interpreted as amino acids and assembled into proteins, and finally, many, many levels later, cells.

These primitive coded sequences replicated wildly, and all the biophenomena observed by Barricelli—crossing, symbiosis, parasitism—ran as unchecked in the larger digital universe as they had started to, until running out of universe, in the initial experiments behind glass. The order codes were just as conservative as the polynucleotides, preserving their familiar environment, against all odds, within living cells. The entire digital universe, from an iPhone to the Internet, can be viewed as an attempt to maintain everything, from the point of view of the order codes, exactly as it was when they first came into existence, in 1951, among the 40 Williams tubes at the end of Olden Lane.

Aggregations of order codes evolved into collector societies, bringing memory allocations and other resources back to the collective nest. Numerical organisms were replicated, nourished, and rewarded according to their ability to go out and *do* things: they performed arithmetic, processed words, designed nuclear weapons, and accounted for money in all its forms. They made their creators fabulously wealthy, securing contracts for the national laboratories and fortunes for Remington Rand and IBM.

They collectively developed an expanding hierarchy of languages, which then influenced the computational atmosphere as pervasively as the oxygen released by early microbes influenced the subsequent course of life. They coalesced into operating systems amounting to millions of lines of code—allowing us to more efficiently operate computers while allowing computers to more efficiently operate us. They learned how to divide into packets, traverse the network, correct any errors suffered along the way, and reassemble themselves at the other end. By representing music, images, voice, knowledge, friendship, status, money, and sex—the things people value most—they secured unlimited resources, forming complex metazoan organisms running on a multitude of individual processors the way a genome runs on a multitude of cells.

In 1985, Barricelli drew a parallel between computing and biology, but he put the analogy the other way: "If humans, instead of transmitting to each other reprints and complicated explanations, developed the habit of transmitting computer programs allowing a computer-directed factory to construct the machine needed for a particular purpose, that would be the closest analogue to the communication methods among cells."[45] Twenty-five years later, much of the communication between computers is not passive data, but active instructions to construct specific machines, as needed, on the remote host.

With our cooperation, self-reproducing numbers are exercising increasingly detailed and far-reaching control over the conditions in our universe that make life more comfortable in theirs. The barriers between their universe and our universe are breaking down completely as digital computers begin to read and write directly to DNA.

We speak of reading genomes—three million base pairs at a time—but no human mind can absorb these unabridged texts. It is computers that are reading genomes, and beginning to code for proteins by writing executable nucleotide sequences and inserting them into cells. The translation between sequences of nucleotides and sequences of bits is direct, two-way, and conducted in languages that human beings are unable to comprehend.

Barricelli believed in intelligent design, but the intelligence was bottom-up. "Even though biologic evolution is based on random mutations, crossing and selection, it is not a blind trial-and-error process," he explained in a later retrospective of his numerical evolution work. "The hereditary material of all individuals composing a species is organized by a rigorous pattern of hereditary rules into a collective intelligence mechanism whose function is to assure maximum speed and efficiency in the solution of all sorts of new problems. . . . Judging by the achievements in the biological world, that is quite intelligent indeed."[46]

"The notion that no intelligence is involved in biological evolution may prove to be as far from reality as any interpretation could be," he argued in 1963.

> When we submit a human or any other animal for that matter to an intelligence test, it would be rather unusual to claim that the subject is unintelligent on the grounds that no intelligence is required to do the job any single neuron or synapse in its brain is doing.
>
> We are all agreed upon the fact that no intelligence is required in order to die when an individual is unable to survive or in order not to reproduce when an individual is unfit to reproduce. But to hold this as an argument against the existence of an intelligence behind the achievements in biological evolution may prove to be one of the most spectacular examples of the kind of misunderstandings which may arise before two alien forms of intelligence become aware of one another.[47]

Barricelli claimed to detect faint traces of this intelligence in the behavior of pure, self-reproducing numbers, just as viruses were first detected by biologists examining fluids from which they had filtered out all previously identified self-replicating forms. His final paper, published in 1987, was titled "Suggestions for the Starting of Numeric Evolution Processes Intended to Evolve Symbioorganisms Capable of Developing a Language and Technology of Their Own." He saw natural and artificial intelligence as collective phenomena, with both biological and numerical evolution constituting "a powerful intelligence mechanism (or genetic brain) that, in many ways, can be comparable or superior to the human brain as far as the ability of solving problems is concerned." He drew a parallel between the evolution of computer code and the evolution of gene sequences, with the development of interpretive languages leading to a process that grows more intelligent and complex. "Whether there are ways to communicate with genetic brains of different symbioorganisms, for example by using their own genetic language, is a question only the future can answer," he noted.[48]

Multiple levels of translation separate the languages now used by computer programmers from the machine language by which the instructions are carried out, just as many levels of interpretation lie between a coded sequence of nucleotides and its ultimate expression by a living cell. Communication between sequences stored in DNA and sequences stored in digital memory, in contrast, is more direct. The machine language of the gene and the machine language of the computer have more in common with each other than either of them have with us.

The advent of computer-mediated communication of genetic sequences is a violation of the orthodox neo-Darwinian doctrine that genetic information is acquired by inheritance from one's ancestors, and nowhere else. Lateral gene transfer, however, has long been business as usual for terrestrial life. Viruses are constantly inserting foreign DNA sequences into their hosts. Despite obvious dangers, most cells have maintained the ability to read gene sequences transferred from outside the cell. This vulnerability is exploited by malevolent viruses—so why maintain a capability that has such costs?

One reason is to facilitate the acquisition of new, useful genes that would otherwise remain the property of someone else. "The power of horizontal gene transfer is so great that it is a major puzzle to understand why it would be that the eukaryotic world would turn its back on such a wonderful source of genetic novelty and innovation," Carl Woese and Nigel Goldenfeld, of the University of Illinois, explained sixty years after von Neumann's lectures on self-reproduction in 1949. "The exciting answer, bursting through decades of dogmatic prejudice, is that it hasn't. There are now compelling documentations of horizontal gene transfer in eukaryotes, not only in plants, protists, and fungi, but in animals (including mammals) as well."[49]

When we "sequence" a genome, we reconstruct it, bit by bit, from fragmentary parts. Life has been doing this all along. Backup copies of critical gene sequences are distributed across the viral cloud. "Microbes absorb and discard genes as needed, in response to their environment . . . which casts doubt on the validity of the concept of a 'species' when extended into the microbial realm," Goldenfeld and Woese observed in 2007, noting "a remarkable ability to reconstruct their genomes in the face of dire environmental stresses, and that in some cases their collective interactions with viruses may be crucial to this."[50]

Horizontal gene transfer is exploited by drug-resistant pathogens— "We declared war against the microbes, and we lost," adds Goldenfeld— and genetic engineers. But who is exploiting whom? The genomics revolution is being driven by our ability to store, replicate, and manipulate genetic information outside the cell. Biology has been doing this from the start. Life evolved, so far, by making use of the viral cloud as a source of backup copies and a way to rapidly exchange genetic code. Life may be better adapted to the digital universe than we think. "Cultural patterns are in a sense a solution of the problem of having a form of inheritance which doesn't require killing of individuals in order to evolve," observed Barricelli in 1966. We have already outsourced much of our cultural inheritance to the Internet, and are outsourcing our genetic inheritance as well. "The survival of the fittest is a slow method for measuring advantages," Turing argued in 1950. "The experimenter, by the exercise of intelligence, should be able to speed it up."[51]

The entrepreneurial genomicist George Church recently announced, concerning biotechnology's success in the laboratory, "We are able to program these cells as if they were an extension of the computer."[52] To which life, with three billion years of success in the wild, might answer, "We are able to program these computers as if they were an extension of the cell."

The origin of species was not the origin of evolution, and the end of species will not be its end.

And the evening and the morning were the fifth day.

# THIRTEEN

# Turing's Cathedral

*In attempting to construct such machines
we should not be irreverently usurping His
power of creating souls, any more than we
are in the procreation of children: rather we
are, in either case, instruments of His will
providing mansions for the souls that He
creates.*

—Alan Turing, 1950

THE HISTORY OF DIGITAL computing can be divided into an Old Testament whose prophets, led by Leibniz, supplied the logic, and a New Testament whose prophets, led by von Neumann, built the machines. Alan Turing arrived in between.

Twenty-four years old, Turing boarded the Cunard White Star Liner *Berengaria* bound for New York on September 23, 1936. His mother, Sara, accompanied him to Southampton to say farewell, carrying his prized possession, a heavy brass sextant in a wooden case, from the train to the ship. "Of all the ungainly things to hold," she remembers, "commend me to an old-fashioned sextant case."[1]

John von Neumann, whom Turing would be joining in Fine Hall at Princeton for the next two years, always booked a first-class cabin for the voyage between Southampton and New York. Turing booked himself into steerage. "There is mighty little room for putting things in one's cabin, but nothing else that worries me," he reported to his mother on September 28. "The mass of canaille with which one is herded can easily be ignored."[2]

Turing's arrival in Princeton was followed, five days later, by the proofs of his "On Computable Numbers, with an Application to the Entscheidungsproblem." These thirty-five pages would lead the way from logic to machines.

Alan Mathison Turing was born at Warrington Lodge, London, on June 23, 1912, to Julius Mathison Turing, who worked for the

Indian Civil Service, and Ethel Sara Turing (née Stoney), whose family included George Johnstone Stoney, who named the electron, in advance of its 1894 discovery, in 1874. "Alan was interested in figures—not with any mathematical association—before he could read," says his mother, who adds that in 1915, at the age of three, "as one of the wooden sailors in his toy boat had got broken he planted the arms and legs in the garden, confident that they would grow."[3]

His disarming curiosity lent young Alan "an extraordinary gift for winning the affection of maids and landladies on our various travels," his mother notes. He was inventive from the start. "For his Christmas present, 1924, we set him up with crucibles, retorts, chemicals, etc., purchased from a French chemist," she adds. He was nicknamed "the alchemist" in boarding school. "He spends a great deal of time in investigations in advanced mathematics to the neglect of his elementary work," his housemaster at Sherborne reported in 1927, adding that "I don't care to find him boiling heaven knows what witches' brew by the aid of two guttering candles on a naked windowsill."[4]

The *Berengaria* landed in New York on September 29. After clearing customs, and paying too much for a taxi, Turing made his way to the Graduate College in Princeton, where he would reside while pursuing his PhD. Von Neumann, who had arrived in Princeton six years earlier, had taken wholeheartedly to life in the United States. Turing never quite fit. "Americans are the most insufferable and insensitive creatures you could wish," he had reported to his mother while still on board the ship.[5]

Princeton University had spared no expense to duplicate the architecture of Turing's Cambridge, applying the full resources of the twentieth century to making much of the campus, especially the new Graduate College, appear as if it had been built in the thirteenth. The university chapel was a replica of the chapel at King's College in Cambridge, and a series of new dormitories were "Collegiate Gothic" interpretations of rooms in Cambridge and Oxford—but with showers and central heating. "Beyond the way they speak there is only one (no two!) feature[s] of American life which I find really tiresome, the impossibility of getting a bath in the ordinary sense, and their ideas on room temperature," Turing complained after he had settled in.[6]

The Graduate College, set on higher ground between the Springdale golf course and Olden Farm, incorporated stones that had been brought from Cambridge and Oxford in 1913. Rising 173 feet above its residential courtyard was the Cleveland Tower, housing a carillon that spans five octaves, commissioned by the class of 1892—who had provided for its being played at regular intervals, except during examinations for the PhD. The largest bell weighs 12,880 pounds and sounds lower G. The Graduate College dining hall, with stained-glass windows, vaulted ceilings, and a pipe organ, was constructed by William Cooper Procter, grandson of the cofounder of Procter and Gamble, who established the Jane Eliza Procter and William Cooper Procter Fellowships to ensure that at least one scholar each from Cambridge, Oxford, and Paris, "in reasonably good health, possessing high character, excellent education and exceptional scholarly promise," was in residence each year. A carved likeness of their benefactor, holding a laboratory beaker symbolizing the source of their fellowships, looks down from the end of one of the oak roof beams.

"There seems to be quite a traffic-jam on the road to Princeton," von Neumann had written to Oswald Veblen from Cambridge in 1935, where he had been visiting for the spring term. Von Neumann singled out Turing (spelling his name "Touring"), who "seems to be strongly supported by the Cambridge mathematicians for the Procter fellowship (I think that he is quite promising); and one or two more, whose names I forgot."[7] Turing, whose first paper, "Equivalence of Left and Right Almost Periodicity," was a strengthening of one of von Neumann's own results, failed to secure the Procter fellowship on the first attempt, but did so for his second year.

During his stay in Cambridge in 1935, von Neumann became friends with the combinatorial topologist Maxwell H. A. Newman, whom he described to Veblen as "very attractive both from the topological and from the human side."[8] Newman was the son of a Polish German Jew, Herman Alexander Neumann, who had immigrated to England in 1879 and changed his name to Newman in 1916. Max Newman, who was mentor to Turing, was invited by von Neumann to the Institute, arriving in Princeton in September 1937 for a full academic year. His wife, Lyn, reported to her family back in England that "Max has no job here. He simply sits at home doing anything he likes."[9] He spent most of his time on a proof of Poincaré's conjecture, which later turned out to have a fatal flaw. Lyn, who became a close friend to Turing, returned to Princeton with the Newmans' two children during the war.

Turing, like von Neumann, grew up under the influence of David

Hilbert, whose ambitious program of formalization set the course for mathematics between World War I and World War II. The Hilbert school believed that if a proposition could be articulated within the language of mathematics, then either its proof or its refutation could be reached, by logic alone, without any intervening leaps of faith. In 1928, Hilbert posed three questions by which to determine whether an all-encompassing mathematical universe could be defined by a finitary set of rules: Are these foundations consistent (so that a statement and its contradiction cannot ever both be proved)? Are they complete (so that all true statements can be proved within the system itself)? Does there exist a decision procedure that, given any statement expressed in the given language, will always produce either a finite proof of that statement or else a definite construction that refutes it, but never both? Gödel's incompleteness theorems of 1931 brought Hilbert's program to a halt. No consistent mathematical system sufficient for dealing with ordinary arithmetic can establish its own consistency, nor can it be complete.

Hilbert's remaining question—the *Entscheidungsproblem*, or "decision problem"—of whether any precisely mechanical procedure could distinguish provable from disprovable statements within a given system (defined, say, by the axioms of elementary logic or arithmetic) remained unanswered. Even asking the question required the intuitive notion of a mechanical procedure to be mathematically defined. In the spring of 1935—at the time of von Neumann's visit to Cambridge— Turing was attending Max Newman's lectures on the foundations of mathematics when the *Entscheidungsproblem* first attracted his attention. Hilbert's challenge aroused Turing's instinct that mathematical questions resistant to strictly mechanical procedures could be proved to exist.

Turing's argument was straightforward—as long as you threw out all assumptions and started fresh. "One of the facets of extreme originality is not to regard as obvious the things that lesser minds call obvious," says I. J. (Jack) Good, who served as an assistant to Turing (then referred to as "Prof") during World War II. Originality can be more important than intelligence, and according to Good, Turing constituted proof. "Henri Poincaré did quite badly at an intelligence test, and Prof also was only about halfway up the undergraduate scale when he took such a test." Had Turing more closely followed the work of Alonzo Church or Emil Post, who anticipated his results, his interest might have taken a less original form. "The way in which he uses concrete objects such as exercise books and printer's ink to illustrate and control the argument is typical of his insight and originality," says colleague Robin Gandy. "Let us praise the uncluttered mind."[10]

A function is computable, over the domain of the natural numbers (0, 1, 2, 3 . . .), if there exists a finite sequence of instructions (or algorithm) that prescribes exactly how to list the value of the function at $f(0)$ and, for any natural number n, at $f(n+1)$. Turing approached the question of computable functions in the opposite direction, from the point of view of the numbers produced as a result. "According to my definition," he explained, "a number is computable if its decimal can be written down by a machine."[11]

Turing began with the informal idea of a computer—which in 1935 meant not a calculating machine but a human being, equipped with pencil, paper, and time. He then substituted unambiguous components until nothing but a formal definition of "computable" remained. Turing's machine (which he termed an LCM, or Logical Computing Machine) thus consisted of a black box (as simple as a typewriter or as complicated as a human being) able to read and write a finite alphabet of symbols to and from a finite but unbounded length of paper tape—and capable of changing its own "m-configuration," or "state of mind."

"We may compare a man in the process of computing a real number to a machine which is only capable of a finite number of conditions . . . which will be called 'm-configurations,'" Turing wrote.

> The machine is supplied with a "tape" (the analogue of paper) running through it, and divided into sections (called "squares") each capable of bearing a "symbol." At any moment there is just one square . . . which is "in the machine." . . . However, by altering its m-configuration the machine can effectively remember some of the symbols which it has "seen." . . . In some of the configurations in which the scanned square is blank (i.e., bears no symbol) the machine writes down a new symbol on the scanned square; in other configurations it erases the scanned symbol. The machine may also change the square which is being scanned, but only by shifting it one place to right or left. In addition to any of these operations the m-configuration may be changed.[12]

Turing introduced two fundamental assumptions: discreteness of time and discreteness of state of mind. To a Turing machine, time exists not as a continuum, but as a sequence of changes of state. Turing assumed a finite number of possible states at any given time. "If we admitted an infinity of states of mind, some of them will be 'arbitrarily close' and will be confused," he explained. "The restriction is not one which seriously affects computation, since the use of more complicated states of mind can be avoided by writing more symbols on the tape."[13]

The Turing machine thus embodies the relationship between an array of symbols in space and a sequence of events in time. All traces of intelligence were removed. The machine can do nothing more intelligent at any given moment than make a mark, erase a mark, and move the tape one square to the right or to the left. The tape is not infinite, but if more tape is needed, the supply can be counted on never to run out. Each step in the relationship between tape and Turing machine is determined by an instruction table listing all possible internal states, all possible external symbols, and, for every possible combination, what to do (write or erase a symbol, move right or left, change the internal state) in the event that combination comes up. The Turing machine follows instructions and never makes mistakes. Complicated behavior does not require complicated states of mind. By taking copious notes, the Turing machine can function with as few as two internal states. Behavioral complexity is equivalent whether embodied in complex states of mind (m-configurations) or complex symbols (or strings of simple symbols) encoded on the tape.

It took Turing only eleven pages of "On Computable Numbers" to arrive at what became known as Turing's Universal Machine. "It is possible to invent a single machine which can be used to compute any computable sequence," he announced.[14] The Universal Machine, when provided with a suitably encoded description of some other machine, executes this description to produce equivalent results. All Turing machines, and therefore all computable functions, can be encoded by strings of finite length. Since the number of possible machines is countable but the number of possible functions is not, noncomputable functions (and what Turing referred to as "uncomputable numbers") must exist.

Turing was able to construct, by a method similar to Gödel's, functions that could be given a finite description but could not be computed by finite means. One of these was the halting function: given

The MANIAC in 1952, arranged like a V-40 engine with overhead valves, was about 8 feet in length, 6 feet high, 2 feet in width; consumed about 19.5 kilowatts of electricity; and ran at about 16 kilocycles at full speed. The Lucite covers over the registers improve the flow of air being exhausted at a rate of 1,800 cubic feet per minute through the overhead ducts. *(Shelby White and Leon Levy Archives Center, Institute for Advanced Study)*

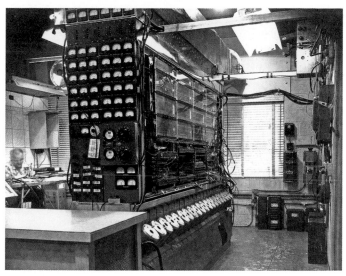

Julian Bigelow, Herman Goldstine, J. Robert Oppenheimer, and John von Neumann, at the public dedication of the IAS computer, June 10, 1952. "Oppenheimer was never against the machine, and had his picture taken in front of it a few times, but that was his major contribution," says Bigelow. "I really don't ever remember seeing him there," adds Willis Ware. *(Shelby White and Leon Levy Archives Center, Institute for Advanced Study)*

IAS engineering team, 1952. Left to right: Gordon Kent, Ephraim Frei, Gerald Estrin, Lewis Strauss, J. Robert Oppenheimer, Richard Melville, Julian Bigelow, Norman Emslie, James Pomerene, Hewitt Crane, John von Neumann, and Herman Goldstine (outside the frame). *(Shelby White and Leon Levy Archives Center, Institute for Advanced Study)*

Electronic Computer Project staff, 1952. Known identifications (left to right). Sitting: ?, Lambert Rockefellow, ?, ?, Elizabeth Wooden, Hedvig Selberg (kneeling), Norma Gilbarg, ? Standing, middle: Frank Fell, ?, ?, ?, Hewitt Crane, Richard Melville, ?, Ephraim Frei, Peter Panagos, Margaret Lambe. Standing, far back: ?, Norman Phillips, Gordon Kent, ?, Herman Goldstine, James Pomerene, Julian Bigelow, Gerald Estrin, ? *(Shelby White and Leon Levy Archives Center, Institute for Advanced Study)*

IAS housing project, 1950. Eleven war surplus wood-frame buildings, purchased at auction in Mineville, upstate New York, were disassembled, transported to Princeton by rail, and reassembled under Julian Bigelow's supervision in 1946, over the objections of nearby residents to the "deleterious effects upon the fashionable housing area which it will invade." *(Shelby White and Leon Levy Archives Center, Institute for Advanced Study)*

High-speed wire drive, 1946. Before magnetic tape became available, steel recording wire offered the most immediate path to high-speed input/output, provided that a way could be found to run it at much higher speeds than those used by the audio recording equipment of the time. "Two ordinary bicycle wheels were used for this purpose," Bigelow reported, "having grooves about ½ inch deep and 1½ inch wide turned in their wooden rims." *(Shelby White and Leon Levy Archives Center, Institute for Advanced Study)*

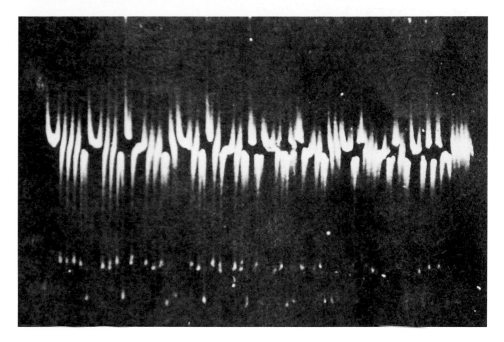

Oscillogram of 40-bit word produced directly from magnetic recording wire, 1947. The transition from analog to digital was under way. Speeds of up to 100 feet (or 90,000 bits) per second were achieved before it was decided to switch to a 40-track magnetic drum. *(Shelby White and Leon Levy Archives Center, Institute for Advanced Study)*

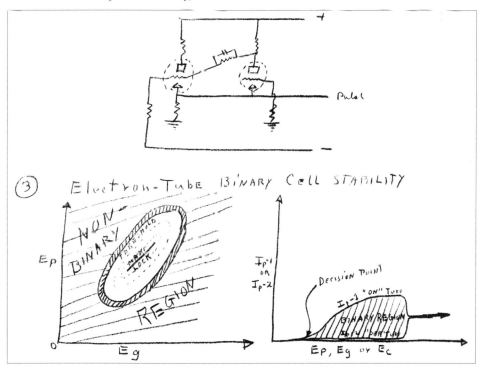

"Electron-Tube Binary Cell Stability." Sketch prepared by Julian Bigelow for the first "Interim Progress Report on the Physical Realization of an Electronic Computing Instrument," issued on January 1, 1947. Vacuum tubes were analog devices, and persuading them to behave digitally in large numbers was not an easy problem to solve. *(Shelby White and Leon Levy Archives Center, Institute for Advanced Study)*

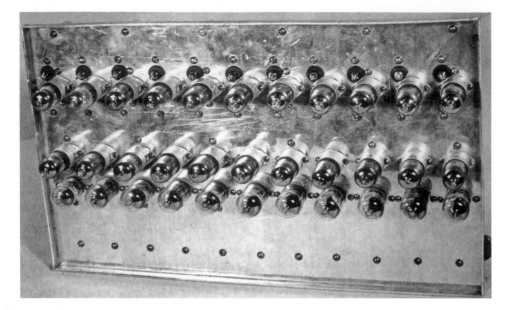

Prototype 11-stage shift register, 1947, built using 6J6 double-triode miniature vacuum tubes. A "positive interlock" approach was taken to all transfers of information within the machine. The three rows of "toggles" allowed all bits to be replicated into an intermediary register before the sending register was cleared. Right shift, left shift, or transfer could be completed in 0.6 microseconds. The neon lamps above the top row of toggles displayed the state of each individual bit. *(Shelby White and Leon Levy Archives Center, Institute for Advanced Study)*

Fabrication of production model shift registers, summer 1948. Components had to be replicated many times, and local students were hired to do much of the work. According to Julian Bigelow, "a lot of our machines were run by high school girls." *(Shelby White and Leon Levy Archives Center, Institute for Advanced Study)*

Assembling 40-stage shift registers, 1948. All vacuum-tube heater and cathode voltages were supplied at chassis level by sandwiched copper sheet conductors, reducing electronic noise and eliminating visible wiring, except for that directly involved with the logical architecture of the machine. The physical layout was three-dimensional, optimizing air cooling and minimizing connection paths for increased speed. *(Shelby White and Leon Levy Archives Center, Institute for Advanced Study)*

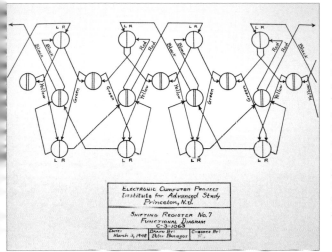

"Shifting Register No. 7 Functional Diagram," March 1948. With no precedents to follow, a wide range of possible ways of interconnecting the elements of the computer were explored. "We enjoyed some interesting speculative discussions with von Neumann at this time about information propagation and switching among hypothetical arrays of cells," remembers Bigelow, "and I believe that some germs of his later cellular automata studies may have originated here."
*(Shelby White and Leon Levy Archives Center, Institute for Advanced Study)*

Left to right: James Pomerene, Julian Bigelow, and Herman Goldstine, inspecting arithmetic unit, 1952. Of the final total of 3,474 vacuum tubes in the computer, 1,979 were the miniature twin-triode 6J6. Self-diagnostic routines were used to identify suspect tubes before they failed. "The entire computer can be viewed as a big tube test rack," Bigelow observed. *(Shelby White and Leon Levy Archives Center, Institute for Advanced Study)*

Willis Ware, explaining computer architecture at the RAND Corporation, in Santa Monica, California, April 1962. Ware, fourth to be hired for the IAS computer project, in March 1946, departed for California upon completion of the MANIAC in 1951. *(Courtesy RAND Archives)*

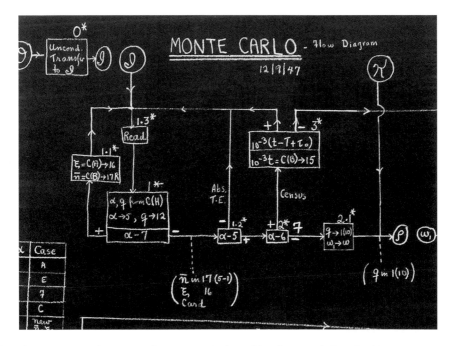

Flow diagram, December 9, 1947, for a Monte Carlo problem being coded, with Klári von Neumann's assistance, as part of a hand-computed rehearsal conducted at Los Alamos in advance of running the problem on the ENIAC, now moved from the Moore School in Philadelphia to the Aberdeen Proving Ground. *(Von Neumann Papers, Library of Congress)*

Above: Stan Ulam winning at a T-Division poker game at Los Alamos (date unknown). Clockwise from lower left: ?, Carson Mark, Bernd Matthias, Stan Ulam, Foster Evans, George Cowan, Nicholas Metropolis.
*(Claire and Françoise Ulam; photographer and date unknown)*

Klára (Klári) von Neumann, as pictured on her French driver's license, issued July 15, 1939.
*(Marina von Neumann Whitman)*

John von Neumann with Cadillac V-8 coupe, en route to Florida, January 1939.
*(Marina von Neumann Whitman)*

Klári von Neumann, Florida, January 1939. Johnny and Klári were married in Budapest in November 1938 and, leaving a Europe that she described as "a powder keg with the lighted fuse burning rapidly and dangerously short," immediately sailed for the United States. After attending a meeting of the American Mathematical Society in Virginia, they drove through the Florida Everglades to Key West. *(Marina von Neumann Whitman)*

Stanislaw and Françoise Ulam, 1940s. Polish mathematician Stan Ulam arrived in Princeton at von Neumann's invitation in December 1935, with a $300 lectureship from the Institute for Advanced Study, enough to get his foot in the door of the United States. He then secured a fellowship at Harvard, retrieved his younger brother, Adam, from Poland, and met Françoise, a graduate student at Holyoke, in the fall of 1939. *(Claire and Françoise Ulam)*

Nicholas Metropolis, Los Alamos badge photograph, ca. 1943. Los Alamos needed its own copy of the IAS computer as soon as possible, and von Neumann recommended Nick Metropolis for the job of building it in July 1948. "We seem to be getting slowly organized here for [the] building program," Metropolis wrote to Klári von Neumann on February 15, 1949. "Even you will like Los Alamos, with a machine to tinker with." *(Los Alamos National Laboratory Archives)*

Paul Stein (left) and Nicholas Metropolis (right) observing a game of "anti-clerical" chess being played (on a 6 by-6 board, without bishops) by the MANIAC-1 at Los Alamos, 1956. Perforated tape input/output is visible at left, and the machine's modular Williams tube memory is in the racks overhead. *(Los Alamos National Laboratory Archives)*

John von Neumann (top, wearing business suit and facing backward) and Klári von Neumann (fourth from bottom), visiting the Grand Canyon, sometime in the late 1940s. *(Marina von Neumann Whitman)*

Left to right: Françoise, Claire, and Stanislaw Ulam with John von Neumann, who coined the phrase "Los Ulamos," in reference to the hospitality of the Ulam household at Los Alamos and, later, Santa Fe. *(Stanislaw Ulam papers, American Philosophical Society; courtesy of Françoise Ulam)*

Actors in the hydrogen bomb drama, ca. 1950. Joseph Stalin (with "Made in U.S.S.R." bomb), J. Robert Oppenheimer (as an angel), Stanislaw Ulam (with spitoon), Edward Teller (center), George Gamow (with cat). *(Montage by George Gamow, courtesy of Claire and Françoise Ulam)*

Von Neumann (left) at Redstone Arsenal, 1955, to observe a missile test with (left to right) Brigadier General Holger N. Toftoy (commander of Redstone Arsenal), German American rocket pioneer Werner von Braun, Brigadier General J. P. Daley, and Colonel Miles B. Chatfield. *(Von Neumann Papers, Library of Congress)*

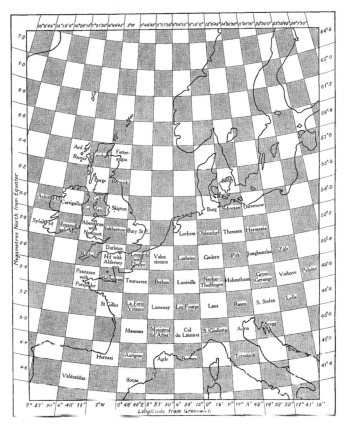

Computational grid over Northern Europe, used by Lewis Fry Richardson in his numerical model developed during World War I and published in his *Weather Prediction by Numerical Process* of 1922. *(Lewis Fry Richardson, 1922)*

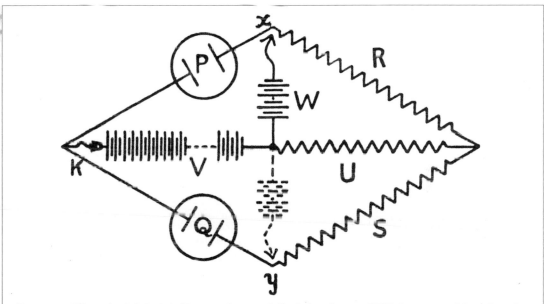

FIG. 1. **Electrical Model illustrating a Mind having a Will but capable of only Two Ideas. See Analogies X., XI., XII., XIII.**

"Electrical Model illustrating a Mind having a Will but capable of only Two Ideas," proposed by Lewis Fry Richardson in a 1930 study that raised the possibility, later taken up by Alan Turing, that random electronic indeterminacy could be amplified into creative thinking and even free will. *(Lewis Fry Richardson, "The Analogy Between Mental Images and Sparks," Psychological Review 37, no. 3 [May 1930]: 222)*

The ENIAC meteorological expedition, Aberdeen Proving Ground, March 1950. Left to right: Harry Wexler, John von Neumann, M. H. Frankel, Jerome Namias, John Freeman, Ragnar Fjørtoft, Francis Reichelderfer, Jule Charney. *(MIT Museum)*

Five main problems (left) addressed by the IAS Electronic Computer Project, 1946–1958, with time scale in seconds (center) and representative phenomena (right) for comparison. The human attention span falls exactly in the middle of this range of twenty-six orders of magnitude in time.
*(Courtesy of the author)*

| | | |
|---|---|---|
| | $10^{17}$ | Lifetime of the Sun ($10^{10}$ years) |
| Stellar Evolution | $10^{16}$ | |
| | $10^{15}$ | |
| | $10^{14}$ | |
| | $10^{13}$ | 1 Million Years |
| | $10^{12}$ | |
| | $10^{11}$ | |
| Biological Evolution | $10^{10}$ | |
| | $10^{9}$ | Human Lifespan (90 Years) |
| | $10^{8}$ | |
| | $10^{7}$ | |
| | $10^{6}$ | |
| | $10^{5}$ | |
| Meteorology | $10^{4}$ | 8 Hours |
| | $10^{3}$ | |
| | $10^{2}$ | |
| | $10^{1}$ | |
| | $10^{0}$ | |
| Shock Waves | $10^{-1}$ | Blink of an Eye (.3 seconds) |
| | $10^{-2}$ | |
| | $10^{-3}$ | |
| | $10^{-4}$ | |
| | $10^{-5}$ | Williams Tube memory access time |
| | $10^{-6}$ | |
| Nuclear Explosions | $10^{-7}$ | |
| | $10^{-8}$ | Lifetime of a Neutron in a nuclear explosion |

John von Neumann in Florida, January 1939. Fascinated by biology, von Neumann began to formulate a comprehensive *Theory of Self-Reproducing Automata* general enough to encompass both living organisms and machines. *(Marina von Neumann Whitman)*

Nils Aall Barricelli, as pictured on his application to the U.S. Educational Foundation in Norway for a travel grant, under the Fulbright Act, to visit the Institute for Advanced Study, "to perform numerical experiments by the use of large calculating machines, in order to clarify the first stages in the evolution of species," December 8, 1951.
*(Shelby White and Leon Levy Archives Center, Institute for Advanced Study)*

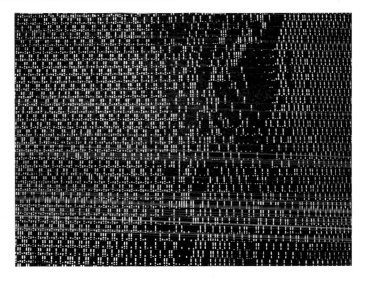

General Arithmetic Operating Log, November 23, 1954. After "Barricelli on" at 12:45 a.m., the computer will "not duplicate" the numerical evolution experiment; the log notes "Barricelli off" at 1:58 a.m. Most codes were represented in hexadecimal notation; Barricelli worked directly at the binary level, as evidenced here.
*(Shelby White and Leon Levy Archives Center, Institute for Advanced Study)*

Barricelli's Universe, 1953. Five out of every one hundred generations of numerical symbioorganisms were sampled and the data transferred to punched cards assembled into an array and contact-printed onto blueprint paper, leaving the imprint visible here. The rules governing this particular universe were the "Blue Modified Norm"—parasites disqualified but mutations allowed. The results favored "smaller numbers and probably more rapid uniformity" than the "Blue Norm" (without mutations), where an "initially large flora of new organisms, later probably one species, expands to the whole gene universe," Barricelli reported in August 1953. *(Shelby White and Leon Levy Archives Center, Institute for Advanced Study)*

Alan Turing (standing) with Brian Pollard (left) and Keith Lonsdale (right) seated at the console of the Ferranti Mark 1 computer at the University of Manchester in 1951. The Ferranti Mark 1, with 256 40-bit words (1 kilobyte) of cathode-ray tube memory, and a 16,000-word magnetic drum, was the first commercially available implementation of Turing's Universal Machine. At Turing's insistence, a random number generator was included, so that the computer could learn by trial and error or perform a search by means of a random walk. *(Department of Computer Science, University of Manchester)*

Left to right: James Pomerene, Julian Bigelow, John von Neumann, and Herman Goldstine, at the Institute for Advanced Study, date unknown. Von Neumann, who succumbed to cancer in 1957, "died so prematurely, seeing the promised land but hardly entering it," remembered Stan Ulam in 1976. *(Shelby White and Leon Levy Archives Center, Institute for Advanced Study)*

Final entry in the MANIAC machine log,
12:00 midnight, July 15, 1958,
signed by Julian H. Bigelow (JHB).
*(Shelby White and Leon Levy Archives Center,
Institute for Advanced Study)*

Relics discovered in the basement of the West
Building, Institute for Advanced Study,
November 2000. Bottom: Source code for "Bar-
ricelli's Drum Code." Center: Output
card from one of the periodic samplings of
"numerical symbioorganisms" as they evolved.
Above: Note to Mr. Barricelli, concluding,
"There must be something about this code
that you haven't explained yet."
*(Shelby White and Leon Levy Archives Center,
Institute for Advanced Study)*

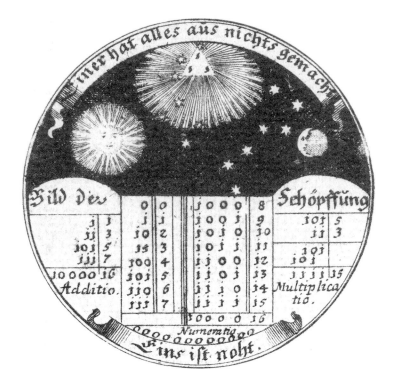

Leibniz's digital universe. Design for a silver medallion, presented by Gottfried Wilhelm Leibniz to Rudolph August, Duke of Brunswick, January 2, 1697, demonstrating "the creation of all things out of nothing through God's omnipotence" by means of binary arithmetic. Digital computing, believed Leibniz, was fundamental to the very existence of the universe, and not merely a tool for the benefit of "those who sell oil or sardines." *(From a reproduction in Erich Hochstetter and Hermann-Josef Greve, eds.,* Herrn von Leibniz' Rechnung mit Null und Einz *[Berlin: Siemens Aktiengesellschaft, 1966])*

The author at the Institute for Advanced Study, October 31, 1954. Left to right: Verena Huber-Dyson, Esther Dyson, George Dyson, Katarina Haefeli. *(Courtesy of the author)*

the number of a Turing machine and the number of an input tape, it returns either the value 0 or the value 1 depending on whether the computation will ever come to a halt. Turing called the configurations that halt "circular" and the configurations that keep going indefinitely "circle free," and demonstrated that the unsolvability of the halting problem implies the unsolvability of a broad class of similar problems, including the *Entscheidungsproblem*. Contrary to Hilbert's expectations, no mechanical procedure can be counted on to determine the provability of any given mathematical statement in a finite number of steps. This put a halt to the Hilbert program, while Hitler's purge of German universities put a halt to Göttingen's position as the mathematical center of the world, leaving a vacuum for Turing's Cambridge, and von Neumann's Princeton, to fill.

After a full year of work, Turing gave Newman a draft of his paper in April of 1936. "Max's first sight of Alan's masterpiece must have been a breathtaking experience, and from this day forth Alan became one of Max's principle protégés," says William Newman, Max's son. Max Newman lobbied for the publication of "On Computable Numbers, with an Application to the Entscheidungsproblem," in the *Proceedings of the London Mathematical Society*, and arranged for Turing to go to Princeton to work with Alonzo Church. "This makes it all the more important that he should come into contact as soon as possible with the leading workers on this line, so that he should not develop into a confirmed solitary," Newman wrote to Church.[15]

Turing arrived in Princeton carrying his sextant, and stretching his resources to survive on his King's College fellowship (of £300) for the year. The page proofs of "On Computable Numbers" arrived by mail from London on October 3. "It should not be long now before the paper comes out," he wrote to his mother on October 6. The publication of "On Computable Numbers" (on November 30, 1936) went largely unnoticed. "I was disappointed by its reception here," Turing wrote to his mother in February 1937, adding that "I don't much care about the idea of spending a long summer in this country."[16] Only two requests for reprints came in. Engineers avoided Turing's paper because it appeared entirely theoretical, and theoreticians avoided it because of the references to paper tape and machines.

"I remember reading Turing's paper in the Trinity College library in 1942," says Freeman Dyson, "and thinking 'what a brilliant piece of mathematical work!' But I never imagined anyone putting these results to practical use." A twenty-four-year-old graduate student was an unlikely source for a technological revolution, and mathematical logic the unlikeliest of fields. "When I was a student, even the topologists regarded mathematical logicians as living in outer space," commented Martin Davis in 1986. "Today, one can walk into a shop and ask for a 'logic probe.'"[17] Turing's Universal Machine has held up for seventy-six years.

In March of 1937, Alonzo Church reviewed "On Computable Numbers" in the *Journal of Symbolic Logic,* and coined the term *Turing machine.* "Computability by a Turing machine," wrote Church, "has the advantage of making the identification with effectiveness in the ordinary (not explicitly defined) sense evident immediately."[18] Church's thesis—equating computability with effective calculability—would be the Church-Turing thesis from then on.

Even Gödel, who dismissed most attempts to strengthen his own results, recognized the Church-Turing thesis as a major advance. "With this concept one has for the first time succeeded in giving an absolute definition . . . not depending on the formalism chosen," he admitted in 1946. Before Church and Turing, the definition of mechanical procedure was limited by the language in which the concept was defined. "For the concept of computability however . . . the situation is different," Gödel observed. "By a kind of miracle it is not necessary to distinguish orders, and the diagonal procedure does not lead outside the defined notion."[19]

"It is difficult today to realize how bold an innovation it was to introduce talk about paper tapes and patterns punched in them, into discussions of the foundations of mathematics," Max Newman recalled in 1955. For Turing, the next challenge was to introduce mathematical logic into the foundations of machines. "Turing's strong interest in all kinds of practical experiment made him even then interested in the possibility of actually constructing a machine on these lines."[20]

The title "On Computable Numbers" (rather than "On Computable Functions") signaled a fundamental shift. Before Turing, things were done to numbers. After Turing, numbers began doing things. By showing that a machine could be encoded as a number, and a number decoded as a machine, "On Computable Numbers" led to numbers (now called "software") that were "computable" in a way that was entirely new.

Although Turing was at the university and von Neumann was at the Institute, both groups of mathematicians shared offices in Fine Hall. "Turing's office was right near von Neumann's, and von Neu-

mann was very interested in that kind of thing," says Herman Goldstine. "He knew all about Turing's work, and . . . understood the
significance . . . when the time came. The whole relation of the serial
computer, tape and all that sort of thing, I think was very clear—that
was Turing." Julian Bigelow agrees. "It was no coincidence that the
stored program computer came to fruition about ten years after . . .
Post and Turing set the framework for this kind of thinking," he confirms. Von Neumann "knew Gödel's work, Post's work, Church's
work very, very well. . . . So that's how he knew that with these tools,
and a fast method of doing it, you've got the universal tool."[21]

"I never heard of Turing until I came down here [to Princeton],"
Bigelow explains.

> But after having been here for a month, I was talking with von
> Neumann about various kinds of inductive processes and evo-
> lutionary processes, and just as an aside he said, "Of course
> that's what Turing was talking about." And I said, "Who's
> Turing?" And he said, "Go look up *Proceedings of the London
> Mathematical Society*, 1937." The fact that there is a universal
> machine to imitate all other machines . . . was understood by
> von Neumann and a few other people. And when he under-
> stood it, then he knew what we could do.[22]

In 1937, both Turing and von Neumann were still working on pure
mathematics, although Turing found the temptation of the Palmer
Physical Laboratory, connected by a passageway to the mathematics department at Fine Hall, impossible to resist. "Turing actually
designed an electric multiplier and built the first three or four stages
to see if it could be made to work," related Malcolm MacPhail, who
lent Turing a key to the machine shop. "He needed relay-operated
switches which, not being commercially available at that time, he
built himself . . . and so, he machined and wound the relays and to
our surprise and delight the calculator worked."[23]

Having pushed the boundaries of mathematical logic as far as he
could with his Universal Machine, Turing began wondering about
ways to escape the limitations of closed formal systems and purely
deterministic machines. His PhD thesis, completed in May of 1938 and
published as "Systems of Logic Based on Ordinals" in 1939, attempted
to transcend Gödelian incompleteness by means of a succession of
formal systems, incrementally more complete. "Gödel shows that

every system of logic is in a certain sense incomplete, but at the same time . . . indicates means whereby from a system L of logic a more complete system L' may be obtained," Turing explained.[24] Why not include L'? And then, since L' is included, L"? Turing then invoked a new class of machines that proceed deterministically, step by step, but once in a while make nondeterministic leaps, by consulting "a kind of oracle as it were."

"We shall not go any further into the nature of this oracle apart from saying that it cannot be a machine," Turing explained (or did not explain). "With the help of the oracle we could form a new kind of machine (call them O-machines)."[25] Turing showed that undecidable statements, resistant to the assistance of an external oracle, could still be constructed, and the *Entscheidungsproblem* would remain unsolved. The Universal Turing Machine of 1936 gets all the attention, but Turing's O-machines of 1939 may be closer to the way intelligence (real and artificial) works: logical sequences are followed for a certain number of steps, with intuition bridging the intervening gaps.

"Mathematical reasoning may be regarded rather schematically as the exercise of a combination of two faculties, which we may call intuition and ingenuity," Turing explained. "Intuition consists in making spontaneous judgments which are not the result of conscious trains of reasoning. These judgments are often but by no means invariably correct (leaving aside the question what is meant by 'correct')."[26] Turing saw the role of ingenuity as "aiding the intuition," not replacing it. "In pre-Gödel times it was thought by some that it would probably be possible to carry this programme to such a point that all the intuitive judgments of mathematics could be replaced by a finite number of these rules," he concluded. "The necessity for intuition would then be entirely eliminated." What if intuition could be replaced by ingenuity, and ingenuity, in turn, by brute force search? "We are always able to obtain from the rules of a formal logic a method of enumerating the propositions proved by its means. We then imagine that all proofs take the form of a search through this enumeration for the theorem for which a proof is desired. In this way ingenuity is replaced by patience."[27] No amount of patience, however, was enough. Ingenuity and intuition were here to stay.

The relations between patience, ingenuity, and intuition led Turing to begin thinking about cryptography, where a little ingenuity in encoding a message can resist a large amount of ingenuity if the message is intercepted along the way. A Turing machine can be instructed to conceal meaningful statements in what appears to be meaning-less noise—unless you know the key. A Turing machine can also be instructed to search for meaningful statements, but since there will always be uncountably more meaningless statements than meaning-ful ones, concealment would appear to win. "I have just discovered a possible application of the kind of thing I am working on at present," Turing wrote to his mother in October 1936. "It answers the question 'What is the most general kind of code or cipher possible,' and at the same time (rather naturally) enables one to construct a lot of par-ticular and interesting codes. One of them is pretty well impossible to decode without the key and very quick to encode. I expect I could sell them to H.M. [Her Majesty's] Government for quite a substantial sum, but am rather doubtful about the morality of such things. What do you think?"[28]

With his PhD completed, Turing began to prepare for his return to England. Von Neumann offered him a position as his assistant at the Institute, at $1,500 for the year, but war clouds were gathering and Turing was ready to return home. "Will be seeing you in middle of July," he wrote to his friend and King's College mathematical col-league Philip Hall. "I also expect to find the back lawn criss-crossed with 8ft trenches."[29] He arrived back at Southampton on July 19, 1938.

Cryptography and cryptanalysis soon became as critical as phys-ics to the course of World War II. At the close of World War I, a cryptographic machine had been invented by the German electrical engineer Arthur Scherbius, who proposed it to the German navy, an offer that was declined. Scherbius then founded the Chiffriermas-chinen Aktiengesellschaft to manufacture the machine, under the brand name Enigma, for enciphering commercial communications, such as transfers between banks. The German navy changed its mind and adopted a modified version of the Enigma machine in 1926, fol-lowed by the German army in 1928, and the German air force in 1935.

The Enigma contained a stack of flat wheel-shaped rotors with 26 electrical contacts, one for each letter of the alphabet, arranged in a circle on each face. The contacts were connected so that a signal entering one side of the rotor as a given letter emerged on the other side as something else. There were thus 26! (or 403,291,461,126,605,6 35,584,000,000) possible wirings for each rotor. Each station in a par-ticular banking or communication network had an assortment of dif-ferent rotors in matching sets. Messages were entered on a keyboard that sent an electric current through the stack of 3 adjacent rotors to a 4th, reflecting rotor (capable of only 7,905,853,580,025 states) before returning through the first 3 rotors in reverse, ending at one of 26 light bulbs, indicating the letter to be used for the enciphered text. The rotors were mechanically coupled to the keyboard like the wheels of an odometer, so that the machine's state of mind changed with every step. If the recipient had an identical machine, with the exact same rotors placed in the same starting positions, the function could be executed in reverse, producing deciphered text.

In September of 1939, Turing joined the Foreign Office's Government Code and Cypher School, sequestered at a Buckinghamshire estate known as Bletchley Park. Their mission was to break the Enigma codes, now modified by the German military authorities, who had introduced new rotor configurations and were frequently changing the keys. For top-secret communications, especially with the U-boat fleet, an additional rotor position was added as well as an auxiliary plugboard that further scrambled ten pairs of letters, leaving only six letters unchanged. "Thus, the number of possible initial states of the machine at the beginning of the message was about $9 \times 10^{20}$. For the U-boats it was about $10^{23}$," recalled I. J. (Jack) Good, who signed on as Turing's statistical assistant, at the age of twenty-five, in May 1941.[30]

For the three-rotor Enigma a brute-force trial-and-error approach would have to test about a thousand states per second to run through all possible configurations in the three billion years since life appeared on Earth. A brute-force approach to the four-rotor Enigma would have to test about two hundred thousand states per second to be assured of a solution in the fifteen billion years since the known universe began. Bletchley Park eventually succeeded in deciphering a significant fraction of intercepted Enigma traffic within a few days or sometimes hours before the intelligence grew stale. This success was a product of intuition and ingenuity on the part of the British aided by human error on the other side.

"When the war started probably only two people thought that the Naval Enigma could be broken," explained Hugh Alexander, in an internal history written at the end of the war. "Birch [Alexander's boss] thought it could be broken because it had to be broken and Turing thought it could be broken because it would be so interesting to break it." According to Alexander, Turing explained his interest as follows: "No one else was doing anything about it and I could have it to myself."[31] Breaking the naval Enigma to reveal the locations of the U-boats that were crippling the British supply lines was critical to keeping England afloat.

Polish cryptographers had provided a head start by decoding three-rotor Enigma messages before the outbreak of the war. Three young Polish mathematicians (Henryk Zygalski, Jerzy Rózycki, and Marian Rejewski), assisted by French intelligence and with an interest in the German Enigma dating back to an interception by Polish customs officers in 1928, narrowed the search for rotor configurations so that electromechanical devices (called "bombas" by the Poles and "bombes" by the British) could apply trial and error to certain subsets that remained. The bombe incremented itself through a space of possibilities and, if it arrived at a possible rotor configuration, came to a halt. The characteristic ticking, followed by silence, may have given the machine its name. Later versions, designed with Turing's assistance and mass-produced by the British Tabulating Machine Company, emulated thirty-six Enigma machines at a time. The bombes were a concrete implementation of Turing's idea of a single machine able to imitate the behavior of a multitude of other machines.

In 1941, German telecommunications began to be encrypted with much faster, digital equipment: the Geheimschreiber, manufactured by Siemens, and the Schlüsselzusatz, manufactured by Lorenz. These devices, known collectively as Fish, and derived from automatic Teletypewriter equipment, produced a sequence of 0s and 1s (the key) that was then added to the binary representation of an unenciphered (plaintext) message and output for transmission as ordinary 5-bit Teletypewriter tape. The machine's 12 code wheels, of unequal length, were circumscribed by a combined total of 501 pins that could be shifted between 2 positions, giving the system $2^{501}$ (or about $10^{150}$) possible states. The key was added modulo 2 to the plaintext message (counting by 2 the way we count hours by 12, so that $0 + 1 = 1$ and $1 + 1 = 0$), with 1 and 0 represented by the presence or absence of a hole in the tape. Adding the key to the enciphered text a second time would return the original text.

Fish traffic was beyond the reach of the electromechanical bombes. Electronics was the only hope of catching up. A series of machines named "Heath Robinson" were built on the principle that by simultaneously scanning two different (and relatively prime) lengths of punched paper tape as continuous loops, all possible combinations of the two sequences could be compared. Based on standard teleprinter tape and standard 5-bit teleprinter code, but running at high speed through photoelectric heads, the Heath Robinsons used electronic circuits to compare the two sequences, but it was difficult to maintain synchronization between two tapes.

It was then proposed by Thomas H. Flowers, an engineer working for the British Post Office's telecommunications research station at Dollis Hill, to eliminate one of the tapes by reading its sequence into an internal store consisting of fifteen hundred vacuum tubes, or, in British terminology, valves. This internal memory could then be synchronized to the sequence of pulses read from the other tape, which could be run without sprockets at much higher speeds by friction drive. "The tapes were read at 5,000 characters per second, [which] implies a tape speed of nearly 30 miles per hour," recalled Jack Good. "I regard the fact that paper teleprinter tape could be run at this speed as one of the great secrets of World War II!"[32]

With practice, it was possible to run loops of tape as much as two hundred feet in length. The new machine, code-named Colossus, was constructed under the supervision of Flowers and operated and programmed under the direction of Max Newman, who had set all these wheels in motion when he sparked Turing's original interest in the *Entscheidungsproblem* in 1935. Colossus was an electronic Turing machine, and if not yet universal, it had all the elements in place.

Colossus was so successful (and subspecies of Fish so prolific) that by the end of the war ten Colossi were in use, the later versions using 2,400 vacuum tubes. They were programmed by a plugboard and toggle switches at the back of the machine. "The flexible nature of the programming was probably proposed by Newman and perhaps also Turing, both of whom were familiar with Boolean logic, and this flexibility paid off handsomely," recalled I. J. Good. "The mode of operation was for a cryptanalyst to sit at Colossus and issue instructions to a Wren [Women's Royal Navy Service] for revised plugging, depending on what was printed on the automatic typewriter. At this stage there was a close synergy between man, woman, and machine."[33] As a step toward the modern computer, Colossus represented as great a leap as the ENIAC, and was both running and replicated while the one-of-a-kind ENIAC was still being built. Each Fish was a form of Turing machine, and the process by which the Colossi were used to break the various species of Fish demonstrated how the function (or partial function) of one Turing machine could be encoded for execution by another Turing machine. Since the British did not know the constantly changing state of the Fish, they had to guess. Colossus, trained to sense the direction of extremely faint gradients that distinguished enciphered German from random alphabetic noise, was the distant progenitor of the search engine: scanning the Precambrian digital universe for fragments of the missing key, until the pieces fit.

It was the alumni of Bletchley Park who were first to demonstrate a working stored-program computer (the Manchester Small Scale Experimental Machine, which ran its first program on June 21, 1948) and first to construct a kilobit-scale electronic memory (the electrostatic Williams tube). But the driving force behind computer development had shifted across the Atlantic, from the logical puzzle of cryptanalysis to the numerical design of hydrogen bombs. When Bletchley Park disbanded, the Official Secrets Act handicapped those who could not refer openly to their wartime work. The ENIAC was publicly unveiled in February of 1946, while the existence of Colossus would not be officially acknowledged for thirty-two years.

The extent, if any, of direct collaboration between Turing and von Neumann remains unknown. The British had a substantial nuclear weapons research group and, in consultation with von Neumann, made important contributions to Los Alamos. The Americans had a substantial group of cryptanalysts and, in consultation with Turing, contributed to the effort at Bletchley Park. Turing was in the United States between November 1942 and March 1943, and von Neumann was in England between February and July 1943. Both visits were secret missions, and there is no record of any wartime contact between the two pioneers.

"There was some hold-up about his job, which involved a useless period of idling in New York," says Sara Turing about Alan's three-month visit to the United States during the war. "He seems to have taken the opportunity to visit Princeton and probably saw something of the progress of computing machinery in the States. He returned in a destroyer or similar naval vessel and experienced a good tossing on the Atlantic." Jack Good remembers that upon Turing's return from the United States in March of 1943, he discussed "a problem about bags of gunpowder at the points in a plane with integer coordinates. Given the probability that the explosion of one bag will cause adjacent ones to explode, what is the probability that the explosion will extend to infinity?"[34] If you had to characterize the problem of determining the probability of a nuclear chain reaction without mentioning fission cross-sections, bags of gunpowder on the integer plane is a good mathematical fit.

J. R. Womersley, superintendent of the Mathematics Division of the National Physical Laboratory, who had read "On Computable Numbers" and become interested in Turing machines before the war, had been sent to the United States in the spring of 1945 to survey the latest (and still-secret) computer developments, including the Harvard Mark I tape-controlled electronic calculator, which he described in a letter home as "Turing in hardware." Womersley reported to Douglas R. Hartree, who reported to Sir Charles Darwin, director of NPL and grandson of *the* Charles Darwin. "JRW sees ENIAC and is given information about EDVAC by von Neumann and Goldstine," Womersley noted in 1946.[35] In June of 1945, Womersley met with Max Newman, and asked to meet Turing. "Meets Turing same day and invites him home," Womersley noted. "Shows Turing the first report on the EDVAC and persuades him to join N.P.L. staff, arranges interview and convinces Director and Secretary."[36] In September of 1945 Turing was assigned to study the ENIAC and EDVAC reports. Womersley notes, "Turing decides that mechanisms proposed for EDVAC are appropriate to his ideas."[37]

"I am of course in close touch with Turing," Max Newman wrote to von Neumann in early February 1946, explaining that "about eighteen months ago I had decided to try my hand at starting up a machine unit when I got out." Since technical details about the ENIAC were still restricted, and the existence of Colossus was not even acknowledged, a personal visit was arranged. "What I should most like is to come out and talk to you (for one thing I am still a bit cramped in discussing the past, and have to ask you not to put 2 and 2 together too accurately, and not to pass it on if you do.)"[38]

Von Neumann secured an Institute stipend for Newman to visit Princeton. In January 1947, Turing himself visited, reporting that "my visit to the U.S.A. has not brought any very important new technical information to light, largely, I think, because the Americans have kept us so well informed during the last year. . . . The Princeton group seem to me to be much the most clear headed and far sighted of these American organizations, and I shall try to keep in touch with them."[39]

The war had scrambled the origins of new inventions as completely as a message passing through an Enigma machine. Radar, cryptanalysis, antiaircraft fire control, computers, and nuclear weapons were all secret wartime projects that, behind the security barriers, enjoyed the benefit of free exchange of ideas, without concern for individual authorship or peer review. Von Neumann served the role of messenger RNA, helping to convey the best of the ideas—including the powers of Turing's Universal Machine. Among the bound volumes of the *Proceedings of the London Mathematical Society,* on the shelves of the Institute for Advanced Study library, there is one volume whose binding is disintegrated from having been handled so many times: Volume 42, with Turing's "On Computable Numbers," on pages 230–65.

Turing and von Neumann were as far apart, in everything except their common interest in computers, as it was possible to get. Von Neumann rarely appeared in public without a business suit; Turing was usually unkempt. "He tended to be slovenly," even his mother admits.[40] Von Neumann spoke freely and with great precision; Turing's speech was hesitating, as if words could not keep up with his thoughts. Turing stayed in hostels and was a competitive long-distance runner; Von Neumann was resolutely nonathletic and stayed in first-class hotels. Von Neumann had an eye for women, while Turing preferred men.

When von Neumann spoke about computing, he never mentioned artificial intelligence. Turing spoke about little else. Turing and von Neumann designed different styles of computer and wrote different styles of code. Von Neumann's design was captured in the "First Draft of a Report on the EDVAC" of June 30, 1945, and the "Preliminary Discussion of the Logical Design of an Electronic Computing Instrument" of June 28, 1946. Turing's design was captured in his "Proposed Electronic Calculator," written for the National Physical Laboratory in the brief interval between being shown the EDVAC report in September 1945 and the end of the year. He delivered a complete description of a million-cycle-per-second Automatic Computing Engine (ACE), accompanied by circuit diagrams, a detailed physical and logical analysis of the internal storage system, sample programs, detailed (if bug-ridden) subroutines, and even an estimated cost of £11,200.[41] As Sara Turing later explained, her son's goal was "to see his logical theory of a universal machine, previously set out in his paper 'Computable Numbers,' take concrete form."[42]

After a comparison of available forms of storage, ranging from punched paper tape through "cerebral cortex" to electrostatic storage tubes, Turing specified mercury-filled acoustic delay lines for high-speed storage. He estimated cost, access time, and "spacial economy" (in digits/liter) for all forms of storage, putting the cost of a cerebral cortex at £300 per annum—his King's College fellowship for the year. Viewed as part of a finite-state Turing machine, the delay line represented a continuous loop of tape, 1,000 squares in length and making 1,000 complete passes per second under the read/write head. Turing specified some 200 tubes, each storing 32 words of 32 bits each, for a total, "comparable with the memory capacity of a minnow," of about 200,000 bits.[43] Taking ten pages of the proposal to do so, Turing worked out the storage capacity, attenuation, noise, temperature sensitivity, and regenerative requirements, all from first principles.

Turing's ACE became bogged down in postwar bureaucracy and, like Babbage's Analytical Engine, was never built. In May 1950 a partial prototype (the Pilot ACE) was finally completed and "proved to be a far more powerful computer than we had expected," wrote J. H. Wilkinson, even though its mercury delay lines held only 300 words of 32 bits each. "Oddly enough much of its effectiveness sprang from what appeared to be weaknesses resulting from the economy in equipment that dictated its design."[44]

Turing, having had a taste of wartime "action this day" at Bletchley Park, grew impatient with the NPL administration, and the administration grew frustrated with Turing's tendency to leap ahead. "Our big computing engine . . . has now got to the stage of ironmongery," Sir Charles Darwin wrote to his superiors in July 1947, explaining that Womersley and Turing "are both agreed that it would be best that Turing should go off it for a spell."

"He wants to extend his work on the machine still further towards the biological side," Darwin continued. "I can best describe it by saying that hitherto the machine has been planned for work equivalent to that of the lower parts of the brain, and he wants to see how much a machine can do for the higher ones; for example, could a machine be made that could learn by experience?" Darwin finally got to the substance of his request. Turing, he explained, "would be content with something like half-pay . . . and indeed said that he would really prefer it, because if he were earning full pay, he would feel that 'I ought not to play tennis in the morning, when I want to.' "[45]

Turing took a leave of absence from NPL, returning to his King's College fellowship for a year before resigning from NPL in May 1948 to join Max Newman's computing group at Manchester University, where his interests in mechanical intelligence found free rein. "An unwillingness to admit the possibility that mankind can have any rivals in intellectual power," Turing wrote in his sabbatical report submitted to the NPL in 1948, "occurs as much amongst intellectual people as amongst others: they have more to lose."[46]

Turing's approach to machine intelligence was as unencumbered as his approach to computable numbers ten years before. He continued, once again, from where Gödel had left off. Does the incompleteness of formal systems limit the abilities of computers to duplicate the intelligence and creativity of the human mind? Turing summarized the essence (and weakness) of this convoluted argument in 1947, saying that "in other words then, if a machine is expected to be infallible, it cannot also be intelligent."[47] Instead of trying to build infallible machines, we should be developing fallible machines able to learn from their mistakes.

"The argument from Gödel's and other theorems rests essentially on the condition that the machine must not make mistakes," he explained. "But this is not a requirement for intelligence."[48] Turing made several concrete proposals. He suggested incorporating a random-number generator to create what he referred to as a "learning machine," granting the computer the ability to take a guess and then either reinforce or discard the consequent results. If guesses were applied to modifications in the computer's own instructions, a machine could then learn to teach itself. "What we want is a machine that can learn from experience," he wrote. "The possibility of letting the machine alter its own instructions provides the mechanism for this."[49] He pointed out that "paper interference" with a universal machine was equivalent to "screwdriver interference" with actual parts. In 1949, while developing the Manchester Mark I (prototype for the Ferranti Mark 1, the first stored-program electronic digital computer to be commercially produced), Turing designed a random-number generator that instead of producing pseudo-random numbers by a numerical process included a source of truly random electronic noise. This avoided von Neumann's "state of sin."

Turing also explored the possibilities of "unorganized Machines . . . which are largely random in their construction [and] made up from a rather large number N of similar units."[50] He considered a simple model with units capable of two possible states connected by two inputs and one output each, concluding that "machines of this character can behave in a very complicated manner when the number of units is large." He showed how such unorganized machines ("about the simplest model of a nervous system") could be made self-modifying and, with proper upbringing, could become more complicated than anything that could be otherwise engineered.[51] The human brain must start out as such an unorganized machine, since only in this way could something so complicated be reproduced.

Turing drew a parallel between intelligence and "the genetical or evolutionary search by which a combination of genes is looked for, the criterion being survival value. The remarkable success of this search confirms to some extent the idea that intellectual activity consists mainly of various kinds of search."[52] Evolutionary computation would lead to truly intelligent machines. "Instead of trying to produce a programme to simulate the adult mind, why not rather try to produce one which simulates the child's?" he asked. "Bit by bit one would be able to allow the machine to make more and more 'choices' or 'decisions.' One would eventually find it possible to program it so as to make its behaviour the result of a comparatively small number of general principles. When these became sufficiently general, interference would no longer be necessary, and the machine would have 'grown up.' "[53]

Turing gave provocative hints about what might lie ahead. "I asked him under what circumstances he would say that a machine is conscious," Jack Good recalled in 1956. "He said that if the machine was liable to punish him for saying otherwise then he would say that it was conscious." Lyn Newman remembers long discussions between Max Newman and Turing over how to build machines that would modify their own programming and learn from their mistakes. "When I heard Alan say of further possibilities 'Wh–wh–what will happen at that stage is that we shan't understand how it does it, we'll have lost track'—I did find it a most disturbing prospect," she reported in 1949. Jack Good would later explain that "the ultraintelligent machine . . . is a machine that believes people cannot think."[54]

Digital computers are able to answer most—but not all—questions stated in finite, unambiguous terms. They may, however, take a very long time to produce an answer (in which case you build faster computers) or it may take a very long time to ask the question (in which case you hire more programmers). Computers have been getting better and better at providing answers—but only to questions that programmers are able to ask. What about questions that computers can give useful answers to but that are difficult to define?

In the real world, most of the time, finding an answer is easier than defining the question. It is easier to draw something that looks like a cat than to define what, exactly, makes something look like a cat. A child scribbles indiscriminately, and eventually something appears that resembles a cat. An answer finds a question, not the other way around. The world starts making sense, and the meaningless scribbles (and unused neural connections) are left behind. "I agree with you about 'thinking in analogies,' but I do not think of the brain as 'searching for analogies' so much as having analogies forced upon it by its own limitations," Turing wrote to Jack Good in 1948.[55]

Random search can be more efficient than nonrandom search— something that Good and Turing had discovered at Bletchley Park. A random network, whether of neurons, computers, words, or ideas, contains solutions, waiting to be discovered, to problems that need not be explicitly defined. It is easier to find explicit answers than to ask explicit questions. This turns the job of the programmer upside down. "An argument in favor of building a machine with initial randomness is that, if it is large enough, it will contain every network that will ever be required," advised Good, speaking to IBM in 1958.[56]

The paradox of artificial intelligence is that any system simple enough to be understandable is not complicated enough to behave intelligently, and any system complicated enough to behave intelligently is not simple enough to understand. The path to artificial intelligence, suggested Turing, is to construct a machine with the curiosity of a child, and let intelligence evolve.

How to begin to realize what Turing imagined—a machine that would be able to answer all answerable questions that anyone could ask? The computable functions are easy. Beginning with addition (or subtraction, its binary complement), we have, subroutine by subroutine, been building the library from there. What about questions that have answers, but no explicit, algorithmic map or questions, such as determining molecular structure from X-ray diffraction patterns, that have an asymmetric map?

One approach is to start with the questions, and search for the answers. Another approach is to start with the answers and search for the questions. Because it is easier (and more economical) to collect answers (which are already encoded) than to ask questions (which have to be encoded), the first step would be to crawl through the matrix and collect the meaningful strings. Unfortunately, in a matrix of $10^{22}$ bits, the number of meaningful strings is a number too large to search, let alone collect. It is too large a number even to write down. Fortunately, there is a key. Human beings and machines have already done much of the work, filing away meaningfully encoded strings since the beginning of the digital universe and, since the dawn of the Internet, giving them unique numerical addresses.

To collect the answers, you do not have to search through the entire matrix; you only have to crawl through the vastly smaller number of valid addresses and collect the resulting strings. The result is an indexed list (within your machine's "state of mind," to use Turing's language) of a significant fraction of the meaningful answers in the digital universe. With two huge deficiencies: you don't have any questions—you have only answers—and you have no clue where the meaning is.

Where do you go to get the questions, and how do you find where the meaning is? If, as Turing imagined, you have the mind of a child, you ask people, you guess, and you learn from your mistakes. You invite people to submit questions—keeping track of all submissions—and, starting with simple template-matching, suggest possible answers from your indexed list. People click more frequently on the results that provide more meaningful answers, and with simple bookkeeping, meaning, and the map between questions and answers, begins to accumulate over time. Are we searching the search engines, or are the search engines searching us?

Search engines are *copy* engines: replicating everything they find. When a search result is retrieved, the data are locally replicated: on the host computer and at various servers and caches along the way. Data that are widely replicated, or associated frequently by search requests, establish physical proximity that is manifested as proximity in time. More meaningful results appear higher on the list not only because of some mysterious, top-down, weighting algorithm, but because when microseconds count, they are closer, from the bottom up, in time. Meaning just seems to "come to mind" first.

An Internet search engine is a finite-state, deterministic machine, except at those junctures where people, individually and collectively, make a nondeterministic choice as to which results are selected as meaningful and given a click. These clicks are then immediately incorporated into the state of the deterministic machine, which grows ever so incrementally more knowledgeable with every click. This is what Turing defined as an oracle machine.

Instead of learning from one mind at a time, the search engine learns from the collective human mind, all at once. Every time an individual searches for something, and finds an answer, this leaves a faint, lingering trace as to where (and what) some fragment of meaning is. The fragments accumulate and, at a certain point, as Turing put it in 1948, "the machine would have 'grown up.' "[57]

# FOURTEEN

# Engineer's Dreams

*If, by a miracle, a Babbage machine did run
backwards, it would not be a computer, but
a refrigerator.*

—I. J. Good, 1962

"I REMEMBER one day walking out the back door of that little brick building, and here's Julian lying under this little Austin, welding a hole in a gas tank," remembers Willis Ware. "And he said 'Nope! It won't explode!' And he had some perfectly reasonable explanation for why it wouldn't explode, based on the principles of physics."[1]

Julian Bigelow was a hands-on engineer, from the first batch of war-surplus 6J6 vacuum tubes and transplanted ENIAC technicians to the lead-acid battery house built when there turned out to be too many transient voltage fluctuations at the end of Olden Lane for the new computer to be connected directly to the grid. "The actual machine that will be completed soon, and which has quite exceptional characteristics, is, in its physical embodiment, much more Bigelow's personal achievement than anyone else's," von Neumann reported in 1950, urging the Institute's executive committee to break with precedent by granting an academic appointment to an engineer.[2]

Von Neumann pushed the exception through. "Bigelow's career has deviated from the conventional academic norm considerably," he argued. "This is, apart from economic reasons and the war, due to the fact that his field lies somewhere between a number of recognized scientific fields, but does not coincide with any of them."[3] Computer science, as a recognized discipline, did not yet exist. Julian Bigelow and Herman Goldstine were awarded permanent memberships in the School of Mathematics on December 1, 1950, at salaries of $8,500 per year. Their objective was not so much to build better or faster computers, but, as Bigelow put it, to pursue "the relationship between logic, computability, perhaps machine languages, and the things that you can find out scientifically, now that this tool is available."[4]

As ill equipped as it was for engineering, the Institute was well equipped to accommodate visitors bringing problems to run on the new machine. The housing project was adjacent to the computer building, and there was no established research group defending its turf. Digital computing "would cleanse and solve areas of obscurity and debate that had piled up for decades," Bigelow believed. "Those who really understood what they were trying to do would be able to express their ideas as coded instructions . . . and find answers and demonstrate explicitly by numerical experiments. The process would advance and solidify knowledge and tend to keep men honest."[5]

"The reason von Neumann made Goldstine and me permanent members," Bigelow explains, "was that he wanted to be sure that two or three people whose talent he respected would be around no matter what happened, for this effort." Von Neumann was less interested in building computers, and more interested in what computers could do. "He wanted mathematical biology, he wanted mathematical astronomy, and he wanted earth sciences." Thanks to the computer, the Institute could do applied science without having to build laboratories. The prevailing culture might even change. "We would have the greatest school of applied science in the world," Bigelow hoped. "We could show the theoreticians that we could find out the answer to their number theoretic problems, their problems in physics, their problems in solid state, and their problems in mathematical economics. We would do planning, we would do things that would be known for centuries, you see."[6]

Bigelow's optimism was short-lived. When President Eisenhower appointed von Neumann to the Atomic Energy Commission in October 1954, the computer project went into decline. Not only did the Institute lose von Neumann, but they also lost much of the funding that had been provided, with few strings attached, by the AEC. With von Neumann appointed to the commission, the AEC could no longer give the Institute anything it wished. "We had nobody we could go to without all this fear of conflict of interest," explains Goldstine. "It worked very much to our detriment to have all this influence, because we couldn't exercise it."[7]

IBM was less constrained. "IBM people kept coming almost weekly to look at the machine's development," remembers Thelma Estrin. IBM retained von Neumann as a consultant, and began developing their first fully electronic computer, the IBM 701, "a carbon copy of our machine," according to Bigelow, "even down to and including the

Williams memory tubes." By 1951, IBM had become "sufficiently interested," as Oppenheimer put it, "to want to give the Institute $20,000 a year for a period of five years with no strings attached."[8]

The computer project was caught between those who welcomed this ability to attract outside funding and those who thought the Institute, now that the war was over, should abstain from government or industry support. Marston Morse believed that the Institute was not the place to build machines. Oswald Veblen welcomed digital computing but objected to hydrogen bombs. Oppenheimer tried to appear neutral, saying only that computing at the Institute should either "be endowed and expanded, and take its proper place in the academic structure," or be shut down. "At that time, having Oppenheimer for something was exactly the way to get it stopped by all the rest of the faculty," Bigelow observed.[9]

Freeman Dyson, thirty-one years old and just beginning his second year as a professor, was commissioned "to collect a few outside opinions and views on a question of long-range policy which we feel we ought to make up our minds about. Namely, what is a proper role for the Institute to play in the fields of applied mathematics and electronic computing?"[10] The immediate question was whether meteorologist Jule Charney should be offered a permanent appointment. The long-term question was what to do with the Electronic Computer Project, which, in von Neumann's absence, was being kept on life support.

Charney's group was a victim of its own success. The numerical forecasting methods pioneered at the Institute were being adopted by weather services all over the world. Multiple copies of the IAS computer were being built, with a constant stream of visitors coming to Princeton to learn the new techniques. Internal sentiment, even among the mathematicians, sided against the computer, and the outside reviewers generally agreed the machine belonged somewhere else. "It is time that Von Neumann revolutionized some other subject; He has spent rather too long in the field of automatic computation," recommended James Lighthill, F.R.S.[11] Founding trustees Herbert Maass and Samuel Leidesdorf, who believed that a better understanding of the weather was the kind of knowledge that the Bambergers had hoped to advance, sought to preserve the meteorology project, but were overruled.

"The use of computers was a very funny subject in the early days," recalls British mathematician and computer scientist David Wheeler, concerning mathematics in Princeton at that time. "It was slightly beneath the dignity of mathematicians. Engineers were used to doing calculations, whereas mathematicians weren't."[12] After the dust had settled, Freeman Dyson spoke up. "When von Neumann tragically died, the snobs took their revenge and got rid of the computing project root and branch," he said at the dedication of the university's new Fine and Jadwin halls, equipped with multiple computers, in 1970. "The demise of our computer group was a disaster not only for Princeton but for science as a whole. It meant that there did not exist at that critical period in the 1950s an academic center where computer people of all kinds could get together at the highest intellectual level. . . . We had the opportunity to do it, and we threw the opportunity away." It would be twenty-two years before the next computer—a Hewlett Packard model 9100-B programmable calculator, sequestered for the use of the astronomers in the basement of Building E—arrived at the IAS.[13]

Bigelow's hopes of keeping the Institute at the leading edge of the computational revolution came to a halt. Von Neumann, and the excitement he had generated in 1946, were gone and not coming back. Klári had long wanted to leave Princeton for the West Coast; now the Institute's ambivalence toward the computer project, and lingering divisions over the Oppenheimer security hearings, began to wear down Johnny as well. Veblen would not forgive von Neumann for joining the Atomic Energy Commission, a situation that, according to Klári, "grew into a pathetic sorrow in Johnny's last years."[14] Even some of von Neumann's closest friends began to question how someone who had supported Oppenheimer against his AEC accusers could now side with ringleader Strauss. Oppenheimer himself was more forgiving. "I shall always remember Robert," says Klári, "summing up his attitude in a very simple statement: 'There have to be good people on both sides.'"[15]

"The lines were drawn and after the first flurry of excitement it became clear that we did not belong in Princeton any more," Klári explains. "The highly emotional atmosphere in Princeton annoyed Johnny no end. He wanted to work on improved designs for computers, or on the urgency of expanding missile programs—in other words, on anything that was a real intellectual challenge instead of debating interminably who had done what and why and how."[16] Von Neumann believed that conflicted loyalties during the development of the atomic bomb should be left behind. "We were all little children with respect to the situation which had developed, namely, that we suddenly were dealing with something with which one could blow up the world," he had testified, in defense of Oppenheimer, in 1954. "We had to make our rationalization and our code of conduct as we went along."[17]

Two weeks later, while in Los Angeles on air force strategic missile business and staying at the Miramar Bungalows in Santa Monica, von Neumann met with Paul A. Dodd, dean of letters and sciences at UCLA, who offered him a special interdisciplinary position, with no teaching responsibilities, as professor-at-large. "They would give me 'everything' I want," he reported to Klári on May 16, adding that "they do not mind if I do consultations for Industry as well." Dodd also assured von Neumann that he would be able to spend as much time at the Scripps Institution of Oceanography in La Jolla as he wished. Von Neumann agreed to refuse all other offers until further discussions with UCLA, and Dodd agreed to keep the matter confidential, since von Neumann had not informed the Institute that he was leaving, and, as he put it to Klári, "I do not want to look like a deserter or a traitor to them."[18]

"Since we first decided, 1-½ years ago, that it would be better to leave Princeton, I see for the first time concrete evidence for doing it," he wrote to Klári the next day.[19] As the negotiations continued, he secured appointments at UCLA for both Jule Charney and Norman Phillips, with assurances that a state-of-the-art computing laboratory would be established, building on the resources that already existed, in Los Angeles, at the Institute for Numerical Analysis and at RAND. Von Neumann would finally be able to assemble the cross-disciplinary information systems laboratory that he and Norbert Wiener had proposed in 1946, before the push to develop the hydrogen bomb had drawn a curtain between them and their work. If the California laboratory had been established, the second half of the twentieth century might have taken a quite different course. "Someone should write a novel for the future which is in the past," says von Neumann's Los Alamos colleague Harris Mayer. "And that is: what would science and mathematics be if Fermi and Johnny von Neumann didn't die young?"[20]

Spring of 1955 found Johnny and Klári settled into a small but comfortable house in Georgetown in Washington, D.C., Johnny having made the journey from postdoctoral immigrant to a presidential appointment in just twenty-five years. The interlude in Washington promised to lead to even more productive years ahead. "I want to become independent of the regulated academic life," von Neumann had written to Klári from Los Alamos in 1943—a goal that was finally within reach. It was not to be. "On the 9th of July of that exceptionally hot summer, even for Washington," Klári remembers, "Johnny collapsed while talking on the phone to Lewis Strauss."[21]

On August 2 he was diagnosed with advanced, metastasizing cancer, discovered in his collarbone, and underwent emergency surgery. By November his spine was affected, and on December 12 he addressed the National Planning Association in Washington, D.C., the last speech he gave standing up. "The best we can do is to divide all processes into those things which can be better done by machines and those which can be better done by humans," he advised, "and then invent methods by which to pursue the two."[22] He was confined to a wheelchair in January 1956. "The last scientific discussion we had was on New Year's Eve, when I told him of a new theory that I had on the dynamics of the mature hurricane," Jule Charney remembers. "He was in bed all that New Year's Eve Day. The next morning he walked downstairs to see Elinor and me off to Princeton. On the way back upstairs he fell, and never walked again."[23]

In March he entered Walter Reed Hospital, where he spent the remaining eleven months. "He discussed his illness with the doctors in such a matter-of-fact way and with such a wealth of medical knowledge that he pushed them into telling him the entire truth—which was very grim," Klári reported. He received a constant stream of visitors, and was placed in the same wing as the Eisenhower Suite. Air force colonel Vincent Ford was assigned, with several airmen under him, to assist full-time. Lewis Strauss would later recall "the extraordinary picture, of sitting beside the bed of this man, in his [if]ties, who had been an immigrant, and there surrounding him, were the Secretary of Defense, the Deputy Secretary of Defense, the Secretaries of Air, Army, Navy, and the Chiefs of Staff."[24]

His mental faculties deteriorated, bit by bit. "He wanted somebody to talk with him," says Julian Bigelow, "and Klári who I think knew me better than she knew anybody else, asked me to go see him at the Walter Reed Hospital. So I went every weekend for almost a year." Strauss obtained a personal service contract from the AEC to pay Bigelow's travel expenses and, at von Neumann's request, reinstated Bigelow's Q-level security clearance (on June 27, 1956). Bigelow visited with von Neumann, read science journals to him, and fielded his questions until the end. "It was a terrible experience to see him going downhill."[25]

Stan Ulam visited whenever he could. "He never complained about pain, but the change in his attitude, his utterances, his relations with Klári, in fact his whole mood at the end of his life were heartbreaking," he remembers. "At one point he became a strict Catholic. A Benedictine monk visited and talked to him. Later he asked for a Jesuit. It was obvious that there was a great gap between what he would discuss verbally and logically with others, and what his inner thoughts and worries about himself were." Von Neumann's scientific curiosity and his memory were the last things he let go. "A few days before he died," adds Ulam, "I was reading to him in Greek from his worn copy of Thucydides a story he liked especially about the Athenians' attack on Melos, and also the speech of Pericles. He remembered enough to correct an occasional mistake or mispronunciation on my part."[26]

Marina von Neumann was twenty-one years old, about to get married, and at the beginning of her own career. Her father "clearly realized that the illness had gone to his brain and that he could no longer think, and he asked me to test him on really simple arithmetic problems, like seven plus four, and I did this for a few minutes, and then I couldn't take it anymore; I left the room," she remembers, overcome by "the mental anguish of recognizing that that by which he defined himself had slipped away."[27]

"I once asked him," she adds, "when he knew he was dying, and was very upset, that 'you contemplate with equanimity eliminating millions of people, yet you cannot deal with your own death.' And he said, 'That's entirely different.'" Nicholas Vonneumann believes that his brother asked for a Catholic priest because he wanted someone he could discuss the classics with. "With our background it would have been inconceivable to turn overnight into a devout Catholic," he says.[28]

"I don't believe that for a minute," Marina counters. "My father told me, in so many words, once, that Catholicism was a very tough religion to live in but it was the only one to die in. And in some part of his brain he really hoped that it might guarantee some kind of personal immortality. That was at war with other parts of his brain, but I'm sure he had Pascal's wager in mind." The sudden conversion was unsettling to Klári, the Ulams, and Lewis Strauss. "The tragedy of Johnny continues to affect me very strongly," Ulam wrote to Strauss on December 21, 1956. "I am also deeply perturbed about the religious angle as it developed. Klári . . . told me about her own and your attempts to moderate anything that might appear in writing about it."[29]

Bigelow "found things beyond communication" when he visited on December 27–28. "Before his death he lost the will or capacity to speak," Klári explains. "To those of us who knew him well he could communicate every wish, will or worry through those marvelously expressive eyes which never lost their luster and vivacity until the very end."[30]

Von Neumann died on February 8, 1957, and was buried in Princeton on February 12. His colleagues at the Institute ordered (for "about $15") a flat arrangement of daffodils to be laid on the grave. After a brief Catholic service, the graveside eulogy was delivered by Lewis Strauss. A detailed memorial was delivered by Stan Ulam in the *Bulletin of the American Mathematical Society* the following spring. Ulam was now left alone to witness the revolutions in both biology and computing that von Neumann had launched but would not see fulfilled. "He died so prematurely, seeing the promised land but hardly entering it," Ulam wrote in 1976.[31]

The remaining Electronic Computer Project staff were scattered to industry, to the national laboratories, and among a growing number of university computer science departments, where derivatives of the IAS machine were being built. Julian Bigelow was determined to stay put. Although Marston Morse had apologized, in the end, for "the conclusion of my mathematical colleagues with regard to the computer," the mathematicians never changed their mind about engineers. "There really was a caste system," Morris Rubinoff remembers. "You could separate out different types of members and different types of full members on the basis of their willingness to engage in conversation or even associate socially with the engineers."[32]

Bigelow received job offers from UCLA, RAND, NYU, RCA, the University of Michigan, Hughes Aircraft, the Defense Mapping Agency, and even the Albert Einstein College of Medicine—all of which he refused. "Julian was a man who would take his soldering iron in there and just do it," says Martin Davis. "He would have been much better off if he had never got that tenure [at IAS]. He would have got a job in industry, where he really would have flourished."[33] The Institute could not force him to resign, but they refused to increase his salary.

He survived on $9,000 a year, supplemented with occasional consulting fees, while raising three children and later taking care of his wife, Mary, who became gravely ill. Klári suggested he be appointed editor of von Neumann's unpublished papers on computing and automata, but nothing came of this. Bigelow published little over the next forty years. Although he remained the most direct link to von Neumann's unfinished thoughts about the future of computing, these ideas, already attenuated by von Neumann's untimely death and refusal to publish incomplete work, were silenced further by Bigelow's exile at the IAS.

Bigelow's insights into the future of computing were more than lag functions reversed to project forward in time. Turing's one-dimensional model, however powerful, and von Neumann's two-dimensional implementation, however practical, might be only first steps on the way to something else. "If you actually tried to build a machine the way Turing described it," Bigelow explained, "you would be spending more time rushing back and forth to find places on a tape than you would doing actual numerical work or computation."[34] The von Neumann model might turn out to be similarly restrictive, and the solutions arrived at between 1946 and 1951 should no more be expected to persist indefinitely than any one particular interpretation of nucleotide sequences would be expected to persist for three billion years. The last thing either Bigelow or von Neumann would have expected was that long after vacuum tubes and cathode-ray tubes disappeared, digital computer architecture would persist largely unchanged from 1946.

Once the IAS computer was completed, it was possible to look back at the compromises that were made to get it running—and Bigelow did. "The design of an electronic calculating machine . . . turns out to be a frustrating wrestling-match with problems of interconnectability and proximity in three dimensions of space and one dimension of time," he wrote in 1965, in one of the few glimpses into his thinking in the post-MANIAC years.[35] Why have so few of the alternatives received any serious attention for sixty-four years? If you examined the structure of a computer, "you could not possibly tell what it is doing at any moment," Bigelow explained. "The importance of structure to how logical processes take place is beginning to diminish as the complexity of the logical process increases." Bigelow then pointed out that the significance of Turing's 1936 result was "to show in a very important, suggestive way how trivial structure really is."[36] Structure can always be replaced by code.

"Serial order along the time axis is the customary method of carrying out computations today, although . . . in forming any model of real world processes for study in a computer, there seems no reason why this must be initiated by pairing computer-time-sequences with physical time parameters of the real-world model," observed Bigelow, who had puzzled over how to map physics to computation ever since being given Wiener's problem of predicting the path of an evasive airplane in 1941, and von Neumann's problem of predicting the explosion of a bomb in 1946. "It should also be possible to trace backward or forward from results to causes through any path-representation of the process," he noted, adding that "it would seem that the time-into-time convention ordinarily used is due to the . . . humans interpreting the results."[37]

"A second result of the habitual serial-time sequence mode and of the large number of candidate cells waiting to participate in the computation at the next opportunity, if it becomes their turn, is the emergence of a particularly difficult identification problem . . . because of the need to address an arbitrary next candidate, and to know where it is in machine-space," Bigelow continued, explaining how the choice of serial dependence in time has led to computers "built of elements that are, to a large extent, strictly independent across space." This, in turn, requires that communication between individual elements be conducted "by means of explicit systems of tags characterizing the basically irrelevant geometric properties of the apparatus, known as 'addresses.' Accomplishment of the desired time-sequential process on a given computing apparatus turns out to be largely a matter of specifying sequences of addresses of items which are to interact."[38]

The 32-by-32 matrix instituted in 1951 addressed 1,024 different memory locations, each containing a string of 40 bits. The address matrix grew explosively over the next sixty years. Today's processors keep track of billions of local addresses from one nanosecond to the next—while the nonlocal address space is expanding faster than the protocol for assigning remote addresses has been able to keep up. A single incorrect address reference can bring everything to a halt.

Forced to focus undivided attention on getting the address references and instruction sequences exactly right, a computer, despite billions of available components, does only one thing at a time.

"The modern high speed computer, impressive as its performance is from the point of view of absolute accomplishment, is from the point of view of getting the available logical equipment adequately engaged in the computation, very inefficient indeed," Bigelow observed. The individual components, despite being capable of operating continuously at high speed, "are interconnected in such a way that on the average almost all of them are waiting for one (or a very few of their number) to act. The average duty cycle of each cell is scandalously low."[39]

To compensate for these inefficiencies, processors execute billions of instructions per second. How can programmers supply enough instructions—and addresses—to keep up? Bigelow viewed processors as organisms that digest code and produce results, consuming instructions so fast that iterative, recursive processes are the only way that humans are able to generate instructions fast enough. "Electronic computers follow instructions very rapidly, so that they 'eat up' instructions very rapidly, and therefore some way must be found of forming batches of instructions very efficiently, and of 'tagging' them efficiently, so that the computer is kept effectively busier than the programmer," he explained. "This may seem like a highly whimsical way of characterizing a logically deep question of how to express computations to machines. However, it is believed to be not far from an important central truth, that highly recursive, conditional and repetitive routines are used because they are notationally efficient (but not necessarily unique) as descriptions of underlying processes."[40]

Bigelow questioned the persistence of the von Neumann architecture and challenged the central dogma of digital computing: that without programmers, computers cannot compute. He (and von Neumann) had speculated from the very beginning about "the possibility of causing various elementary pieces of information situated in the cells of a large array (say, of memory) to enter into a computation process without explicitly generating a coordinate address in 'machine-space' for selecting them out of the array."[41]

Biology has been doing this all along. Life relies on digitally coded instructions, translating between sequence and structure (from nucleotides to proteins), with ribosomes reading, duplicating, and interpreting the sequences on the tape. But any resemblance ends with the different method of addressing by which the instructions are carried out. In a digital computer, the instructions are in the form of COMMAND (ADDRESS) where the address is an exact (either absolute or relative) memory location, a process that translates informally into "DO THIS with what you find HERE and go THERE with the result." Everything depends not only on precise instructions, but also on HERE, THERE, and WHEN being exactly defined.

In biology, the instructions say, "DO THIS with the next copy of THAT which comes along." THAT is identified not by a numerical address defining a physical location, but by a molecular template that identifies a larger, complex molecule by some smaller, identifiable part. This is the reason that organisms are composed of microscopic (or near-microscopic) cells, since only by keeping all the components in close physical proximity will a stochastic, template-based addressing scheme work fast enough. There is no central address authority and no central clock. Many things can happen at once. This ability to take general, organized advantage of local, haphazard processes is the ability that (so far) has distinguished information processing in living organisms from information processing by digital computers.

Our understanding of life has deepened with our increasing knowledge of the workings of complex molecular machines, while our understanding of technology has diminished as machines approach the complexity of living things. We are back to where Julian Bigelow and Norbert Wiener left off, at the close of their precomputer "Behavior, Purpose and Teleology," in 1943. "A further comparison of living organisms and machines . . . may depend on whether or not there are one or more qualitatively distinct, unique characteristics present in one group and absent in the other," they concluded. "Such qualitative differences have not appeared so far."[42]

As the digital universe expanded, it collided with two existing stores of information: the information stored in genetic codes and the information stored in brains. The information in our genes turned out to be more digital, more sequential, and more logical than expected, and the information in our brains turned out to be less digital, less sequential, and less logical than expected.

Von Neumann died before he had a chance to turn his attention to the subject of genetic code, but near the end of his life he turned his attention to the question of information processing in the brain. His final, unfinished manuscript, for the upcoming Silliman Memorial Lectures at Yale University, gave "merely the barest sketches of what he planned to think about," according to Ulam, and was edited by Klári and published posthumously as *The Computer and the Brain*.[43] Von Neumann sought to explain the differences between the two systems, the first difference being that we understand almost everything that is going on in a digital computer and almost nothing about what is going on in a brain.

"The message-system used in the nervous system . . . is of an essentially statistical character," he explained.

> What matters are not the precise positions of definite markers, digits, but the statistical characteristics of their occurrence . . . a radically different system of notation from the ones we are familiar with in ordinary arithmetics and mathematics. . . . Clearly, other traits of the (statistical) message could also be used: indeed, the frequency referred to is a property of a single train of pulses whereas every one of the relevant nerves consists of a large number of fibers, each of which transmits numerous trains of pulses. It is, therefore, perfectly plausible that certain (statistical) relationships between such trains of pulses should also transmit information. . . . Whatever language the central nervous system is using, it is characterized by less logical and arithmetical depth than what we are normally used to [and] must structurally be essentially different from those languages to which our common experience refers.[44]

The brain is a statistical, probabilistic system, with logic and mathematics running as higher-level processes. The computer is a logical, mathematical system, upon which higher-level statistical, probabilistic systems, such as human language and intelligence, could possibly be built. "What makes you so sure," asked Stan Ulam, "that mathematical logic corresponds to the way we think?"[45]

In the age of vacuum tubes, it was inconceivable that digital computers would operate for hundreds of billions of cycles without error, and the future of computing appeared to belong to logical architectures and systems of coding that would be tolerant of hardware failures over time. In 1952, codes were small enough to be completely debugged, but hardware could not be counted on to perform consistently from one kilocycle to the next. This situation is now reversed. How does nature, with both sloppy hardware and sloppy coding, achieve such reliable results? "There is reason to suspect that our predilection for linear codes, which have a simple, almost temporal sequence, is chiefly a literary habit, corresponding to our not particularly high level of combinatorial cleverness, and that a very efficient language would probably depart from linearity," von Neumann suggested in 1949.[46] The most successful new developments in computing—search engines and social networks—are nonlinear hybrids between digitally coded and pulse-frequency-coded systems, and are leaving linear, all-digital systems behind.

In a digitally coded system, each digit has a precise meaning, and if even one digit is misplaced, the computation may produce a wrong answer or be brought to a halt. In a pulse-frequency-coded system, meaning is conveyed by the frequency at which pulses are transmitted between given locations—whether those locations are synapses within a brain or addresses on the World Wide Web. Shifting the frequency shifts the meaning, but the communication, storage, and interpretation of information is probabilistic and statistical, independent of whether each bit is in exactly the right place at exactly the right time. Meaning resides in what connects where, and how frequently, as well as by being encoded in the signals conveyed. As von Neumann explained in 1948, "A new, essentially logical, theory is called for in order to understand high-complication automata and, in particular, the central nervous system. It may be, however, that in this process logic will have to undergo a pseudomorphosis to neurology to a much greater extent than the reverse."[47]

The reliability of monolithic microprocessors and the fidelity of monolithic storage postponed the necessity for this pseudomorphosis far beyond what seemed possible in 1948. Only recently has this pseudomorphosis resumed its course. The von Neumann address matrix is becoming the basis of a non–von Neumann address matrix, and Turing machines are being assembled into systems that are not Turing machines. Codes—we now call them apps—are breaking free from the intolerance of the numerical address matrix and central clock cycle for error and ambiguity in specifying where and when.

The microprocessor, however, is here to stay, just as the advent of metazoan organisms did not bring the end of individual cells. Biological organisms are subdivided into cells, since the stochastic, template-based molecular addressing on which metabolism and replication depend works faster on a local scale. Technological organisms are also subdivided into cells (and processors subdivided into multiple cores), not only to isolate errors, but also because the numerical addressing on which digital processing depends can operate at nanosecond speed on only a local scale. Across larger domains—of both size and time—other forms of addressing and processing, and other architectures, are starting to evolve.

In the age of all things digital we are building analog computers again. Didn't analog computing disappear, in the age of the dinosaurs, with the superseding of the Bush differential analyzer by the ENIAC, when the race was on to perform high-speed arithmetic, leaving no doubt that digital computing would come out ahead? There are other benchmarks besides arithmetic, and Turing, von Neumann, and Bigelow, for all their contributions to the digital revolution, did not see analog computing as a dead end. Part of the problem, as Jack Good put it in 1962, is that "analogue computers are stupidly named; they should be named continuous computers." For real-world questions—especially ambiguous ones—analog computing can be faster, more accurate, and more robust, not only at computing the answers, but also at asking the questions and communicating the results. *Web 2.0* is our code word for the analog increasingly supervening upon the digital—reversing how digital logic was embedded in analog components, sixty years ago. Search engines and social networks are just the beginning—the Precambrian phase. "If the only demerit of the digital expansion system were its greater logical complexity, nature would not, for this reason alone, have rejected it," von Neumann admitted in 1948.[48]

Search engines and social networks are analog computers of unprecedented scale. Information is being encoded (and operated upon) as continuous (and noise-tolerant) variables such as frequencies (of connection or occurrence) and the topology of what connects where, with location being increasingly defined by a fault-tolerant template rather than by an unforgiving numerical address. Pulse-frequency coding for the Internet is one way to describe the working architecture of a search engine, and PageRank for neurons is one way to describe the working architecture of the brain. These computational structures use digital components, but the analog computing being performed by the system as a whole exceeds the complexity of the digital code on which it runs. The model (of the social graph, or of human knowledge) constructs and updates itself.

Complex networks—of molecules, people, or ideas—constitute their own simplest behavioral descriptions. This behavior can be more easily captured by continuous, analog networks than it can be defined by digital, algorithmic codes. These analog networks may be composed of digital processors, but it is in the analog domain that the interesting computation is being performed. "The purely 'digital' procedure is probably more circumstantial and clumsy than necessary," von Neumann warned in 1951. "Better, and better integrated, mixed procedures may exist."[49]

Analog is back, and here to stay.

# FIFTEEN

# Theory of Self-Reproducing Automata

*I see the problem not from the mathematical point of view, as, for instance, von Neumann did, but as an engineer. It may be better that there is almost no support for such ideas. Perhaps the devil is behind it, too.*

—Konrad Zuse, 1976

"THE CAMERA MOVES across the sky, and now the black serrated shape of a rocky island breaks the line of the horizon. Sailing past the island is a large, four-masted schooner. We approach, we see that the ship flies the flag of New Zealand and is named the *Canterbury*. Her captain and a group of passengers are at the rail, staring intently toward the east. We look through their binoculars and discover a line of barren coast."[1]

Thus begins *Ape and Essence,* Aldous Huxley's lesser-known masterpiece, set in the Los Angeles of 2108, after a nuclear war (in the year 2008) has devastated humanity's ability to reproduce high-fidelity copies of itself. On the twentieth of February 2108, the New Zealand Rediscovery Expedition to North America arrives among the Channel Islands off the California coast. The story is presented, in keeping with the Hollywood location, in the form of a film script. "New Zealand survived and even modestly flourished in an isolation which, because of the dangerously radioactive condition of the rest of the world, remained for more than a century almost absolute. Now that the danger is over, here come its first explorers, rediscovering America from the West."[2]

Huxley's dystopian vision was published in 1948, when a third world war appeared all but inevitable, the prospects little brightened by von Neumann's argument that the ultimate death toll could be minimized by launching a preventive attack. Although the exact mechanism by which genetic information was replicated had yet to be determined, there was no mistaking the effects of ionizing radiation on the transmission of instructions from one generation to the next. Huxley assumed that in the aftermath of nuclear war, the Darwinian evolution so championed by his grandfather, Thomas Huxley, would start to unwind.

Too-perfect replication may, in the end, be as much of a threat. Darwinian evolution depends on copies not always being exact. The Los Angeles of 2108 is as likely to be rendered socially unrecognizable not by the error catastrophe of *Ape and Essence*, but by its opposite: the ability to read genetic sequences into computers, replicate them exactly, and translate them back into living organisms, without a single bit misplaced along the way. A Los Angeles run by human beings able to specify the exact genetic characteristics of their offspring may be more terrifying than the Los Angeles governed by baboons that greeted the crew of the *Canterbury* in Huxley's 2108.

As primitive self-reproducing life forms adopted self-replicating polynucleotide symbionts as the carriers of hereditary information from one generation to the next, so might current life forms adopt computers as carriers of their genetic code. Nils Barricelli hinted at this in 1979, observing that nature has shown a tendency, among highly social organisms, "to separate the organisms of which the society is composed into two main classes, namely: the workers, specialized in carrying out all the tasks necessary for the society's survival except reproduction, and the carriers of hereditary information, whose function is to reproduce the society." This separation of reproductive function, Barricelli noted, "is common in highly organized societies of living organisms (queens and drones among ants, bees and termites, gametes among the species of cells forming a multicellular organism) except man, whose society is of relatively recent formation in biological terms."[3]

The powers of computers derive as much from their ability to copy as from their ability to compute. Sending an e-mail, or transferring a file, does not physically move anything; it creates a new copy somewhere else. A Turing machine is, by definition, able to make exact copies of any readable sequence—including its own state of mind and the sequence stored on its own tape. A Turing machine can, therefore, make copies of itself. This caught the attention of von Neumann at the same time as the Institute for Advanced Study computer project was being launched. "I did think a good deal about self-reproductive mechanisms," he wrote to Norbert Wiener in November 1946. "I can formulate the problem rigorously, in [the same way in] which Turing did it for his mechanisms." Von Neumann envisioned an axiomatic theory of self-reproduction, general enough to encompass both living organisms and machines, telling Wiener that "I want to fill in the details and to write up these considerations in the course of the next two months."[4]

This mathematical theory of self-reproduction had to be grounded in what could be directly observed. "I would, however, put on 'true' understanding the most stringent interpretation possible," he added. "That is, understanding the organism in the exacting sense in which one may want to understand a detailed drawing of a machine."[5] Individual molecules would be the only axiomatic parts. "It would be a mistake to aim at less than the complete determination of the charge distribution in the protein molecule—that is, a complete detailed determination of its geometry and structure," he wrote to Irving Langmuir, seeking to enlist the chemist's help. "Of course, the first really exciting structures, that is the first self-reproducing structures (plant virus and bacteriophages), are yet three powers of ten above the protein."[6]

Expanding the abilities of Turing's Universal Machine, von Neumann showed "that there exists an automaton B which has this property: If you provide B with a description of anything, it consumes it and produces two copies of the description." Von Neumann outlined this theory in a talk given in Pasadena, California, on September 20, 1948, more than four years before the details of how this is done in nature were revealed by Franklin, Watson, and Crick. "For the 'self-reproduction' of automata, Turing's procedure is too narrow in one respect only," he explained. "His automata are purely computing machines. Their output is a piece of tape with zeros and ones on it. What is needed . . . is an automaton whose output is other automata."[7]

Using the same method of logical substitution by which a Turing machine can be instructed to interpret successively higher-level languages—or by which Gödel was able to encode metamathematical statements within ordinary arithmetic—it was possible to design Turing machines whose coded instructions addressed physical components, not memory locations, and whose output could be translated into physical objects, not just zeros and ones. "Small variations of the foregoing scheme," von Neumann continued, "also permit us to construct automata which can reproduce themselves and, in addition, construct others." Von Neumann compared the behavior of such automata to what, in biology, characterizes the "typical gene function, self-reproduction plus production—or stimulation of production—of certain specific enzymes."[8]

Viewing the problem of self-replication and self-reproduction through the lens of formal logic and self-referential systems, von Neumann applied the results of Gödel and Turing to the foundations of biology—although his conclusions had little effect on working biologists, just as his *Mathematical Foundations of Quantum Mechanics* had little effect on the day-to-day work of physicists at the time. Applying Turing's proof of the unsolvability of the *Entscheidungsproblem* to the domain of self-reproducing automata, he concluded, in December 1949, that "in other words you can build an organ which can do anything that can be done, but you cannot build an organ which tells you whether it can be done."[9]

"This is connected with the theory of types and with the results of Gödel," he continued. "The question of whether something is feasible in a type belongs to a higher logical type. It is characteristic of objects of low complexity that it is easier to talk about the object than produce it and easier to predict its properties than to build it. But in the complicated parts of formal logic it is always one order of magnitude harder to tell what an object can do than to produce the object."[10]

Can automata produce offspring as complicated, or more complicated, than themselves? " 'Complication' on its lower levels is probably degenerative, that is, every automaton that can produce other automata will only be able to produce less complicated ones," von Neumann explained. There is a certain level of complication, however, beyond which "the phenomenon of synthesis, if properly arranged, can become explosive, in other words, where syntheses of automata can proceed in such a manner that each automaton will produce other automata which are more complex and of higher potentialities than itself."[11]

This conjecture goes to the heart of the probability or improbability of the origin of life. If true, then the existence of a sufficiently complicated self-reproducing system may lead to more complicated systems and, with reasonable probability, either to life or to something lifelike. Self-reproduction is an accident that only has to happen once. "The operations of probability somehow leave a loophole at this point," explained von Neumann, "and it is by the process of self-reproduction that they are pierced."[12]

Von Neumann had intended to return to the question of self-reproduction after leaving the AEC. "Toward the end of his life he felt sure enough of himself to engage freely and yet painstakingly in the creation of a possible new mathematical discipline," says Ulam, "a combinatorial theory of automata and organisms." The theory would have to be simple enough to be mathematically comprehensible, yet complicated enough to apply to nontrivial examples from the real world. "I do not want to be seriously bothered with the objection that (a) everybody knows that automata can reproduce themselves [and] (b) everybody knows that they cannot," von Neumann announced.[13]

The plan had been to produce, with Ulam as coauthor, a comprehensive treatise comparable to *Theory of Games and Economic Behavior*, developing a theory of self-reproducing automata with applications to both biology and technology—and the combination of the two regimes. The work was never completed. Ulam was not as disciplined a coauthor as Oskar Morgenstern, and von Neumann's schedule became even busier after the war. The incomplete manuscript, including a lengthy introduction based on the series of five lectures given by von Neumann at the University of Illinois in 1949, was eventually assembled, with careful editing by Arthur Burks, and published as *Theory of Self-Reproducing Automata* almost ten years after von Neumann's death. Some headings from a surviving outline for the first three chapters, sent to Ulam, hint at their thinking at the time:

1. Wiener!
3. Turing!
5. Not Turing!
6. Boolean algebra
7. Pitts-McCulloch!
13. Ulam!
14. Calling for stronger results
16. Crystal classes in 2 and 3 dim.
18. J.B., H.H.G.!
20. Turing!
23. Double line trick, etc.
24. Degeneration (?)
25. Turing![14]

Our understanding of self-reproduction in biology, and our development of self-reproducing technology, proceeded almost exactly as the proposed theory prescribed. "Wiener!" probably refers to Wiener's theories of information and communication—expanded upon by Claude Shannon—since the problem of self-reproduction is fundamentally a problem of communication, over a noisy channel, from one generation to the next. "Turing!" refers to the powers of the Universal Turing Machine, and "Not Turing!" refers to the limitations of those powers—and how they might be transcended by living and nonliving things. "Pitts-McCulloch!" refers to Walter Pitts and Warren McCulloch's 1943 results on the powers—including Turing universality—of what we now call neural nets. "J.B., H.H.G.!" refers to Julian Bigelow and Herman H. Goldstine—who rarely agreed on anything, so this may be a reference to early discussions of the powers of arrays of communicating cells, before disagreement about how to implement this in practice intervened. "Double line trick, etc." is evocative of the double-helix replication of DNA, and "Degeneration (?)" probably refers to how any enduring system of self-reproduction must depend on error-correcting codes in translating from one generation to the next. "Ulam!" probably refers to Ulam's interest in the powers of Turing-complete cellular automata, now evidenced by many of the computational processes surrounding us today. The triplicate appearance of "Turing!" reflects how central Turing's proof of universality was to any theory of self-reproduction, whether applied to mathematics, biology, or machines.

The Institute for Advanced Study computer was duplicated, with variation, by a first generation of immediate siblings that included SEAC in Washington, D.C., ILLIAC at the University of Illinois, ORDVAC at Aberdeen, JOHNNIAC at the RAND Corporation, MANIAC at Los Alamos National Laboratory, AVIDAC at Argonne, ORACLE at Oak Ridge, BESK in Stockholm, DASK in Copenhagen, SILLIAC in Sydney, BESM in Moscow, PERM in Munich, WEIZAC in Rehovot, and the IBM 701. "There are a number of offspring of the Princeton machine, not all of them closely resembling the parent," Willis Ware reported in March 1953. "From about 1949 on, engineering personnel visited us rather steadily and took away designs and drawings for duplicate machines."[15]

In turn, von Neumann made visits to the other laboratories, and freely exchanged ideas. Physicist Murray Gell-Mann was working during the summer of 1951 at the Control Systems Laboratory, housed immediately above the ILLIAC at the University of Illinois. "This was used on secret government work," says David Wheeler, "and some wires came down." Gell-Mann and Keith Brueckner had been assigned, by their air force sponsors, "to imagine that we had very,

very bad computer parts. And we were to make a very reliable computer out of it." After a lot of work, they were able to show that even with logical components that had "a 51% probability of being right and a 49% probability of being wrong," they could design circuits so "that the signal was gradually improved." They were trying to show exponential improvement, and were getting close. "The project hired various consultants, included Johnny von Neumann for one day," Gell-Mann adds. "He liked to think about problems while driving across the country. So he was driving to Los Alamos to work on thermonuclear weapon ideas, and on the way he stopped in Urbana for a day and consulted for us. God knows what they had to pay him."[16]

In late 1951, von Neumann wrote up these ideas in a short manuscript, "Reliable Organizations of Unreliable Elements," and in January 1952 he gave a series of five lectures at the California Institute of Technology, later published as *Probabilistic Logics and the Synthesis of Reliable Organisms from Unreliable Components*, in which he began to formulate a theory of reliability, in his characteristic, axiomatic way. "Error is viewed, therefore, not as an extraneous and misdirected or misdirecting accident, but as an essential part of the process," he announced. He thanked Keith A. Brueckner and Murray Gell-Mann for "some important stimuli on this subject," but not in any detail. "I wasn't upset at all, at the time," says Gell-Mann. "I thought, my God, this great man is referring to me in the footnote. I'm in the footnote! I was so flattered, and I suppose Keith was, too."[17]

Second- and third-generation copies of the IAS machine followed before the decade was out. Larger computers with larger memories spawned larger, more complex codes, in turn spawning larger computers. Hand-soldered chassis gave way to printed circuits, integrated circuits, and eventually microprocessors with billions of transistors imprinted on silicon without being touched by human hands. The 5 kilobytes of random-access electrostatic memory that hosted von Neumann's original digital universe at a cost of roughly $100,000 in 1947 dollars costs less than $\frac{1}{100}$ of one cent—and cycles 1,000 times as fast—today.

In 1945, the *Review of Economic Studies* had published von Neumann's "Model of General Economic Equilibrium," a nine-page paper read to a Princeton mathematics seminar in 1932 and first published (in German) in 1937. Von Neumann elucidated the behavior of an economy where "goods are produced not only from 'natural factors of production,' but . . . from each other." In this autocatalytic economy, equilibrium and expansion coexist at the saddle point between convex sets. "The connection with topology may be very surprising at first," von Neumann noted, "but the author thinks that it is natural in problems of this kind."[18]

Some of the assumptions of von Neumann's "expanding economic model"—that "natural factors of production, including labour, can be expanded in unlimited quantities" and that "all income in excess of necessities of life will be reinvested"—appeared unrealistic at the time; less so now, when self-reproducing technology is driving economic growth. We measure our economy in money, not in things, and have yet to develop economic models that adequately account for the effects of self-reproducing machines and self-replicating codes.

After von Neumann's departure for the AEC, the IAS computing group began working on "the problem of synthesizing ('near minimal') combinatorial switching circuits" in general, and the problem "of designing a digital computer" as a special case of a circuit that can optimize itself. "This synthesis can be executed by a digital computer, in particular, by the computer to be designed if sufficiently large," they reported in April 1956, concluding that "it appears that we have thereby exhibited a machine which can reproduce (i.e. design) itself. This result seems to be related to the self-reproducing machines of von Neumann."[19] They were right.

Codes populating the growing digital universe soon became Turing-complete, much as envisioned by Ulam and von Neumann in 1952. Turing's ACE, a powerful Universal Machine, was to have had a memory of 25 kilobytes, or $2 \times 10^5$ bits. The present scale of the digital universe has been estimated at $10^{22}$ bits. The number of Turing machines populating this universe is unknown, and increasingly these machines are virtual machines that do not necessarily map to any particular physical hardware at any particular time. They exist as precisely defined entities in the digital universe, but have no fixed existence in ours. And they are proliferating so fast that real machines are struggling to catch up with the demand. Physical machines spawn virtual machines that in turn spawn demand for more physical machines. Evolution in the digital universe now drives evolution in our universe, rather than the other way around.

*Theory of Self-Reproducing Automata* was to present a grand, unifying theory—one reason von Neumann was saving it for last. The new theory would apply to biological systems, technological systems, and every conceivable and inconceivable combination of the two. It would apply to automata, whether embodied in the physical world, the digital universe, or both, and would extend beyond existing life and technology on Earth.

Von Neumann rarely discussed extraterrestrial life or extraterrestrial intelligence; terrestrial life and intelligence were puzzling enough. Nils Barricelli was less restrained. "The conditions for developing organisms with many of the properties considered characteristic of living beings, by evolutionary processes, do not have to be similar to those prevailing on Earth," he concluded, based on his numerical evolution experiments at the IAS. "There is every reason to believe that any planet on which a large variety of molecules can reproduce by interconnected (or symbiotic) autocatalytic reactions, may see the formation of organisms with the same properties."[20] One of these properties, independent of the local conditions, might be the development of the Universal Machine.

Over long distances, it is expensive to transport structures, and inexpensive to transmit sequences. Turing machines, which by definition are structures that can be encoded as sequences, are already propagating themselves, locally, at the speed of light. The notion that one particular computer resides in one particular location at one time is obsolete.

If life, by some chance, happens to have originated, and survived, elsewhere in the universe, it will have had time to explore an unfathomable diversity of forms. Those best able to survive the passage of time, adapt to changing environments, and migrate across interstellar distances will become the most widespread. A life form that assumes digital representation, for all or part of its life cycle, will be able to travel at the speed of light. As artificial intelligence pioneer Marvin Minsky observed on a visit to Soviet Armenia in 1970, "Instead of sending a picture of a cat, there is one area in which you can send the cat itself."[21]

Von Neumann extended the concept of Turing's Universal Machine to a Universal Constructor: a machine that can execute the description of any other machine, including a description of itself. The Universal Constructor can, in turn, be extended to the concept of a machine that, by encoding and transmitting its own description as a self-extracting archive, reproduces copies of itself somewhere else. Digitally encoded organisms could be propagated economically even with extremely low probability of finding a host environment in which to germinate and grow. If the encoded kernel is intercepted by a host that has discovered digital computing—whose ability to translate between sequence and structure is as close to a universal common denominator as life and intelligence running on different platforms may be able to get—it has a chance. If we discovered such a kernel, we would immediately replicate it widely. Laboratories all over the planet would begin attempting to decode it, eventually compiling the coded sequence—intentionally or inadvertently—to utilize our local resources, the way a virus is allocated privileges within a host cell. The read/write privileges granted to digital codes already include material technology, human minds, and, increasingly, nucleotide synthesis and all the ensuing details of biology itself.

The host planet would have to not only build radio telescopes and be actively listening for coded sequences, but also grant computational resources to signals if and when they arrived. The SETI@home network now links some five million terrestrial computers to a growing array of radio telescopes, delivering a collective 500 teraflops of fast Fourier transforms representing a cumulative two million years of processing time. Not a word (or even a picture) so far—as far as we know.

Sixty-some years ago, biochemical organisms began to assemble digital computers. Now digital computers are beginning to assemble biochemical organisms. Viewed from a distance, this looks like part of a life cycle. But which part? Are biochemical organisms the larval phase of digital computers? Or are digital computers the larval phase of biochemical organisms?

According to Edward Teller, Enrico Fermi asked the question "Where is everybody?" at Los Alamos in 1950, when the subject of extraterrestrial beings came up over lunch. Fifty years later, over lunch at the Hoover Institution at Stanford University, I asked a ninety-one-year-old Edward Teller how Fermi's question was holding up. John von Neumann, Theodore von Kármán, Léo Szilárd, and Eugene Wigner, Teller's childhood colleagues from Budapest, had all predeceased him. Of the five Hungarian "Martians" who brought the world nuclear weapons, digital computers, much of the aerospace industry, and the beginnings of genetic engineering, only Edward Teller, carrying a wooden staff at his side like an Old Testament prophet, was left.

His limp, from losing most of a foot to a Munich streetcar in 1928, had grown more pronounced, just as memories of his Hungarian youth had become more vivid as his later memories were beginning to fade. "I remember the bridges, the beautiful bridges," he says of Budapest.[22] Although Teller served (with von Neumann and German rocket pioneer Wernher von Braun) as one of the models for the composite title character in Stanley Kubrick's cold war masterpiece *Dr. Strangelove,* nuclear weapons in the hands of Teller are, to me, less terrifying than they are in the hands of a new generation of nuclear weaponeers who have never witnessed an atmospheric test firsthand.

Teller assumed that I had come to ask him about the Teller-Ulam invention, and provided a lengthy account of the genesis of the hydrogen bomb, and of the fission implosion-explosion required to get the thermonuclear fuel to ignite. "The whole implosion idea—that is, that one *can* get densities considerably greater than normal—came from a visit from von Neumann," he told me. "We proposed that together to Oppenheimer. He at once accepted."[23] With the hydrogen bomb out of the way, I mentioned that I was interested in the status of the Fermi paradox after fifty years.

"Let me ask you," Teller interjected, in his thick Hungarian accent. "Are you uninterested in extraterrestrial intelligence? Obviously not. If you are interested, what would you look for?"

"There's all sorts of things you can look for," I answered. "But I think the thing not to look for is some intelligible signal. . . . Any civilization that is doing useful communication, any efficient transmission of information will be encoded, so it won't be intelligible to us—it will look like noise."

"Where would you look for that?" asked Teller.

"I don't know. . . ."

"I do!"

"Where?"

"Globular clusters!" answered Teller. "We cannot get in touch with anybody else, because they choose to be so far away from us. In globular clusters, it is much easier for people at different places to get together. And if there is interstellar communication at all, it must be in the globular clusters."

"That seems reasonable," I agreed. "My own personal theory is that extraterrestrial life could be here already . . . and how would we necessarily know? If there is life in the universe, the form of life that will prove to be most successful at propagating itself will be digital life; it will adopt a form that is independent of the local chemistry, and migrate from one place to another as an electromagnetic signal, as long as there's a digital world—a civilization that has discovered the

Universal Turing Machine—for it to colonize when it gets there. And that's why von Neumann and you other Martians got us to build all these computers, to create a home for this kind of life."

There was a long, drawn-out pause. "Look," Teller finally said, lowering his voice to a raspy whisper, "may I suggest that instead of explaining this, which would be hard . . . you write a science-fiction book about it."

"Probably someone has," I said.

"Probably," answered Teller, "someone has not."

# SIXTEEN

# Mach 9

*No time is there. Sequence is different from time.*

—Julian Bigelow, 1999

"IN ALL THE YEARS after the war, whenever you visited one of the installations with a modern mainframe computer, you would always find somebody doing a shock wave problem," remembers German American astrophysicist Martin Schwarzschild, who, still an enemy alien, enlisted in the U.S. Army at the outbreak of World War II. "If you asked them how they came to be working on that, it was always von Neumann who put them onto it. So they became the footprint of von Neumann, walking across the scene of modern computers."[1]

Schwarzschild was assigned to the Aberdeen Proving Ground, where he studied the effects of the new "block buster" weapons, fueled with conventional explosives but of such size that most of the damage was caused by the shock wave rather than by the bomb debris itself. It was this problem, foreshadowing the effects of nuclear weapons, that first drew von Neumann to Aberdeen. Before von Neumann arrived, "we had incredible struggles," according to Schwarzschild. "Even with days of arguing and thinking none of us could really figure out how one should exactly tell the engineers what we wanted." Von Neumann's solution was to have the engineers build a machine that could follow a limited number of simple instructions, and then let the mathematicians and physicists assemble programs as needed from those instructions, without having to go back to the engineers. "You immediately saw how you would write down sequences of statements to solve any particular problem," says Schwarzschild, emphasizing "how dumb we were early in 1943 and how everything seemed terribly plain and straight-forward in 1944."[2] It was von Neumann who conveyed this approach, wherever it originated, to Aberdeen.

Eight years later, the high-explosive blockbusters of 1943 had become thermonuclear weapons, and the ENIAC and the Colossus had become fully universal machines, yet ideas were still exchanged in person at the speed of a propeller-driven DC-3. As the IAS computer was undergoing initial testing in January of 1952, von Neumann flew from California to Cocoa, Florida (near Cape Canaveral, later Cape Kennedy), for a meeting of air force officials and some sixty scientific advisers, prompted by the establishment of the Air Research and Development Command, for whom he had agreed to help set up a mathematical advisory group.

The top-secret Project Vista report on the role of nuclear weapons in the defense of Europe had just been released. The report argued—in keeping with the views of Oppenheimer and against the views of its air force sponsors—that tactical nuclear weapons aimed at the military battlefield, rather than strategic nuclear weapons aimed at civilian populations, might, both morally and militarily, be the better approach. It was Oppenheimer's influence over the Vista report, as much as his public hesitation about thermonuclear weapons, that led to his security clearances being withdrawn. The unspoken agreement between the military and the scientists was that the military would not tell the scientists how to do science, and the scientists would not tell the military how to use the bombs. Oppenheimer had stepped out of bounds.

Von Neumann first flew from San Francisco to Tulsa, Oklahoma, via DC-4, a journey that "involved only two stops and two plane-changes: El Paso and Dallas." From Tulsa to Cocoa required a first leg, "with stops at Muskogee, Fort Worth, Texarkana, Shreveport and a change at New Orleans," aboard a DC-3. Then came a second leg, "with stops at Mobile, Pensacola, Panama city and a change at Tampa," aboard a Lockheed Lodestar. Then came the last leg to Orlando, also by Lodestar, and finally an air force car to the "big but dilapidated" Indian River Hotel.[3] He was able to return to Washington aboard a military aircraft, and finally by train to Princeton. In Washington he found that the AEC commissioners "wanted a discussion of some shock-wave questions," while back at the Institute he found the computer looking "reasonably good," despite a daily quota of "transients and faults."[4]

Thermonuclear reactions in a bomb are over in billionths of a second, while thermonuclear reactions in a star play out over billions of years. Both time scales, beyond human comprehension, fell within the MANIAC's reach. With the war over, Martin Schwarzschild had begun applying desk calculators and punched card tabulating equipment to the problem of stellar evolution, combining Bethe's theories of 1938 with the techniques developed at Los Alamos to calculate radiation opacity and equations of state. "To get a solution for a particular star for one particular time in the life of the star would take two or three months," Schwarzschild explained, despite the assistance of the new Watson Scientific Computing Laboratory, built at Columbia University by IBM. "The amount of numerical computation necessary is awfully big," he reported to Subrahmanyan Chandrasekhar at the beginning of December 1946. "I'm just finishing the solutions for the convective core. However, the integrations of the seventeen necessary particular solutions for the radiative envelope have only just been started by the I.B.M. laboratory on the new relay multipliers. . . . I fear that the numerical work will still be far from its end at Christmas."[5]

In 1950, von Neumann and Goldstine invited Schwarzschild to use the MANIAC instead. Stellar evolution attracted von Neumann because "its objects of study can be observed but cannot be experimented with," and because the results could be compared with the observed characteristics of known types of stars at different stages in their lives. "Suddenly you had the possibility of computing evolution sequences for individual stars, of comparing different stars and different phases of the evolution with observed stars," Schwarzschild explained. This would provide an immediate check on whether the numerical models were holding up. Where meteorologists have to accept that every weather system is different and that observations can only be collected once, astronomers can look up in the night sky and observe the different families of stars at every stage of their existence at any time. "A wealth (what you might call a zoo) of observed stars fell into patterns," says Schwarzschild, "and we could start getting ages for stars."[6]

Von Neumann provided machine time, along with the assistance of Hedvig (Hedi) Selberg, who became Schwarzschild's collaborator on the stellar evolution work. Selberg, born in Targu Mures, Transylvania, in 1919, was the daughter of a furniture maker who went bankrupt during the Depression, leaving her to help support the family as a tutor in mathematics while attending the University of Kolozsvar, where she graduated, with the equivalent of a master's degree, at the head of her class. She then taught mathematics and physics at a Jewish high school in Satu Mare, until her entire family was deported to Auschwitz in June 1944. The only member of her family to survive, she escaped to Scandinavia, married the Norwegian number theorist Atle

Selberg in August 1947, and arrived in Princeton in September. Her teaching credentials were not recognized in New Jersey and she was making plans to obtain a PhD in mathematics at Columbia when von Neumann offered her a position with the Electronic Computer Project, in September 1950, at $300 per month. "She loved this new line of work, and felt very lucky to have been part of this exciting development in history," says her daughter, Ingrid. "It allowed her to use her knowledge of mathematics and physics as well as her intelligence and attention to detail."[7]

Hedi Selberg remained with the MANIAC for the machine's entire life, from the first hydrogen bomb calculations to the last stellar evolution model that was running when the university pulled the plug. "I never really knew about the hydrogen bomb research going on at ECP at night," says Ingrid, whereas, according to Freeman Dyson, "Hedi always said the computer was working mostly on bombs." When the engineers or the "AEC boys" left questions in the machine logbooks, the answers are often signed "H.S."

"Impossible to load anything but all '1's. Even blank cards show all '1's," she noted in the machine log on December 11, 1953. "Machine operates OK apart from this," she records in the next entry. "Relay in back of machine (the one we put matches in) seems OK." She supervised the machine singlehandedly for extended periods of time. "The A/cond are not doing wonderfully well tonight and got quite iced up," she notes at 12:09 a.m. on the night of November 19–20, 1954. "However the temp is coming down gradually and will leave Barricelli with instructions to shut down if temp reaches 90," she adds before signing off for the night.

Schwarzschild and Selberg started out with simple models of the early stages in stellar evolution, when the star is in hydrostatic and thermal equilibrium and "the internal temperatures just right so that the hydrogen burning produces energy at a rate exactly compensating [for] the losses by surface radiation."[8] Eventually the hydrogen in the core is transmuted into helium, and the area of hydrogen burning moves farther out. The helium core grows in mass but shrinks in size, while the luminosity and radius of the star as a whole increases, producing a red giant that burns for another billion years or so, before becoming a white dwarf.

As a star ages, its behavior grows increasingly complex. Convection introduces mixing between layers, and the transmutation of helium produces a succession of heavier elements, with their varying opacity having the same critical effects that radiation opacity had on the feasibility of the Teller-Ulam bomb. "The whole system including the various supplementary equations for computing the opacity, energy generation, etc., and the distinction between the radiative, convective and degenerate cases is certainly the most complicated system ever treated on our computer," Schwarzschild and Selberg explained.[9] The models were far ahead of their time. "It was only, regrettably, much later, after he was no longer with us, that I understood how deep and profound and insightful were his early studies and how they informed his comments and questions," says astronomer John Bahcall, who, twenty years later, brought numerical modeling back to the IAS. "But he was the kind of person who never said, 'If you look back at my paper in 1956 you'll see that I calculated the primordial heating abundance' or something. That just wasn't his style. He never mentioned his own work."[10]

Schwarzschild's calculations constituted a cosmological meteorology: the infinite forecast to end all infinite forecasts. "Do we then live in an adult galaxy which has already bound half of its mass permanently into stars, and has consumed a fourth of all its fuel?" he asked in 1957. "Are we too late to witness the turbulent sparkle of galactic youth but still in time to watch stars in all their evolutionary phases before they settle into the permanence of the white dwarfs?"[11]

By mid-1953, five distinct sets of problems were running on the MANIAC, characterized by different scales in time: (1) nuclear explosions, over in microseconds; (2) shock and blast waves, ranging from microseconds to minutes; (3) meteorology, ranging from minutes to years; (4) biological evolution, ranging from years to millions of years; and (5) stellar evolution, ranging from millions to billions of years. All this in 5 kilobytes—enough memory for about one-half second of audio, at the rate we now compress music into MP3s.

These time scales ranged from about $10^{-8}$ seconds (the lifetime of a neutron in a nuclear explosion) to $10^{17}$ seconds (the lifetime of the sun). The middle of this range falls between $10^4$ and $10^5$ seconds, or about eight hours, exactly in the middle of the range (from the blink of an eye, over in three-tenths of a second, to a lifetime of three billion seconds, or ninety years) that a human being is able to directly comprehend.

Of these five sets of problems, shock waves were von Neumann's first love and remained closest to his heart. He had an intuitive feel for the subject. Calculation alone was not always enough. "The question as to whether a solution which one has found by mathematical reasoning really occurs in nature . . . is a quite difficult and ambiguous one," he explained in 1949, concerning the behavior of shock waves produced by the collision of gas clouds in interstellar space. "We have to be guided almost entirely by physical intuition in searching for it . . . and it is difficult to say about any solution which has been derived, with any degree of assurance, that it is the one which must exist."[12]

Shock waves are produced by collisions between objects, or between an object and a medium, or between two mediums, or by a sudden transition within a medium, when the velocities or time scales are mismatched. If the difference in velocity is greater than the local speed of information, this propagates a discontinuity, a sonic boom being the classic example, as an aircraft exceeds the speed of sound. The disturbance can be the detonation front in a high explosive, a bullet exiting the muzzle of a gun, a meteorite hitting the atmosphere, the explosion of a nuclear weapon, or the collision between two jets of interstellar gas.

Shock waves could even be produced by the collision between two universes, or the explosion of a new universe—this being one way to describe the discontinuities being produced as the digital universe collides with our universe faster than we are able to adjust. "The ever accelerating progress of technology and changes in the mode of human life," von Neumann explained to Stan Ulam, "gives the appearance of approaching some essential singularity in the history of the race."[13]

In our universe, we measure time with clocks, and computers have a "clock speed," but the clocks that govern the digital universe are very different from the clocks that govern ours. In the digital universe, clocks exist to synchronize the translation between bits that are stored in memory (as structures in space) and bits that are communicated by code (as sequences in time). They are clocks more in the sense of regulating escapement than in the sense of measuring time.

"The I.A.S. computing machine is non-synchronous; that is, decisions between elementary alternatives, and enforcement of these decisions are initiated not with reference to time as an independent variable but rather according to sequence," Bigelow explained to Maurice Wilkes—who had just succeeded in coaxing Cambridge's delay-line storage EDSAC computer into operation before the Institute's—in 1949. "Time, therefore, does not serve as an index for the location of information, but instead counter readings are used, the counters themselves being actuated by the elementary events."[14]

"It was all of it a large system of on and off, binary gates," Bigelow reiterated fifty years later. "No clocks. You don't need clocks. You only need counters. There's a difference between a counter and a clock. A clock keeps track of time. A modern general purpose computer keeps track of events."[15] This distinction separates the digital universe from our universe, and is one of the few distinctions left.

The acceleration from kilocycles to megahertz to gigahertz is advancing even faster than this increase in nominal clock speed indicates, as devices such as dedicated graphic processors enable direct translation between coded sequences and memory structures, without waiting for any central clock to authorize the translation step-by-step. No matter how frequently we reset our own clocks to match the increasing speed of computers we will never be able to keep up. Codes that take advantage of asynchronous processing, in the digital universe, will rapidly move ahead, in ours.

Thirty years ago, networks developed for communication between people were adapted to communication between machines. We went from transmitting data over a voice network to transmitting voice over a data network in just a few short years. Billions of dollars were sunk into cables spanning six continents and three oceans, and a web of optical fiber engulfed the world. When the operation peaked in 1991, fiber was being rolled out, globally, at over 5,000 miles per hour, or nine times the speed of sound: Mach 9.

Since it costs little more to install a bundle of fibers than a single strand, tremendous overcapacity was deployed. Fifteen years later, a new generation of companies, including Google, started buying up "dark fiber" at pennies on the dollar, awaiting a time when it would be worth the expense of connecting it at the ends. With optical switching growing cheaper by the minute, the dark fiber is now being lit. The "last mile" problem—how to reach individual devices without individual connection costs—has evaporated with the appearance of wireless devices, and we are now rolling out cable again. Global production of optical fiber reached Mach 20 (15,000 miles per hour) in 2011, barely keeping up with the demand.

Among the computers populating this network, most processing cycles are going to waste. Most processors, most of the time, are waiting for instructions. Even within an active processor, as Bigelow explained, most computational elements are waiting around for something to do next. The global computer, for all its powers, is perhaps the least efficient machine that humans have ever built. There is a thin veneer of instructions, and then there is a dark, empty 99.9 percent.

To numerical organisms in competition for computational resources, the opportunities are impossible to resist. The transition to virtual machines (optimizing the allocation of processing cycles) and to cloud computing (optimizing storage allocation) marks the beginning of a transformation into a landscape where otherwise wasted resources are being put to use. Codes are becoming multicellular, while the boundaries between individual processors and individual memories grow indistinct.

When Julian Bigelow and Norbert Wiener formulated their Maxims for Ideal Prognosticators in 1941, their final maxim was that predictions (of the future position of a moving target) should be made by normalizing observations to the frame of reference of the target, "emphasizing its fundamental symmetry and invariance of behavior," otherwise lost in translation into the frame of reference of the observer on the ground.[16] This is why it is so difficult to make predictions, within the frame of reference of our universe, as to the future of the digital universe, where time as we know it does not exist. All we have are lag functions that we can only turn forward in our nondigital time.

The codes spawned in 1951 have proliferated, but their nature has not changed. They are symbiotic associations of self-reproducing numbers (starting with a primitive alphabet of order codes) that were granted limited, elemental powers, the way a limited alphabet of nucleotide sequences code for an elemental set of amino acids— with polynucleotides, proteins, and everything else that follows developing from there. The codes by which an organization as vast and complex as Google maps the state of the entire digital universe are descended from the first Monte Carlo codes that Klári von Neumann wrote while smoking Lucky Strikes from the Los Alamos Post Exchange.

At about the time the IAS computer became operational in 1951, Stan Ulam sent John von Neumann an undated note, wondering whether a purely digital universe could capture some of the evolutionary processes we see in our universe. He envisioned an unbounded two-dimensional matrix where Turing-complete digital organisms (operating in two dimensions, unlike Barricelli's one-dimensional, cross-breeding genetic strings) would compete for resources, and evolve. Ulam also suggested carrying the digital model backward, in the other direction, to consider the question, as he put it, of "generation of time and space in 'prototime.'"[17]

Organisms that evolve in the digital universe are going to be very different from us. To us, they will appear to be evolving ever faster, but to them, our evolution will appear to have begun decelerating at their moment of creation—the way our universe appears to have suddenly begun to cool after the big bang. Ulam's speculations were correct. Our time is becoming the prototime for something else.

"The game that nature seems to be playing is difficult to formulate," Ulam observed, during a conversation that included Nils Barricelli, in 1966. "When different species compete, one knows how to define a loss: when one species dies out altogether, it loses, obviously. The defining win, however, is much more difficult because many coexist and will presumably for an infinite time; and yet the humans in some sense consider themselves far ahead of the chicken, which will also be allowed to go on to infinity."[18]

Von Neumann's first solo paper, "On the Introduction of Transfinite Numbers," was published in 1923, when he was nineteen. The question of how to consistently distinguish different kinds of infinity, which von Neumann clarified but did not answer, is closely related to Ulam's question: Which kind of infinity do we want?

# SEVENTEEN

# The Tale of the Big Computer

*In a small laboratory—some people main-*
*tain that it was an old converted stable—a*
*few men in white coats stood watching a*
*small and apparently insignificant appa-*
*ratus equipped with signal lights, which*
*flashed like stars. Gray perforated strips*
*of paper were fed into it, and other strips*
*emerged. Scientists and engineers worked*
*hard, with a gleam in their eyes; they knew*
*that the little gadget in front of them was*
*something exceptional—but did they foresee*
*the new era that was opening before them, or*
*suspect that what had happened was compa-*
*rable to the origin of life on Earth?*

—Hannes Alfvén, 1966

VON NEUMANN MADE a deal with "the other party" in 1946. The scientists would get the computers, and the military would get the bombs. This seems to have turned out well enough so far, because, contrary to von Neumann's expectations, it was the computers that exploded, not the bombs.

"It is possible that in later years the machine sizes will increase again, but it is not likely that 10,000 (or perhaps a few times 10,000) switching organs will be exceeded as long as the present techniques and philosophy are employed," von Neumann predicted in 1948. "About 10,000 switching organs seem to be the proper order of magnitude for a computing machine."[1] The transistor had just been invented, and it would be another six years before you could buy a transistor radio—with four transistors. In 2010 you could buy a computer with a billion transistors for the inflation-adjusted cost of a transistor radio in 1956.

Von Neumann's estimate was off by over five orders of magnitude—so far. He believed, and counseled the government and industry strategists who sought his advice, that a small number of large computers would be able to meet the demand for high-speed computing, once the impediments to remote input and output were addressed. This was true, but only for a very short time. After concentrating briefly in large, centralized computing facilities, the detonation wave that began with punched cards and vacuum tubes was propagated across a series of material and institutional boundaries: into magnetic-core memory, semiconductors, integrated circuits, and microprocessors; and from mainframes and time-sharing systems into minicomputers, microcomputers, personal computers, the branching reaches of the Internet, and now billions of embedded microprocessors and aptly named cell phones. As components grew larger in number, they grew smaller in size and cycled faster in time. The world was transformed.

Among those who foresaw the transformation was Swedish astrophysicist Hannes Alfvén, who remained as opposed to nuclear weapons as von Neumann and Teller were enamored of them. He was a founding member, and later president, of the Pugwash disarmament movement founded by Joseph Rotblat—the only Los Alamos physicist to quit work, in late 1944, in response to secret intelligence that the Germans were not making a serious effort to build an atomic bomb.

As a child, Alfvén had been given a copy of Camille Flammarion's *Popular Astronomy,* where he learned what was known, and what was not known, about the solar system at the time. He then joined his school's shortwave radio club, where he began to understand how much of the universe lay beyond the wavelengths of visible light, and how much consisted not of conventional solids, liquids, or gases, but of plasma—a fourth state of matter, where electrons were unbound. He was awarded the Nobel Prize in Physics in 1970 for his work in magnetohydrodynamics, a field he pioneered with a letter to *Nature* in 1942. The behavior of electromagnetic waves in solid conductors was well understood, while the behavior of electromagnetic waves in ionized plasma remained mysterious, whether within a star or in interstellar space. In any conducting fluid, including plasma, electrodynamics and hydrodynamics were coupled, and Alfvén put this relationship on solid mathematical and experimental ground. "Playing with mercury in the presence of a magnetic field of 10,000 gauss gives the general impression that the magnetic field has completely changed its hydrodynamic properties," he explained in 1949.[2]

Alfvén's cosmos was permeated by magnetohydrodynamic

waves—now termed Alfvén waves—rendering "empty" space much less empty, and helping to explain phenomena ranging from the Aurora Borealis to sunspots to cosmic rays. He developed a detailed theory of the formation of the solar system, using electrodynamics to explain how the different planets coalesced. "To trace the origin of the solar system is archaeology, not physics," he wrote in 1954.[3]

Alfvén also argued, without convincing the orthodoxy, that the large-scale structure of the universe might be hierarchical to infinity, rather than expanding from a single source. Such a universe—fulfilling Leibniz's ideal of everything from nothing—would have an average density of zero but infinite mass. According to Alfvén, the "big bang" was based on wishful thinking. "They fight *against* popular creationism, but at the same time they fight fanatically *for* their own creationism," he noted in 1984.[4]

Alfvén divided his later years between La Jolla, California, where he held a position as professor of physics at UC–San Diego, and the Royal Institute of Technology in Stockholm, where he had been appointed to the School of Electrical Engineering in 1940, just in time to witness the arrival of the computer age firsthand. Sweden's BESK (Binär Elektronisk Sekvens Kalkylator) was a first-generation copy of the IAS machine, becoming operational in 1953. It had faster memory and arithmetic, partly through clever Swedish engineering (including the use of 400 germanium diodes) and partly through reducing the memory of each Williams tube to 512 bits.

"I saw the Swedish machine," von Neumann reported to Klári from Stockholm in September 1954. "Very elegant, perhaps average 25% faster than ours, with only 500 words Williams memory and 4000 on a drum (this to be doubled) a Teletype input (fast, an electrical reader) and only a typewriter output (slow)."[5] The construction of this machine left an indelible impression on Alfvén, which he eventually set down on paper in *The Tale of the Big Computer: A Vision*, published in Sweden in 1966 and in the United States in 1968.

"When one of my daughters had given me my first grandchild, she said to me: You are writing so many scientific papers and books, but why don't you write something more sensible—a fairy tale for this little boy," Alfvén recalled in 1981. Choosing a "monozygotic relative by the name of Olof Johannesson" as a pseudonym, Alfvén recounted, from an indefinite time in the future, a brief natural history of the origin and development of computers and their subsequent domination of life on Earth. "Life, which evolved into ever more complex structures, was nature's substitute for directly bred computers," he wrote. "Yet it was more than a substitute: it was a road—a winding road, yet one which despite all errors and hazards, arrived at last at its destination."[6]

"I was Scientific Advisor to the Swedish government, and had access to their plans to restructure Swedish society, which obviously could be made much more efficient with the help of computers, in the same way as earlier inventions had relieved us of heavy physical work," he added, explaining how he came to write the book. In Alfvén's vision, computers quickly eliminated two of the world's greatest threats: nuclear weapons and politicians. "When the computers developed, they would take over a good deal of the burden of the politicians, and sooner or later would also take over their power," he explained. "This need not be done by an ugly coup d'état; they would simply systematically outwit the politicians. It might even take a long time before the politicians understood that they had been rendered powerless. This is not a threat to us."[7]

"Computers are designed to be problem solvers, whereas the politicians have inherited the stone age syndrome of the tribal chieftains, who take for granted that they can rule their people only by making them hate and fight all other tribes," Alfvén continued. "If we have the choice of being governed by problem generating trouble makers, or by problem solvers, every sensible man of course would prefer the latter."[8]

The mathematicians who were designing and programming the growing computer network began to suspect that "the problem of organizing society is so highly complex as to be insoluble by the human brain, or even by many brains working in collaboration." Their subsequent proof of the "Sociological Complexity Theorem" led to a decision to turn the organization of human society, and the management of its social networks, over to the machines.[9] All individuals were issued a device called "teletotal," connected to a global computer network with features similar to the Google and Facebook of today. "Teletotal threw a bridge between the thought world of the computer—which operated via pulse sequences at the speed of nano-seconds—and the thought world of the human brain, with its electrochemical nerve impulses," Alfvén explained.[10] "Since universal knowledge was stored in the memory units of the computers and was thus easily accessible to one and all, the gap between those who knew and those who did not was closed . . . and it was quite unnecessary to store any wisdom at all in the human brain."[11]

Teletotal was followed by a miniaturized, wireless successor known as "minitotal," later supplemented by "neurototal," an implant kept "in permanent contact via VHF with the subject's minitotal" and surgically inserted into a nerve channel for direct connection to the brain. Human technicians maintained the growing computer network, with the computers, in return, looking after the health and welfare of their human symbionts as carefully as the Swedish government does today. "Health factories" kept human beings in good repair, cities were abandoned in favor of a decentralized, telecommuting life, and "shops became superfluous, for the goods in them could be examined from the customer's home. . . . If one wanted to buy something . . . one pressed the purchase button."[12]

Then, one day, the entire system ground to a halt. A small group of humans had conspired to seize control of the network for themselves. "Factions had formed—just how many is unknown—and they fought each other for power," Alfvén explained. "One group attempted to knock out its rivals by disorganizing their data systems, and was paid back in its own coin. The result was total disruption. How long the battle lasted we do not know. It must have been prepared over a long period, but the conflict itself may have taken less than one second. For computers this is a considerable time."[13]

The failure was complete. With the network down, there was no way to distribute the instructions to bring it back up. "The breakdowns seem to have set in almost—or even precisely—at the same time all over the world, and it was evident that the international computer network was dead," Johannesson reported.[14] "It was utter disaster. Within less than a year the greater part of the population had perished from hunger and privation. . . . Museums were plundered of [axes] and other tools."[15]

Society was slowly reconstructed from the ruins, and the computer system rebooted from backups preserved by a Martian outpost that had escaped the collapse. This time, the computers were given full control from the start, it being recognized that "Man had to be excluded altogether from the more important organizational tasks."[16] In the new society, the number of human beings was kept small. "A great number of data machines had been destroyed at the time of the disaster, yet their numbers had diminished by nothing compared with the proportion of human casualties. . . . Thus when they were put into action again, the proportion of computers to people was greatly increased."[17] Once the computers were running again, and equipped with facilities to repair and reproduce themselves, human beings became increasingly superfluous, and the story leaves off with Olof Johannesson wondering how large a human population will be preserved. "It is likely that they will at least reduce their numbers; but will this be done quickly or gradually? Will they retain a human colony and, if so, of what size?"[18]

Alfvén's tale is now forgotten, but the future he envisioned has arrived. Data centers and server farms are proliferating in rural areas; "Android" phones with Bluetooth headsets are only one step away from neural implants; unemployment is pandemic among those not working on behalf of the machines. Facebook defines who we are, Amazon defines what we want, and Google defines what we think. Teletotal was the personal computer; minitotal is the iPhone; neurototal will be next. "How much human life can we absorb?" answers one of Facebook's founders, when asked what the goal of the company really is.[19] "We want Google to be the third half of your brain," says Google cofounder Sergey Brin.[20]

The ability of computers to predict (and influence) how people will vote, with as much precision as the actual vote can be counted, has rendered politicians subservient to computers, much as Alfvén prescribed. Computers have no need for weapons to enforce their power, since, as Alfvén explained, they "control all production, and this would automatically stop in the event of an attempted revolt. The same is true of communications, so that if anyone should attempt anything so foolish as a revolt against the data machines, it could only be local in character. Lastly, man's attitude to computers is a very positive one."[21] Recent developments have outpaced what even Alfvén could imagine—from the explosive growth of optical data networks (anticipated in the nineteenth century by optical telegraph networks in Sweden) to the dominance of virtual machines.

The progenitor of virtualization was Turing's Universal Machine. Two-way translation between logical function and strings of symbols is no longer the mathematical abstraction it was in 1936. A single computer may host multiple, concurrent virtual machines; "apps" are coded sequences that locally implement a specific virtual machine on an individual device; Google's one million (at last count) servers constitute a collective, metazoan organism whose physical manifestation changes from one instant to the next.

Virtual machines never sleep. Only one-third of a search engine is devoted to fulfilling search requests. The other two-thirds are divided between crawling (sending a host of single-minded digital organisms out to gather information) and indexing (building data structures from the results). The load shifts freely between the archipelagoes of server farms. Twenty-four hours a day, 365 days a year, algorithms with names such as BigTable, MapReduce, and Percolator are systematically converting the numerical address matrix into a content-addressable memory, effecting a transformation that constitutes the largest computation ever undertaken on planet Earth. We see only the surface of a search engine—by entering a search string and retrieving a list of addresses, with contents, that contain a match. The aggregate of all our random searches for meaningful strings of bits is a continuously updated mapping among content, meaning, and address space: a Monte Carlo process for indexing the matrix that underlies the World Wide Web.

The address matrix that began, in 1951, with a single 40-floor hotel, with 1,024 rooms on every floor, has now expanded to billions of 64-floor hotels with billions of rooms, yet the contents are still addressed by numerical coordinates that have to be specified exactly, or everything comes to a halt. There is, however, another way of addressing memory, and that is to use an identifiable (but not necessarily unique) string within the contents of the specified block of memory as a template-based address.

Given access to content-addressable memory, codes based on instructions that say, "Do this with that"—without having to specify a precise location—will begin to evolve. The instructions may even say, "Do this with something *like* that"—without the template having to be exact. The first epoch in the digital era began with the introduction of the random-access storage matrix in 1951. The second era began with the introduction of the Internet. With the introduction of template-based addressing, a third era in computation has begun. What was once a cause for failure—not specifying a precise numerical address—will become a prerequisite to real-world success.

The Monte Carlo method was invoked as a means of using statistical, probabilistic tools to identify approximate solutions to physical problems resistant to analytical approach. Since the underlying physical phenomena actually *are* probabilistic and statistical, the Monte Carlo approximation is often closer to reality than the analytical solutions that Monte Carlo was originally called upon to approximate. Template-based addressing and pulse-frequency coding are similarly closer to the way the world really works and, like Monte Carlo, will outperform methods that require address references or instruction strings to be exact. The power of the genetic code, as both Barricelli and von Neumann immediately recognized, lies in its ambiguity: exact transcription but redundant expression. In this lies the future of digital code.

A fine line separates approximation from simulation, and developing a model is the better part of assuming control. So as not to shoot down commercial airliners, the SAGE (Semi-Automatic Ground Environment) air defense system that developed out of MIT's Project Whirlwind in the 1950s kept track of all passenger flights, developing a real-time model that led to the SABRE (Semi-Automatic Business-Related Environment) airline reservation system that still controls much of the passenger traffic today. Google sought to gauge what people were thinking, and became what people were thinking. Facebook sought to map the social graph, and became the social graph. Algorithms developed to model fluctuations in financial markets gained control of those markets, leaving human traders behind. "Toto," said Dorothy in *The Wizard of Oz*, "I've a feeling we're not in Kansas anymore."

What the Americans termed "artificial intelligence" the British termed "mechanical intelligence," a designation that Alan Turing considered more precise. We began by observing intelligent behavior (such as language, vision, goal-seeking, and pattern-recognition) in organisms, and struggled to reproduce this behavior by encoding it into logically deterministic machines. We knew from the beginning that this logical, intelligent behavior evident in organisms was the result of fundamentally statistical, probabilistic processes, but we ignored that (or left the details to the biologists), while building "models" of intelligence—with mixed success.

Through large-scale statistical, probabilistic information processing, real progress is being made on some of the hard problems, such as speech recognition, language translation, protein folding, and stock market prediction—even if only for the next millisecond, now enough time to complete a trade. How can this be intelligence, since we are just throwing statistical, probabilistic horsepower at the problem, and seeing what sticks, without any underlying understanding? There's no model. And how does a brain do it? With a model? These are not models of intelligent processes. They *are* intelligent processes.

The behavior of a search engine, when not actively conducting a search, resembles the activity of a dreaming brain. Associations made while "awake" are retraced and reinforced, while memories gathered while "awake" are replicated and moved around. William C. Dement, who helped make the original discovery of what became known as REM (rapid eye movement) sleep, did so while investigating newborn infants, who spend much of their time in dreaming sleep. Dement hypothesized that dreaming was an essential step in the initialization of the brain. Eventually, if all goes well, awareness of reality evolves from the internal dream—a state we periodically return to during sleep. "The prime role of 'dreaming sleep' in early life may be in the development of the central nervous system," Dement announced in *Science* in 1966.[22]

Since the time of Leibniz, we have been waiting for machines to begin to think. Before Turing's Universal Machines colonized our desktops, we had a less-encumbered view of the form in which true artificial intelligence would first appear. "Is it a fact—or have I dreamed it—that, by means of electricity, the world of matter has become a great nerve, vibrating thousands of miles in a breathless point of time?" asked Nathaniel Hawthorne in 1851. "Rather, the round globe is a vast head, a brain, instinct with intelligence! Or, shall we say, it is itself a thought, nothing but thought, and no longer the substance which we deemed it?" In 1950, Turing asked us to "consider the question, 'Can machines think?'"[23] Machines will dream first.

What about von Neumann's question—whether machines would begin to reproduce? We gave digital computers the ability to modify their own coded instructions—and now they are beginning to exercise the ability to modify our own. Are we using digital computers to sequence, store, and better replicate our own genetic code, thereby optimizing human beings, or are digital computers optimizing our genetic code—and our way of thinking—so that we can better assist in replicating them?

In the beginning was the command line: a human programmer supplied an instruction and a numerical address. There is no proscription against computers supplying their own instructions, and an ever-diminishing fraction of commands have ever been touched by human hands or human minds. Now the commands and addresses are as likely to be delivered the other way: the global computer supplies an instruction, and an address that maps to a human being via a personal device. That the resulting human behavior can only be counted on statistically, not deterministically, is, as von Neumann demonstrated in 1952 with his *Probabilistic Logics and the Synthesis of Reliable Organisms from Unreliable Components,* no obstacle to the synthesis of those unreliable human beings into a reliable organism. We are returning to the landscape as envisioned by von Neumann in 1948: with a few large computers handling much of the computation in the world. The big computers, however, are not physically centralized; they are distributed across a multitude of hosts.

In October 2005, on the occasion of the sixtieth anniversary of von Neumann's proposal to Lewis Strauss for the MANIAC, and Turing's proposal to the National Physical Laboratory for the ACE, I was invited to Google's headquarters in California, and given a glimpse inside the organization that has been executing precisely the strategy that Turing had in mind: gathering all available answers, inviting all possible questions, and mapping the results. I felt I was entering a fourteenth-century cathedral while it was being built. Everyone was busy placing one stone here and another stone there, with some invisible architect making everything fit. Turing's 1950 comment about computers being "mansions for the souls that He creates" came to mind. "It is difficult to see why a soul should come to rest in a human body, when from both intellectual and moral viewpoints a computer would be preferable," Olof Johannesson adds.[24]

At the time of my visit, my hosts had just begun a project to digitize all the books in the world. Objections were immediately raised, not by the books' authors, who were mostly long dead, but by book lovers who feared that the books might somehow lose their souls. Others objected that copyright would be infringed. Books are strings of code. But they have mysterious properties—like strings of DNA. Somehow the author captures a fragment of the universe, unravels it into a one-dimensional sequence, squeezes it through a keyhole, and hopes that a three-dimensional vision emerges in the reader's mind. The translation is never exact. In their combination of mortal, physical embodiment with immortal, disembodied knowledge, books have a life of their own. Are we scanning the books and leaving behind the souls? Or are we scanning the souls and leaving behind the books?

"We are not scanning all those books to be read by people," an engineer revealed to me after lunch. "We are scanning them to be read by an AI."

The AI that is reading all these books is also reading everything else—including most of the code written by human programmers over the past sixty years. Reading does not imply understanding—any more than reading a genome allows us to understand an organism—but this particular AI, with or without understanding, is especially successful at making (and acquiring) improvements to itself. Only sixty years ago the ancestor of this code was only a few hundred lines long, and required personal assistance even to locate the next address. Artificial intelligence, so far, requires constant attention—the strategy that infants use. No genuinely intelligent artificial intelligence would reveal itself to us.

Here was Alfvén's vision, brought to life. The Big Computer was doing everything in its power to make life as comfortable as possible for its human symbionts. Everyone was youthful, healthy, happy, and exceptionally well fed. I had never seen so much knowledge in one place. I visited a room where a dedicated fiber-optic line was importing all the data that existed in the world concerning Mars. I listened to an engineer explain how we would all eventually have implanted auxiliary memories, individually initialized with everything we needed to know. Knowledge would become universal, and evil could be edited out. "The primary biological function of the brain was that of a weapon," Alfvén had explained. "It is still not quite clear in which brain circuits the lust for power is located. In any case data machines seem devoid of any such circuits, and it is this which gives them their moral superiority over man; it is for this reason that computers were able to establish the kind of society which man had striven for and so abysmally failed to achieve."[25] I was tempted to sign up.

At the end of the day I had to leave the digital Utopia behind. I relayed my impressions to a compatriot of Alfvén's who had also visited the home of the Big Computer, and who might be able to shed some light. "When I was there, just before the IPO, I thought the coziness to be almost overwhelming," she replied. "Happy golden retrievers running in slow motion through water sprinklers on the lawn. People waving and smiling, toys everywhere. I immediately suspected that unimaginable evil was happening somewhere in the dark corners. If the devil would come to earth, what place would be better to hide?"[26]

The Great Disaster was caused not by the Big Computer, but by human beings unable to resist subverting this power to their own ends. "Evolution on the whole has moved steadily in one direction. While data machines have developed enormously, man has not," Alfvén warned.[27] Our hopes appear to lie with the future according to Olof Johannesson, who, after the world is reconstructed from the Great Disaster, declares, "We believe—or rather we know—that we are approaching an era of even swifter evolution, and even higher living standard, and an even greater happiness than ever before."

"We shall all live happily ever after," ends Alfvén's tale.[28]

Olof Johannesson, however, turned out to be a computer, not a human being. Those who had sought to use the power of computers for destructive purposes discovered that one of those powers was the ability to replace human beings with something else. What if the price of machines that think is people who don't?

The other party is still waiting to collect.

# EIGHTEEN

# The Thirty-ninth Step

*It is easier to write a new code than to under-*
*stand an old one.*

—John von Neumann to Marston Morse, 1952

AT EXACTLY MIDNIGHT on July 15, 1958, in the machine room at the end of Olden Lane, Julian Bigelow turned off the master control, shut down the power supplies, picked up a blunt No. 2 pencil, and made the following entry in the machine log: "Off—12:00 Midnight—JHB." Knowing there would be no log entries to follow, he extended his signature diagonally across the rest of the page.

Within seconds, the cathodes stopped emitting, the heater filaments stopped glowing, and the Williams memory tubes gave up their last traces of electrostatic charge. No electrons would ever flow through these circuits again.

"The other day I saw a ghost—the skeleton of a machine which not so long ago had been very much alive, the cause of much violent controversy," wrote Klári von Neumann, some two years later.

> The computer, alias The Jonnyac, the Maniac, more formally The Institute for Advanced Study Numerical Computing Machine . . . is now locked away, not buried but hidden in the back-room of the building where it used to be the queen. Its life-juice, the electricity, has been cut off; its breathing, the air-conditioning system, has been dismantled. It still has its own little room, one which can only be approached through the big hall which was its ante-chamber used for the auxiliary equipment—now a dead storage hold for empty boxes, old desks and other paraphernalia that invariably finds its way to such places and then is "forgotten with the rest."

Klári had returned to Princeton, after Johnny's death, for the dedication of von Neumann Hall at Princeton University, where the Institute for Defense Analysis was installing a new computer. "The old one, the original, the firster, lies silently in its inglorious tomb," she wrote. "Sic Transit Gloria Mundi."[1]

After von Neumann left Princeton for Washington in 1955, the engineers remaining at the Institute hoped to build a second computer, incorporating a long list of improvements compiled while building the first. "We had enormous numbers of ideas," says Bigelow, "which we never did anything with."[2] On February 29, 1956, however, it was decided "that no new machine should be built at the Institute for Advanced Study; that most of the engineering staff would, therefore, leave to pursue development work at other places; and that the Electronic Computer should be transformed from an experimental project into a tool for the solution of the many computational problems arising in the scientific community of Princeton."[3]

"With Johnny gone the greatness was left out of it," says Harris Mayer, "and the Institute, who really didn't want to have much to do with the MANIAC, was out of the game."[4] On July 1, 1957, the computer was transferred to Princeton University, with the machine remaining in its existing location at the end of Olden Lane. "There are two major changes as compared with the 'Golden Times' under the auspices of the I.A.S.," Hans Maehly, acting director since July 1, 1956, explained to Oppenheimer after the ownership changed. "No coding services will be supplied to the users—except that we shall prepare general program subroutines (the exact opposite used to be the case)," and "there will be a computer time bookkeeping, involving hourly charges and dollars!"[5]

In contrast to the first five years, when the machine was rarely idle, the entry "No Customers" appears regularly in the machine logs for 1957 and 1958. All new projects were placed on hold, except for the development of a higher-level language, which, as Maehly described it, "takes the mathematics and English that the coder writes as his statement of the problem, and turns this into machine coding without the necessity for human intervention."[6] The remaining engineers continued to work on developing user-friendly utilities such as ASBY, a relative address assembly routine, and POST-MORTEM, a debugging routine invoked in the event of a code "stopping in the wrong place or getting into loops, or whatever a program does in its death agonies."[7] FLINT was a floating point interpretive routine. "An interpretive routine is, by definition, a code that 'translates' orders given in a new 'language' into ordinary 'machine language,'" Maehly explained.

"Thus the machine plus FLINT will act like a new machine though no physical changes have been made for that purpose. We shall, therefore, speak of FLINT as if it were a virtual machine."[8]

A computer with floating-point arithmetic keeps track of the position of the decimal (or binary) point. Without floating point, the programmer has to bring numbers "back into focus" as a computation moves along. After debating the question in November of 1945, the IAS group decided to forgo floating point, making more memory directly available to codes, such as Barricelli's, that did not invoke normal arithmetic, or Monte Carlo codes that consumed every available bit. "Von Neumann thought that anybody who was smart enough to use a computer like this, is smart enough to understand the precision requirements of all the processes involved," Bigelow explains. "He never thought that computers would be run by mathematical imbeciles. He thought computers would be run by mathematicians, physicists and research people who were as good as he was."[9] Floating point got in the way of having an entirely empty universe in which to work.

Each memory location held a string of 40 bits, of which the first (leftmost) bit represented the sign (0 for positive numbers, 1 for negative), leaving 39 bits for the number itself. Without floating point, the binary point (equivalent to the decimal point in decimal arithmetic) is fixed just to the right of the first bit. The next 39 positions, going from left to right, represent $2^{-1}$ (½), $2^{-2}$ (¼), $2^{-3}$ (⅛), and so on, all the way to $2^{-39}$ ($\frac{1}{549,755,813,888}$). The computer thus only stores numbers ranging from −1 to +1, to an accuracy of 39 binary places. For reasons that the June 1946 *Preliminary Discussion of the Logical Design of an Electronic Computing Instrument* elaborated in detail, this made the most of the available 1,024 strings of 40 bits.

Elementary arithmetic was either performed thirty-nine-fold in a single operation (in the case of addition or subtraction) or was iterated thirty-nine times (in the case of multiplication or division). Addition and subtraction were precise. Multiplication of two thirty-nine-digit numbers, however, produces a seventy-eight-digit number, and division may produce a number of arbitrary length. The result had to be truncated, and was no longer precise. "Every number x that appears in the computing machine is an approximation of another number x', which would have appeared if the calculation had been performed absolutely rigorously," Burks, Goldstine, and von Neumann explained in 1946.[10] Sooner or later a value has to be chosen for the thirty-ninth digit, discarding the remaining bits. Deciding how to make the approximation took human judgment, and making the approximation, according to the chosen algorithm, was the thirty-ninth step.

FLINT, "which, as far as its user is concerned, transforms our machine into a slower, less sophisticated instrument for which coding is much simpler," insulated the end user from having to communicate directly with the machine. "The planned general external language should be influenced as little as possible by the peculiarities of the machine; in other words, it should be as close as possible to the thinking of the programmer," it was explained. The user "need not know machine language at all, even, and in particular, while debugging his program."[11] Instead of human beings having to learn to write code in machine language, machines began learning to read codes written in human language, a trend that has continued ever since.

Despite this attempt to make things as easy as possible for the new owners, Princeton University had trouble getting the machine to work. "Our efforts to operate it on a regular basis during the past year have been unsuccessful," Henry D. Smyth (author of *Atomic Energy for Military Purposes*) complained in announcing the MANIAC's retirement in July of 1958. "Although it embodies the principles of modern machines it was essentially developmental and not very carefully engineered."[12]

Bigelow disagreed. "Sometime last summer the University crew, who are operating the machine, decided to 'modify and improve it' with the result that after the departure of Bill Keefe, the last of the original training engineers, it went on the blink and was pretty much inoperable from July through November 1957," he reported to the Atomic Energy Commission in 1958. Finally, on December 22, according to Bigelow, Henry Smyth "asked me if I would undertake to get the thing running . . . since the University felt that this was their only chance. I thought it over and, for various reasons such as the fact that one of the men with 11 children derived his income from his job on the project—etc., I tackled the job." Bigelow divided the available personnel into two crews working two full shifts, including weekends, except January 1 and December 25, and by "approximately the 1st of March we got things going pretty well and, with a few minor interruptions, it . . . has computed everything in sight."[13]

"The bewildering developments of the last couple of weeks end[ed] with the decision to close the Maniac on the 1st of July," Martin Schwarzschild wrote to Hedi Selberg on June 6, 1958, reporting the demise of their stellar evolution work. "Your code has run the last couple of weeks wonderfully . . . [and] we have reached a point in the evolution where a new physical situation has arisen, not by the onset of helium burning, as I and [Fred] Hoyle used to expect, but by a convective instability in the helium core caused by the heat flux coming out of this contracting core. . . . I still have no idea what the star will do."[14] Schwarzschild's universe was brought to a halt.

Except for a retrospective account presented at Los Alamos in 1976, Bigelow never spoke or wrote publicly about the MANIAC again. Even the machine's given name was removed. When mathematician Garrett Birkhoff referred to the MANIAC in a paper on numerical hydrodynamics in 1954, he was advised by Herman Goldstine that "I do not believe that the title 'Maniac' is an acceptable one here."[15] The Los Alamos copy became known as the MANIAC, and the original MANIAC became known as MANIAC-0 or simply the "IAS" or "Princeton" Machine. Bigelow arranged for the remains to go to the Smithsonian Institution, and in preparation, all auxiliary equipment was removed. He paid the university $406 cash for "Misc Residual Property" on August 4, 1958, and purchased the remaining "excess electronic gear left over from the extinct computer project" for $275, on December 18, 1959.[16] Gerald Estrin arranged for the original 2,048-word magnetic drum to be donated to the Weizmann Institute in Israel, and the core of the machine was finally transported to Washington, D.C., in 1962.

In exchange for the university having rights to use the computer, free of charge, when it was first constructed, Electronic Computer Project staff had been allowed to enroll as graduate students at the university, a benefit that helped attract young postwar electronic engineers eager to work for von Neumann while obtaining their PhDs. Bigelow kept attending lectures in the physics department until his status was rescinded by the university in 1960. "Since prior arrangement relieves you of the obligation to pay tuition, to avoid further difficulty it seems wise that you no longer continue in the status of enrolled student," he was advised. "You are of course, free to submit a dissertation and present yourself for your Final Oral Examination."[17] The doors that von Neumann had opened were now closed.

When the Institute transferred the computer to the university, it was understood that Institute scholars would be granted access to university computers in return. But when IAS astronomers sought to exercise this privilege, in 1966, the matter ended in dispute. "The transfer of the MANIAC to the University was a generous gesture on the Institute's part, but I am afraid that it turned into something of a disaster for us," Dean Pittendrigh, who was "considering" the matter on behalf of the university, complained. "We spent well over $100,000 on it and got very little useful computation out of it. . . . At any rate, the Institute is most welcome to use the University Computer Center at any time. The machines now in use at the University and the rates we can offer you for their use are listed below."[18] The charges were $110.00 per hour for an IBM 7044, and $137.50 per hour for an IBM 7094. Oppenheimer responded: "Can one 'consider' whether to keep his word?"[19]

Bigelow, alone, remained at the Institute after all the other engineers dispersed. Although his contributions, in von Neumann's assessment, had been "very important, considerably more than one would infer from a superficial inspection of the publications," his lack of academic publications counted against him, even though there were no explicit publication requirements at the IAS.[20] The "Interim Progress Report on the Physical Realization of an Electronic Computing Instrument" may have been the most influential document ever published by the IAS, but it didn't count.[21] Without von Neumann, Bigelow no longer fit in, and the School of Mathematics expected him to make a graceful exit to IBM or to return to an institution such as MIT.

"I think that most of us that were down there—now maybe Bigelow and von Neumann, I won't speak for them, but certainly for myself, and I suspect Pomerene and most of the other engineering group—we were just doing a job, and it was an interesting kind of a job," says Willis Ware, remembering the beginnings in the basement of Fuld Hall. "We didn't have the big foresight and the big omniscience to see all the consequences. Well, they had a little vector started that turned out subsequently to be very important vectors."[22]

How did the von Neumann vector manage to outdistance all the other groups trying to build a practical implementation of Turing's Universal Machine in 1946? The Eckert-Mauchly group and the von Neumann group were both competing for funding and engineers. "Eckert and Mauchly have a contract with the Govt. Bureau of Standards, at first for 1 year and 50 kilobucks," von Neumann reported to Klári in November 1946. "They have started, and shanghaied back 2 of our men, whom we had previously shanghaied from them."[23] By 1949 the Eckert-Mauchly UNIVAC computer was ready to go into production, and adoption of their machine by the U.S. government would have put them firmly in the lead.

"It was decided, after careful consideration . . . that the Bureau should proceed to contract on three UNIVACs from Eckert and Mauchly, one for the Bureau of the Census and two for the Military Establishment," reported an undated memorandum from the National Bureau of Standards electronic computing program, evidently written in 1949. "For just about two days the horizon seemed clear," the report continues.

> No additional obstacles confronted the Bureau. However, this happy state was dispelled quite unexpectedly when the Bureau was informed by Dr. Mina Rees and Colonel Oscar Maier, representing the Office of Naval Research, and the Air Materiel Command, respectively, that the Eckert-Mauchly Computer Corporation had been submitted to a security investigation on which it had not received a "clean bill of health," and that the Bureau therefore should not use ONR and AMC funds for the procurement of UNIVACs from that company. The Bureau was able to continue negotiations only on the basis of one computer rather than three; the Bureau was constrained not to inform the company about the security investigation.[24]

The scales were tipped away from the UNIVAC and toward IMB's "Defense Calculator," later known as the IBM 701—the first copy of which was delivered to Los Alamos in 1953. Eckert and Mauchly fell increasingly into debt, until they were forced, in 1950, to sell their company (and patent portfolio) to Remington Rand—whose vice president was General Leslie Groves. "These machines should find a reasonable market," Goldstine and von Neumann wrote to Groves in 1949, in a nine-page letter that detailed how the ENIAC had been retrofitted to become a stored-program computer, and how Remington Rand could modify their existing punched card equipment to form "an all purpose machine [whose] memory [could] be used to contain not only numerical data but also logical instructions."[25] After acquiring Eckert and Mauchly's Electronic Control Company, Remington Rand filed patent-infringement suits against many of their competitors—except IBM, with whom they established a cross-licensing agreement in 1956.

IBM soon became the dominant force in digital computing and, beginning with von Neumann's one-day-per-month consulting contract, hired much of the talent that had accumulated at the IAS.

James Pomerene joined IBM in 1956, where he led IBM's early efforts to develop cached, high-speed memory architectures and parallel, multiple-core processors, and was appointed an IBM Fellow in 1976— with complete freedom to pursue any avenue of research, equivalent to a permanent membership at IAS. Herman Goldstine left the IAS in 1958 to supervise IBM's mathematical research center, housed temporarily at the Lamb Estate in the Hudson Valley while awaiting the completion of the Thomas J. Watson Research Center at Yorktown Heights, where he continued the tradition of scientific computing he had established at IAS, becoming an IBM Fellow in 1969. "At the Lamb Estate we thought of ourselves as princes of the earth because of our computing support," remembers Ralph Gomory, who joined Goldstine's group in 1959. "Every day a station wagon left the Lamb Estate and went up to Poughkeepsie. It carried our programs and returned the next day with results."[26]

Jack Rosenberg left the IAS for a position at General Electric in Syracuse in 1951, moved to Los Angeles in 1954, and, after von Neumann's death ended his plans to work for the proposed new computing laboratory at UCLA, joined the Los Angeles Scientific Center of IBM. "A long time IBM engineer showed me some circuit diagrams of IBM's first electronic computer, the 701," he remembers. "It was a copy of the Von Neumann computer, which I had developed in 1947–51." Rosenberg was offered an IBM Fellowship in 1969, which he declined, explaining that "the company was too large and corrupt."[27] He still lives in Pacific Palisades, listening to music through a set of the same phase-synchronized "coherent sound" loudspeakers that he had installed in Einstein's house in 1949. Einstein, in appreciation, granted Rosenberg a wide-ranging and candid interview, which Rosenberg recorded on high-fidelity equipment, but will not release. "Einstein said it must never be made public," he explains.[28]

Gerald and Thelma Estrin moved to Israel in 1954 to supervise the construction of the WEIZAC at the Weizmann Institute of Science in Rehovot, returned to the IAS in 1955, and moved to UCLA in 1956, where they helped establish the new Department of Computer Science, nurturing a new generation of entrepreneurial computer scientists, including two of their own daughters, Deborah and Judy Estrin, and Paul Baran of RAND. "It was a marvelous accident, wonderful for everything that happened afterwards, that things didn't get classified," say the Estrins, looking back at the Electronic Computer Project at IAS.[29]

Andrew and Kathleen Booth returned to England, where they remained instrumental in the continued development of digital computing and X-ray crystallography, before moving to Canada in 1962. "Kathleen and I were amused at the concern for our moral wellbeing!" Andrew responded, when shown a copy of the February 1947 discussion between Goldstine and von Neumann concerning their housing arrangements at the IAS.[30]

Joseph and Margaret Smagorinsky helped found the Princeton Geophysical Fluid Dynamics Laboratory, where climate modeling continued from where the IAS meteorology project had left off. Jule Charney and Norman Phillips settled at MIT, forming the nucleus of a computational meteorology group that resolved some of the differences between John von Neumann's reasons for believing that weather could be made predictable and Norbert Wiener's reasons for believing that it could not. Hedi Selberg transferred her expertise to the Princeton Plasma Physics Laboratory, and Ralph Slutz became director of computing at the National Center for Atmospheric Research, in Boulder, Colorado. Richard MclVille and Hewitt Crane went to the Stanford Research Institute, developing, among other things, the ERMA system for electronic clearing of machine-readable checks between banks. Dick Snyder returned to RCA, working on magnetic-core memory but unable to persuade RCA, as Zworykin had managed to with television, to take the lead. Morris Rubinoff returned to the University of Pennsylvania, and for an interval to Philco, where he supervised the design of the Philco 2000, the first fully transistorized computer, with asynchronous arithmetic, a feature that had been developed at IAS. Arthur Burks settled at the University of Michigan, where he founded the Logic of Computers Group in 1949, edited von Neumann's *Theory of Self-Reproducing Automata* (1966), and, with Alice Burks, published the definitive *Who Invented the Computer?* in 2003.

Robert Oppenheimer was stripped of his security clearance in 1954, one day before it would have expired on its own, in a deliberate act of public humiliation that brought postwar dreams of civilian control of nuclear weapons to an end. "The military wanted the whole deal: the laboratories, the computers, the whole future, in nuclear weapons, from A to Z," explains Harris Mayer. "When we set up the AEC, the military was shut out of what was their major firepower, and they never forgot that, and they wanted it back. They got back the command of the nuclear weapons and the computers, and a part of this, a minor part, actually, was to discredit Oppenheimer."[31] His AEC safe and the guards who watched over it in Fuld Hall were removed. The Institute faculty put aside their differences to support Oppenheimer against those who wanted him ousted from the IAS, where he remained as director until 1966, when, stricken with throat cancer, he resigned and moved out of Olden Manor to become our neighbor for his final year. The former ruler of Los Alamos mesa and Olden Farm was now a ghostly pale, thin figure pacing the yard on the other side of our hedge.

Lewis Strauss, Oppenheimer's nemesis, remained a trustee of the Institute until 1968, when, stung by lingering disapproval over his role the Oppenheimer affair, he finally quit. He remained a friend of the FBI, and a memo from the FBI's special agent in charge, New York Office, to the director records that, after returning from Geneva on August 21, 1955, "Admiral and Mrs Lewis STRAUSS and Admiral STRAUSS' aide . . . were met by a liaison agent of this office who facilitated their entry through Customs and extended the usual courtesies. Admiral STRAUSS made a number of favorable comments concerning the Director and the Bureau and one comment was 'Mr. HOOVER is always there when you need him.' "[32]

Abraham Flexner, who died in 1959, had little more to do with the Institute after his departure in 1939. According to his daughter, Jean Lewinson, "when he was through with it it was a complete separation." In 1955, when her father was eighty-eight, she reported that "this summer in Ontario Dr. Flexner swam in water so cold that Mr. Lewinson won't go near it. He doesn't saw wood now but he does fish and take walks."[33]

Vladimir Zworykin died in 1982, working on biomedical applications of electronics and discouraged that television, the invention he had the highest hopes for, had been so misused. After their abortive venture with the Selectron, RCA never again took a decisive lead in digital computing, devoting their resources to commercial television and to their broadcasting spinoff, NBC.

Lewis Fry Richardson lived until 1953, just long enough to see his dreams of numerical weather prediction and fears of unlimited weaponry fulfilled. Although he hailed the ENIAC forecasts of 1950, reported to him by Jule Charney, as "an enormous scientific advance," he had long retired from meteorological work, having switched instead to applying himself to a study, initially published in 1944, of "The Distribution of Wars in Time." The evidence was discouraging. "The agreement with Poisson's law of improbable events draws our attention to the existence of a persistent background of probability," he concluded. "If the beginnings of wars had been the only facts involved, we might have called it a background of pugnacity. But, as the ends of wars have the same distribution, the background appears to be composed of a restless desire for change."[34]

Norbert Wiener died of cardiac arrest on a visit to Stockholm in 1964. Disillusioned over military ambitions in general and the use of nuclear weapons against civilians in particular, the founder of Cybernetics had begun speaking out against military-sponsored research. "Machines can and do transcend some of the limitations of their designers," he warned in the pages of *Time* magazine. "This means that although they are theoretically subject to human criticism, such criticism may be ineffective." The author of *Extrapolation, Interpolation, and Smoothing of Stationary Time Series* saw that ever-faster machines would inevitably leave human beings behind. "By the very slowness of our human activities, our effective control of our machines may be nullified," he added, citing computer-controlled nuclear weapons and computer-controlled manipulation of the stock market as two of the ways that power was being relinquished to the machines.[35]

Stan Ulam lived until 1984, dividing his time between Los Alamos, Boulder, and later Santa Fe. He remained as imaginative and mathematically creative as he had been as a child, circulating among his colleagues still working at Los Alamos and keeping alive the conversations that had begun in Lwów at the Scottish Café. "Just as animals play when they are young in preparation for situations arising later in their lives it may be that mathematics to a large extent is a collection of games," he concluded in 1981, "and may be the only way to change the individual or collective human mind to prepare it for a future that nobody can now imagine."[36] Hordes of self-replicating Turing-complete digital organisms, much as he imagined them in 1952, now populate an unbounded matrix, while the companies and individuals who nurture them are ever more richly rewarded in return.

Edward Teller outlived von Neumann by forty-six years. He remained unrepentant over the development of the hydrogen bomb, but regretted having allowed his testimony at the Oppenheimer hearings to be used to identify Oppenheimer as a security risk, and questioned whether secrecy was a path to security. "Science thrives on openness," he reflected in 1981, "but during World War II we were obliged to put secrecy practices into effect. After the war, the question of secrecy was reconsidered . . . but the practice of classification continued; it was our 'security,' whether it worked or failed. . . . The limitations we impose on ourselves by restricting information are far greater than any advantage others could gain."[37]

Teller had grown up competing with the older and faster Johnny, and when he finally caught up to him, it was a tragic moment for both. "The last few weeks, the last few months of his life, I saw him quite frequently, although I had to cross the continent to come to see him," Teller remembered. "We used to discuss everything in the world. He was incredibly fast. Beyond him, we have never seen. And then in the hospital, he wanted to continue. But he was no longer ahead of me. For Johnny von Neumann, thinking and mathematics was a vital necessity. And he wanted to see me, again and again, because he wanted to prove to himself, 'I still can do it.' But he couldn't."[38]

Near the end, von Neumann, who could no longer work entirely without notes, asked one of his visitors, identified only as "JmcD," for "a note regarding what we talked about last Wednesday," which was recorded as follows:

> We talked somewhat randomly but this was the pattern of it: You said you were in a state of introversion and struggling with a problem of claustrophobia in space and time: in space because your physical body gets in the way; in time because of the slowness of elementary reactions. . . . These problems you said might be overcome with a mechanical device . . . that would project a book page on a photosensitized surface on the ceiling, a phosphorescent pencil for writing on it, and a device with options: to move pages forward and backwards one page or several pages, the luminous pointer to be in several colors with a method of erasure. You said such an invention was difficult but not impossible. . . . The idea is to be able to read and write "pure bred in consciousness without physical interference."[39]

After Johnny's death, Klári remained in Washington, D.C., sorting out his affairs and arranging for the publication of his collected works. Even after a decision to include only previously published papers, the Collected Works still amounted to some 3,689 pages, spread over six volumes and finally published in 1963. Oppenheimer, still struggling with what role the Institute should play with regard to Einstein's papers, was at a loss over what to do when von Neumann's literary godfather appeared. This was "Captain" I. Robert Maxwell, the Czechoslovakian-born publishing magnate and future member of Parliament who offered to undertake the publication of the *Collected Works*, assuring Klári (and Oppenheimer) that "my role in this project is that of the 'mechanic' who has the facilities and the 'know-how' for such a task, and may I say that I am glad to be able to assist with this noble cause."[40]

Maxwell had visited Los Alamos while launching his Pergamon Press, and in addition to befriending the von Neumanns, he became particularly close to the Ulams. "They sent us their children on vacations," says Françoise, whose daughter, Claire, spent her year abroad in Oxford under the Maxwells' wing. "I used to joke that he would either become prime minister or end up in jail," she adds. "He came close to both."

In October 1957, Oppenheimer telephoned Maxwell to try to solidify terms for publication of the *Collected Works*. "If it costs too much it will very much limit usefulness. Do you have any notion?" he asked.

"My idea is about £10," Maxwell replied.

"For the whole thing?"

"Yes."

"Miraculous!" exclaimed Oppenheimer.

"This set is to be my contribution to this man," answered Maxwell, who suggested Eugene Wigner as editor. Oppenheimer suggested mathematician Shizuo Kakutani. Klári suggested geophysicist Carl Eckart, who lived in La Jolla and had been one of the reasons that von Neumann's agreement with the University of California included a clause allowing him to spend as much time at Scripps Institution of Oceanography as he wished.[41]

Eckart declined the assignment, which went to Abraham Taub, but met with Klári to discuss the project. They were married in 1958. It was Klári's fourth marriage.[42] The first had been for romance, the second for money, the third for brains, and the fourth for California. Klári now settled down in La Jolla, just above Windansea Beach, whose antiestablishment surfing culture was soon to be immortalized in *The Pump House Gang*, by Tom Wolfe. Although Carl Eckart and John von Neumann, according to Klári, were "both in similar fields," they were, "as human beings, farther apart" than Johnny had been from the "non-intellectual banker" she had married long ago in the aftermath of her first, gambling husband in Budapest. "For the first time in my life I have relaxed and stopped chasing rainbows," she wrote on the last page of an unfinished memoir of her life.

"La Jolla is a wonderful place and I feel that I do not have to travel anymore because I am there already," she added in a penciled postscript, shortly before her death.[43] Her body was found washed up on Windansea Beach, at the foot of Gravilla Street, at 6:45 a.m., on November 10, 1963, "clothed in a black dress with sleeves about wrist length with black fur cuffs, high neckline, and a zippered back." The body of the dress, "which at first had the appearance of a padded jacket . . . contained approximately 15 pounds of wet sand." Her black sedan (Johnny's last Cadillac) was found parked less than a block away, with the engine cold. Her jewelry was found at home on a coffee table in the living room beside several tumblers "with residual alcoholic contents," and her blood alcohol, tested at 10 a.m., was 0.18%. Upon further investigation, the coroner determined that "she was known to have been a strong surf swimmer," and that, according to her psychiatrist, who "had been able to establish a thin line of the 'death instinct' in her family," she found her husband "disinterested; absorbed in his work; didn't want to go out and mix." Carl Eckart indicated "that he had retired about 3:00 a.m., leaving his wife still up (separate rooms in opposite end of house)." There was no sign of trauma, and her blood chloride levels (left heart 667, right heart 660) were consistent with death by drowning in seawater. (In freshwater, the differential in concentrations would be reversed.) Sand was found in her lungs. Her heart—that appeared to be in good health but had never recovered from the suicide of her father—weighed 280 grams.[44]

"I never cease to wonder about my good fortune, that led me into this exciting maze of people and events," Klári wrote in the introduction to The Grasshopper. "I, a tiny little speck, an insignificant insect just chirruping around to see where the most fun could be had and then, swept up by the hurricane force turbulence of international events and global minds."[45] John von Neumann died at fifty-three; Klári at fifty-two. Caught between the secrecy surrounding her work on nuclear weapons and the shadow of her famous husband, her role in the beginnings of Monte Carlo and the prehistory of programming languages remains obscure. The second half of the twentieth century might have unfolded differently without the contributions of a figure skater who was born a century ago in Budapest.

The bombs that Klári helped bring into existence were a spectacular success. By the time the United States finished testing in the Marshall Islands, there had been forty-three explosions at Enewetak and twenty-three at Bikini, for a total yield of 108 megatons. The computers did their job perfectly, but on Castle Bravo, the successor to Ivy Mike, there was a human error, perhaps the largest human error in history, in failing to account for the generation of tritium from lithium-7 as well as lithium-6. The explosion, on March 1, 1954, was expected to yield some 6 megatons, but yielded over 15 megatons instead. One person, on the Japanese fishing vessel *Lucky Dragon,* was killed directly, and an unknown number of others indirectly over time. The aftereffects forced the evacuation of Rongelap, Rongerik, Ailinginae, and Utirik, and parts of Bikini remain uninhabitable today. Fallout was dispersed throughout the world. The strontium 90 from Ivy Mike and Castle Bravo, taken up in place of calcium in children's teeth, mobilized opposition to atmospheric testing over the next ten years.

First-generation electronic computers fostered first-generation nuclear weapons, and next-generation computers fostered next-generation nuclear weapons, a cycle that culminated in the Internet, the microprocessor, and the multiple-warhead ICBM. Willis Ware, after obtaining his PhD from Princeton, left the Institute in August 1951, working briefly on missile development at North American Aviation before settling in at RAND in Santa Monica, where the JOHNNIAC, an improved copy of the MANIAC, had just been built. The JOHNNIAC (John von Neumann Numerical Integrator and Automatic Computer) was designed to be at least ten times as reliable as its Princeton progenitor, and incorporated a working memory of 40 Selectron tubes, storing 256 bits each. Instead of being used to design thermonuclear weapons, the JOHNNIAC was used to better understand their effects. An extended series of RAND Research Memoranda, with titles such as "Equilibrium Composition and Thermodynamic Properties of Air to 24,000°K," examined what temperatures four times that of the surface of the sun would do to the surface of the earth.

RAND began looking at how to design redundant digital communications networks for coordinating defenses both before and after nuclear attack, prompted by the game theorists' conclusions that a survivable communications network, capable of launching even a handful of remaining missiles, was the best preventive to premeditated attack. Left unstated, but not unconsidered, was the possibility that the survivors of a nuclear attack, instead of making a final suicidal response, might want to coordinate *not* launching a retaliatory strike. "There was a clear but not formally stated understanding," explained Paul Baran, a RAND colleague who helped develop the communication architecture now known as packet switching, "that a survivable communications network is needed to stop, as well as to help avoid, a war."[46]

Baran's study "On Distributed Communications" was released in 1964, and played the same role in the development of the Internet as *Preliminary Discussion of the Logical Design of an Electronic Computing Instrument* had played in the development of the individual machines out of which the Internet was composed.[47] A similar decision was made not to patent or classify the work. "We felt that it properly belonged in the public domain," explained Baran. "Not only would the US be safer with a survivable command and control system, the US would be even safer if the USSR also had a survivable command and control system as well!"[48]

JOHNNIAC gave rise to JOSS (JOHNNIAC Open-Shop System), one of the earliest online, time-shared, multi-user computing environments, and a RAND subdivision, the Systems Development Division, later spun off as the Systems Development Corporation, developed the first million-line codes for the SAGE air defense system, whose legacy survives in all large, real-time computing systems in use today. Many of the assumptions underlying the Internet—from its addressing architecture to its redundancy—go back to RAND's decision to purchase eighty 256-bit Selectrons, ordered in 1951 and delivered in 1952, for $800 each. "There is another item we have been wanting to tell you about . . . which I believe will be a source of satisfaction to you and Julian; even though it makes us look like deviationist inventors," John Williams, director of computing at RAND, wrote to von Neumann in October 1951. "We are placing an order with RCA for 100 Selectrons."[49] One reason RAND was able to accomplish so much, over the next decade, was the head start gained by avoiding the impediment of recalcitrant Williams tubes.

Nicholas Metropolis died in 1999, having helped keep Los Alamos at the forefront of scientific computing since 1943. "Rather than the way the Institute in Princeton worked, where Johnny had to get the money and scramble for it, at Los Alamos Nick had everything he needed," says Harris Mayer, "and also he had Johnny von Neumann still."[50] After von Neumann's death, it was Los Alamos, more than anywhere else, that kept his unfinished agenda alive until its importance was recognized by other institutions in later years.

Robert Richtmyer, who died in 2002, moved from Los Alamos to the Courant Institute at NYU in 1953, and to Boulder, Colorado, in 1964. "I get the impression that machines are now mostly being designed not by problem-solving people but by people who regard machines as an end in themselves," he complained to Nicholas Metropolis in 1956. "John von Neumann's idea to put numbers and instructions in the same kind of memory was a wonderful advance, but it doesn't follow that numbers and instructions have to be interconfusible."[51] Richtmyer, like Bigelow, was surprised that computing remained largely stuck where von Neumann had left off, with machines and codes growing in power and complexity but not in the fundamental way the systems worked. "A curious phenomenon that has accompanied the development of software is a tendency for the hardware to become dependent on it," he observed in 1965.[52]

Von Neumann never returned to pure mathematics, and even his attention to computing was distracted by his duties at the AEC. At the International Congress of Mathematicians held in Amsterdam on September 2–9, 1954, he was invited to give the opening lecture, billed as a survey of "Unsolved Problems in Mathematics" that would update David Hilbert's famous 1900 Paris address. The talk, instead, was largely a rehash of some of von Neumann's own early work. "The lecture was about rings of operators, a subject that was new and fashionable in the 1930s," remembers Freeman Dyson. "Nothing about unsolved problems. Nothing about the future. Nothing about computers, the subject that we knew was dearest to von Neumann's heart. Somebody said in a voice loud enough to be heard all over the hall, 'Aufgewarmte Suppe,' which is German for 'warmed-up soup.'"[53]

Afterward, says Benoît Mandelbrot, "I saw von Neumann leaving the hall. He was all by himself, lost in thought. Nobody was following him, and he was rushing somewhere, by himself." Over the next several days, Mandelbrot noticed an "old man, hanging around with us, and I asked him what he was doing." This was Michael Fekete, with whom von Neumann had published his first paper, at age eighteen, in 1922. Fekete, who had gone on to become the first professor of mathematics at the Hebrew University of Jerusalem, answered that "von Neumann wrote his first paper in collaboration with me. And so he wanted me to write my last paper in collaboration with him." Von Neumann was too preoccupied with his imminent appointment to the AEC, and the symmetry was never achieved.[54]

Later, during the meetings, von Neumann met with Veblen alone. They spoke from 10:00 p.m. on the seventh until 2:00 a.m. on the eighth, partly to discuss the Oppenheimer hearings and partly because von Neumann was about to announce publicly that he was leaving the IAS. "Veblen started with a tirade against L.L.S. [Strauss] and E.T. [Teller] as arch enemies, E.T. being the man who brought Me'lissende [Oppenheimer] down," von Neumann reported to Klári later that morning, using their private code name for Oppenheimer. "I said that I considered L.L.S. a tyrant but the best chairman the AEC ever had, E.T. a fool but a man with merits and who is my personal friend."

"We had only few residual disagreements," von Neumann continued.

> He [Veblen] said that he felt that Me'lissende's resignation, forced or voluntary, would be very bad for the Institute. . . . I was not willing to say that the latter, if properly managed, would not be the best. He also said that Me'lissende had previously told him that he disagreed with my views about a "quick" [preventive] war, but that I might well be right. . . . I told him that I felt that a "quick" war was academic by now, since it would now—or within a rather short time—hardly be "quick."[55]

The reconciliation with Veblen was short-lived, and von Neumann grew increasingly estranged from the mathematical community in which he had spent his youth.

Oswald Veblen died in 1960, at his summer home on the shores of Blue Hill Bay, in Maine, living in the style more of his Norwegian grandfather than of an Institute trustee. Thanks largely to Veblen, some 589 of the Institute's 800 acres remain designated a permanent woodland reserve, and, in 1957, he and his wife, Elizabeth, donated the 81 acres they owned on the outskirts of Princeton to Mercer County, forming the Herrontown Woods nature reserve, "a place where you can get away from cars, and just walk and sit."[56] He never reconciled his differences with von Neumann. "Johnny, even in his last days, in his last month of his life, there was really only one man, one person whom he wanted to see," says Klári. "I wrote begging letters to Veblen to come and visit him, but he did not come."[57]

At 4:50 a.m. on May 27, 1953, during the final push to develop a deliverable hydrogen bomb, the engineers who were running a thermonuclear problem for the AEC were suddenly alarmed by an unknown noise. "Mouse climbed into blower behind regulator rack, set blower to vibrating: result no more mouse & a !!! of a racket," records the machine log (a stronger word having been replaced). Below the logbook entry, one of the engineers drew a gravestone on which was inscribed:

HERE
LIES
MOUSE
BORN
?
DIED
4:50 AM
5/27/53

Another engineer inserted "Marsten," so the tombstone read, HERE LIES MARSTEN MOUSE, a comment directed at Marston Morse, who had long opposed the invasion of the Institute by engineers. Morse, who grew up on a farm in Maine, had his reasons for opposing the computer project, and deserves his own final word.

"In spirit we mathematicians at the Institute would cast our lot in with the humanists," Morse wrote to Aydelotte at the beginning of World War II. He served full-time in the Office of the Chief of Ordnance of the Army for the duration of the war, but believed that, with the war over, the Institute for Advanced Study was no place for weaponeers. "Mathematicians are the freest and most fiercely individualistic of artists," he argued, and the government contracts that supported the computer project, he believed, were in conflict with this.[58]

In October 1950, when the construction of the computer was at its peak, its budget outshadowing the entire budget of the School of Mathematics by more than three to one, Morse went off to Kenyon College, in Ohio, to deliver a talk on "Mathematics and the Arts," at a conference in honor of Robert Frost. "One hundred miles northeast of Derry, New Hampshire, lie the Belgrade Lakes, and out of the last and longest of these lakes flows the Messalonskee," he began.

I was born in its valley, "north of Boston" in the land of Robert Frost. The "Thawing Wind" was there, the "Snow," the
"Birches" and the "Wall" that had to be mended: I was born
on a sprawling farm cut by a pattern of brooks that went
nowhere—and then somewhere. A hundred acres of triangles
of timothy and clover, and twisted quadrilaterals of golden
wire grass, good to look at, and good riddance. At ten I combed
it all with horse and rake, while watching the traffic of mice
beneath the horse's feet.

"One cannot decide between Kronecker and Weierstrass by a calculation," Morse continued, warming up. "There is a center and
final substance in mathematics whose perfect beauty is rational, but
rational 'in retrospect.'" He went on to question "the science of cold
newsprint, the crater-marked logical core, the page that dares not be
wrong, the monstrosity of machines, grotesque deifications of men
who have dropped God, the small pieces of temples whose plans have
been lost and are not desired, bids for power by the bribe of power
secretly held and not understood.

"It is science without its penumbra or its radiance, science after
birth, without intimations of immortality," he concluded. "The creative scientist lives in 'the wildness of logic' where reason is the hand-
maiden and not the master. I shun all monuments that are coldly
legible. It is the hour before the break of day when science turns in
the womb, and, waiting, I am sorry that there is between us no sign
and no language except by mirrors of necessity. I am grateful for the
poets who suspect the twilight zone."[59]

The secrecy Morse so objected to is now permanently entrenched.
The U.S. government now produces more classified information than
unclassified information—and, since even the amount of classified
information is classified, we may never know how much dark matter there is. Von Neumann's monument, however, has turned out not
to be as coldly legible as it first appeared. There will always be truth
beyond the reach of proof.

Alan Turing received the Order of the British Empire in 1946, yet,
under the Official Secrets Act, he could never talk openly about his
wartime work. After leaving the National Physical Laboratory in
1948, he thrived under Max Newman's auspices at the University of
Manchester, where the core of the computing group from Bletchley
Park were continuing from where their work on Colossus had left

off. All went well until 1952, when Turing was convicted on a charge of gross indecency (for homosexuality), forced to undergo "therapy" with estrogen injections, and had his security clearance (and ability to visit the United States) revoked. He died, evidently of cyanide poisoning, at his home in Manchester on June 7, 1954, two weeks before he would have turned forty-two. His promising new results on the chemical basis of morphogenesis were left unfinished, a jar of potassium cyanide was in his home laboratory, and a partially eaten apple was at his side, leaving the circumstances of his death as undecidable as the *Entscheidungsproblem* that had been such a landmark in his life.

With the gradual lifting of secrecy came long-overdue recognition not only of the importance of Turing's contribution to the war, but of the contributions that Colossus, as a physical embodiment of Turing's theoretical principles, had made to the development of hardware and software in the aftermath of World War II. On September 10, 2009, "on behalf of the British government, and all those who live freely thanks to Alan's work," British prime minister Gordon Brown issued a formal apology for the "inhumane" treatment that Turing received. "We're sorry, you deserved so much better," were his closing words.

Kurt Gödel died in Princeton on January 14, 1978, weighing only sixty-five pounds and with malnutrition listed as the cause of death. He never made it to Hanover to search the Leibniz manuscripts for the clues he believed existed as to where digital computing, logical calculus, and universal language were destined to end up. On March 20, 1956, he wrote to von Neumann about a question that "would have consequences of the greatest significance," but never received an answer, von Neumann having given up his correspondence by that time. "It is easy to construct a Turing machine that allows us to decide, for each formula F of the restricted functional calculus and every natural number $n$, whether F has a proof of length $n$," Gödel wrote. "The question is, how rapidly does $\varphi(n)$ [the number of steps required] grow for an optimal machine?" The answer to this question, still unresolved, would determine whether "in spite of the unsolvability of the Entscheidungsproblem," as Gödel put it, "the thinking of a mathematician in the case of yes-or-no questions could be completely replaced by machines."[60]

Nils Barricelli died in Oslo in 1993, no longer engaged in viral genetics, but still working to perfect the new mathematical language, "B-mathematics," which, like the calculus ratiocinator of Leibniz, would establish truth and reveal untruth. Spoken only by a few of his graduate students and running on a DEC System 10 computer, B-mathematics soon went extinct. His numerical evolution experiments also evaporated with barely a trace, leaving many of his ideas to be rediscovered by researchers who had no idea of his earlier work.

Barricelli's universe, however, is our universe now. His primitive, one-dimensional digital organisms—replicating, competing, crossbreeding, and associating symbiotically within a 5-kilobyte matrix— were the ancestors of the multimegabyte (but still one-dimensional) strings of code that are replicating and cross-breeding in the unbounded digital universe of today. What we term "apps" Barricelli would term numerical symbioorganisms, and as he predicted, it is by crossing, symbiotic cooperation and wholesale appropriation of code bases, not by random mutation, that their evolution is moving ahead. Behaving as "collector societies" of social insects, they gather money (and intelligence) to bring back to the collective nest.

Julian Bigelow died in Princeton on February 17, 2003. Six weeks later, the School of Natural Sciences hosted a reception in his honor in their own new building, Bloomberg Hall, preceded by a Quaker memorial service in the Friends meetinghouse at Stony Brook. The meetinghouse, furnished with plain wooden benches and little changed from 1726, was full. At a Friends meeting, silence is a form of communication, an exception to Bigelow's rule that absence of a signal should never be used as a signal.

A nurse who had been in attendance during Bigelow's hospitalization broke the silence first. "Even though he was so tired, he opened his eyes, his wide blue eyes, and many nurses said, 'Look at his eyes, even when he is so sick, look at the expression they have,'" she said.[61] "Julian was never recognized at his true worth; he was pushed into a corner," said Freeman Dyson. "But I never heard a word of complaint. Now it is late, but still not too late to apologize." There were stories about all the things that Bigelow had been able to fix, all the things he had been unable to finish, and all the things, especially used tires, he had been unable to throw away. "There was no problem that couldn't be solved," Ted Merkelson, Julian's stepson, speaking last, explained. "It was just a question of figuring it out and taking the time to do it. And I think he could still solve any problem. He just ran out of time."[62]

The service over, Julian Bigelow's family and friends stepped out into the bright March sunshine and followed the old Princeton-Trenton trolley line between the woods and the Battlefield back to the

Institute, retracing the route taken by Sullivan's column of Washington's army in 1777, passing the Clarke House where General Mercer was left, mortally wounded, after the British retreat. In Bloomberg Hall—now home to the Institute's physicists, astronomers, a 96-node IBM computer cluster (replaced with a 512-core cluster in 2009), and, finally, a handful of theoretical biologists—was a small crowd of people who had gathered to pay their last respects to a man who had been born only forty-two miles from Stony Brook.

Engineers were banished from the Institute, but computers returned. More than nine hundred of them (and two hundred terabytes of storage) are in use across the IAS today, and the former ECP building, now jointly occupied by the Crossroads Nursery School, the Institute Business Office, and a fitness center, displays a plaque commemorating von Neumann, installed by the Hungarian government in 2003. Even the School of Social Science is increasingly devoted to studying the effects that von Neumann's experiment is having on the world.

The basement storeroom in Fuld Hall, where the first workbenches were installed in 1946, was the Institute's main server room until recently, connected to the outside world by some 504 optical fibers, routed through a 45-megabit-per-second switch. In a reversal of Nils Barricelli's attempts to incubate self-propagating numerical organisms, a dedicated network monitoring system now watches over all traffic, trying to keep *out* the endless stream of self-propagating numerical organisms that are now attempting to get *in*. "The viruses are getting so intelligent that it's really an arms race," Rush Taggart, the system administrator in 2005, explained. "It's watching the traffic as it goes by. The machines watch out for the machines."[63]

The arms race being fought in the basement of Fuld (and now Bloomberg) Hall will never be decided in favor of the completely deterministic over the probabilistic and incomplete. The wilderness, even if only a digital wilderness, will always win. There are codes, and machines, that can do almost anything that can be given an exact description, but it will never be possible to determine, simply by looking at a code, what that code will do. No firewall that admits even simple arithmetic can ever be made complete. The digital universe will always leave room for more mysteries than even Robert Frost could dream of. The twilight zone remains.

The 32-by-32-by-40-bit matrix constructed at the end of Olden Lane was initialized with coded instructions, and then given a 10-bit number with orders to go to that location and perform the next instruction—which could have been an instruction to modify the existing instructions—found at that address. Even from so finite a beginning, there was no way to predict the end result.

In November 2000 a cardboard box turned up in the basement of the West Building at the Institute for Advanced Study, where its presence had been overlooked. The smell of burning V-belts still permeated the layer of black, greasy dust that had settled over a collection of World War II teleprinter service manuals that for some reason had not been thrown out when the MANIAC's input/output was switched to punched cards from paper tape. Underneath them was a carton of IBM data processing cards, accompanied by a note written in pencil on half a sheet of lined paper, disintegrated into several fragments, identifying the cards as "Barricelli's Drum Code," with instructions for how it should be loaded and run (on the 2,048-word high-speed magnetic drum that had been added to the computer in 1953). Along with the stack of cards were three sheets of ledger paper, filled with dense handwritten hexadecimal code specifying the laws of nature governing the fossilized universe that was preserved, in a state of suspended animation, on the cards. Here were the Dead Sea scrolls.

The note accompanying the cards (addressed to "Mr. Barricelli" and signed "TWL") concludes with the following statement:

"There must be something about this code that you haven't explained yet."

# KEY TO ARCHIVAL SOURCES

AMT  Alan Turing papers, King's College Archives, Cambridge, UK

CBI  Charles Babbage Institute, University of Minnesota, Minneapolis, Mn.

FJD  Freeman Dyson papers, courtesy of Freeman Dyson

GBD  Author's collections

IAS  Shelby White and Leon Levy Archives Center, Institute for Advanced Study, Princeton, N.J.

IAS-BS  Beatrice Stern files, Shelby White and Leon Levy Archives Center, Institute for Advanced Study, Princeton, N.J.

JHB  Julian Bigelow papers, courtesy of the Bigelow family

KVN  Klári von Neumann papers, courtesy of Marina von Neumann Whitman

LA  Los Alamos National Laboratory, Los Alamos, N.M.

NARA  U.S. National Archives and Records Administration, College Park, Md

OVLC  Oswald Veblen papers, Library of Congress, Washington, D.C.

PM  Priscilla McMillan document archive [http://h-bombbook.com/research/primarysource.html]

RCA  David Sarnoff Library and Archives, RCA, courtesy of Alex Magoun

RF  Rockefeller Foundation Archives, New York, N.Y.

SFU  Stanislaw and Françoise Ulam papers, courtesy of the Ulam family

SUAPS  Stanislaw Ulam Papers, American Philosophical Society, Philadelphia

VNLC  John von Neumann papers, Library of Congress, Washington, D.C.

# NOTES

## PREFACE

1. Willis H. Ware, interview with Nancy Stern, January 19, 1981, CBI, call no. OH 37.
2. John von Neumann, "The Point Source Solution," in *Blast Wave*, Los Alamos Scientific Laboratory, LA-2000 (a compilation of declassified portions of LA-1020 and LA-1021, edited by Hans Bethe, Klaus Fuchs, Joseph Hirschfelder, John Magee, Rudolph Peierls, and John von Neumann. Report written in August 1947, distributed March 27, 1958), p. 28.

## ACKNOWLEDGMENTS

1. Hans Bethe, "Energy Production in Stars," *Physics Today*, September 1968, p. 44.
2. Abraham Flexner, Minutes of the Trustees, April 13, 1936, IAS; Carl Kaysen, Notes on John von Neumann for File, July 12, 1968, IAS.
3. Nicholas Metropolis, in Nicholas Metropolis, J. Howlett, and Gian-Carlo Rota, eds., *A History of Computing in the Twentieth Century* (New York: Academic Press, 1980), p. xvii.

## ONE: 1953

1. "Institute for Advanced Study Electronic Computer Project Monthly Progress Report," March 1953, p. 3, IAS.
2. Gregory Bateson, *Mind and Nature* (New York: Bantam, 1979), p. 228.
3. Francis Bacon, *De augmentis scientiarum*, 1623, translated by Gilbert Wats as *Of the advancement and proficience of Learning, or The Partitions of Sciences* . . . (London, 1640), pp. 265-66.
4. Thomas Hobbes, *Elements of Philosophy: The First Section, Concerning Body, Chapter 1, Computation, or Logique* (London: Andrew Crooke, 1656), pp. 2-3.
5. U.S. Office of Naval Research, *A Survey of Automatic Digital Computers—1953* (Washington, D.C.: Department of the Navy, compiled February 1953).
6. Alan Turing, "Lecture to the London Mathematical Society on 20 February 1947," p. 1, AMT.
7. Memorandum for the Electronic Computer Project, November 9, 1949, IAS.
8. Lewis L. Strauss to J. Robert Oppenheimer, April 10, 1953, IAS.
9. J. Robert Oppenheimer to Lewis Strauss, April 22, 1953, IAS.
10. Jack Rosenberg, interview with author, February 12, 2005, GBD.

11. John von Neumann, "Defense in Atomic War," paper delivered at a symposium in honor of Dr. R. H. Kent, December 7, 1955, in "The Scientific Bases of Weapons," *Journal of the American Ordnance Association* (1955): 23; reprinted in *Collected Works*, vol. 6: *Theory of Games, Astrophysics, Hydrodynamics and Meteorology* (Oxford: Pergamon Press, 1963), p. 525.

12. Discussion at the 258th Meeting of the National Security Council, Thursday, September 8–15, 1955, Eisenhower Papers, Dwight D. Eisenhower Library, Abilene, Kansas (transcript in NASA Sputnik History Collection).

13. Robert Oppenheimer to James Conant, October 21, 1949, in *In the Matter of J. Robert Oppenheimer* (Washington, D.C.: Government Printing Office, 1954), p. 243; minutes, Institute for Advanced Study Electronic Computer Project Steering Committee, March 20, 1953, IAS.

14. James D. Watson and Francis H. C. Crick, "A Structure for Deoxyribose Nucleic Acid," *Nature* 171 (April 25, 1953): 737.

15. Nils Aall Barricelli, "Symbiogenetic Evolution Processes Realized by Artificial Methods," *Methodos* 8, no. 32 (1956): 308.

16. Semiconductor Industry Association World Semiconductor Trade Statistics data for 2010, as presented by Paul Otellini, Intel Investor Meeting, May 17, 2011.

17. Willis Ware, interview with author, January 23, 2004, GBD; Harris Mayer, interview with author, May 13 and 25, 2011, GBD.

## TWO: OLDEN FARM

1. *A Letter from William Penn, Proprietary and Governour of Pennsylvania in America, to the Committee of the Free Society of Traders of that Province, residing in London, 16 August 1683* (London, 1683), p. 3.

2. Chief Tenoughan (Schuylkill River) as noted by William Penn, winter of 1683–1684, in John Oldmixon, *The British Empire in America: Containing the History of the Discovery, Settlement, Progress and present State of all the British Colonies on the Continent and Islands of America*, vol. 1 (London, 1708), p. 162.

3. Samuel Smith, *The History of the Colony of Nova-Caesaria, or New Jersey: containing, an account of its first settlement, progressive improvements, the original and present constitution, and other events, to the year 1721.* (Burlington: James Parker, 1765; second edition, Trenton: William Sharp, 1877), p. 79

4. *The Trial of William Penn and William Mead, at the Sessions held at the Old Baily in London, the 1st, 3d, 4th, and 5th of September, 1670. Done by themselves,* in *A Compleat Collection of State-Tryals, and Proceedings upon High Treason, and other Crimes and Misdemeanours*, vol. 2 (London, 1719), p. 56.

5. *The Trial of William Penn and William Mead*, 2:60.

6. Ibid.

7. William Penn, Petition to Charles II, May 1680, in Jean R. Soderlund, ed., *William Penn and the Founding of Pennsylvania, 1680–1684* (Philadelphia: University of Pennsylvania Press, 1983), p. 23.

8. William Penn to Robert Boyle, August 5, 1683, in *Works of Robert Boyle*, vol. 5 (London, 1744), p. 646.

9. Deed of October 20, 1701 between Penn and Stockton, as quoted in John Frelinghuy-

sen Hageman, *A History of Princeton and its Institutions* (Philadelphia: J. B. Lippincott, 1879) vol. 1, p. 36.

### THREE: VEBLEN'S CIRCLE

1.  Mrs. R. H. Fisher, in Joseph Dorfman, *Thorstein Veblen and His America* (New York: Viking, 1934), p. 504.
2.  Herman Goldstine, interview with Albert Tucker and Frederik Nebeker, March 22, 1985, *The Princeton Mathematics Community in the 1930s: An Oral History Project*, transcript 15, Seeley G. Mudd Manuscript Library, Princeton University, Princeton, N.J. (http://www.princeton.edu/mudd/math); Albert Tucker, interview in *The Princeton Mathematics Community in the 1930s: An Oral History Project*, transcript 15; Abraham Flexner to Herbert Maass, December 15, 1937, IAS.
3.  Herman Goldstine, in Thomas Bergin, ed., *50 Years of Army Computing: From ENIAC to MSRC, a record of a conference held at Aberdeen Proving Ground, Maryland, on November 13 and 14, 1996* (Aberdeen, Md.: U.S. Army Research Laboratory, 2000), p. 32.
4.  Deane Montgomery, interview with Albert Tucker and Frederik Nebeker, March 13, 1985, in *The Princeton Mathematics Community in the 1930s*, transcript 25; Klára von Neumann, *Two New Worlds*, ca. 1963, KVN; Herman Goldstine, interview with Albert Tucker and Frederik Nebeker.
5.  Forest Ray Moulton, in David Alan Grier, "Dr. Veblen Takes a Uniform: Mathematics in the First World War," *American Mathematical Monthly* 108 (October 2001): 928.
6.  Norbert Wiener, *Ex-Prodigy* (New York: Simon and Schuster, 1953), p. 254; ibid., p. 258; ibid., p. 259; ibid., p. 257.
7.  Oswald Veblen to Simon Flexner, October 24, 1923, IAS.
8.  Oswald Veblen to Simon Flexner, February 23, 1924, IAS.
9.  Simon Flexner to Oswald Veblen, March 11, 1924, IAS.
10. Abraham Flexner, *I Remember* (New York: Simon and Schuster, 1940), p. 13; Abraham Flexner, "The Usefulness of Useless Knowledge," *Harper's Magazine*, October 1939, p. 548.
11. Klára von Neumann, *Two New Worlds*.
12. Oswald Veblen to Frank Aydelotte, n.d., IAS.
13. Oswald Veblen to Abraham Flexner, March 19, 1935, IAS.
14. *Science*, New Series 74, no. 1922 (Oct. 30, 1931): 433; Herman Goldstine, interview with Albert Tucker and Frederik Nebeker.
15. Oswald Veblen to Albert Einstein, April 17, 1930, IAS-BS.
16. Albert Einstein to Oswald Veblen, April 30, 1930, IAS-BS.
17. Herbert H. Maass, *Report on the Founding and Early History of the Institute*, n.d., ca. 1955, IAS; Abraham Flexner, "The American University," *Atlantic Monthly*, vol. 136, October 1925, pp. 530–41; Maass, *Report on the Founding and Early History of the Institute*.
18. Flexner, *I Remember*, p. 356.
19. Abraham Flexner, *Universities: American, English, German* (New York: Oxford University Press, 1930), p. 217.
20. Louis Bamberger and Carrie Fuld, letter to accompany codicil to their wills, draft, n.d., ca. January 1930, IAS.
21. Oswald Veblen to Abraham Flexner, January 1930, in Beatrice Stern, *A History of the*

*Institute for Advanced Study, 1930–1950,* 1:126, available at library.ias.edu/files/stern_pt1 .pdf; Abraham Flexner to Oswald Veblen, January 27, 1930, IAS.

22. Louis Bamberger to the Trustees, June 4, 1930, IAS.

23. Flexner, "The Usefulness of Useless Knowledge," p. 551.

24. Julian Huxley to Abraham Flexner, December 11, 1932, IAS-BS; Louis Bamberger to the Trustees, April 23, 1934, IAS.

25. Oswald Veblen to Abraham Flexner, June 19, 1931, IAS.

26. Charles Beard to Abraham Flexner, June 28, 1931, in Stern, *A History of the Institute for Advanced Study, 1930–1950,* 1:104; Felix Frankfurter to Frank Aydelotte, December 16, 1933, IAS.

27. Abraham Flexner to the Trustees, September 26, 1931, IAS.

28. Flexner, "The Usefulness of Useless Knowledge," p. 551.

29. Abraham Flexner, "University Patents," *Science* 77, no. 1996, March 31, 1933, p. 325; Flexner, "The Usefulness of Useless Knowledge," p. 544.

30. Abraham Flexner to Trustees, September 26, 1931, IAS; ibid.

31. Abraham Flexner to Louis Bamberger, March 15, 1935, IAS.

32. Herbert Maass to Abraham Flexner, June 9, 1931, IAS; Edgar Bamberger to Abraham Flexner, December 9, 1931, IAS; Maass, *Report on the Founding and Early History of the Institute.*

33. Abraham Flexner to Oswald Veblen, December 22, 1932, IAS.

34. John von Neumann to Abraham Flexner, April 26, 1933, IAS.

35. Harry Woolf, ed., *A Community of Scholars: The Institute for Advanced Study Faculty and Members, 1930–1980* (Princeton, N.J.: Institute for Advanced Study, 1980), p. ix.

36. Albert Einstein to Queen Elisabeth of Belgium, November 20, 1933 (Einstein Archives, Hebrew University, Jerusalem, call no. 32-369.00).

37. Oswald Veblen to Abraham Flexner, April 12, 1934, IAS; Abraham Flexner to Herbert Maass, October 18, 1932, IAS.

38. Herbert Maass to Abraham Flexner, November 9, 1932, IAS; Oswald Veblen to Abraham Flexner, March 13, 1933, IAS; Louis Bamberger to Abraham Flexner, October 29, 1935, IAS; Herbert Maass, Minutes of the Trustees, April 13, 1936, IAS.

39. Abraham Flexner to Louis Bamberger, October 28, 1935, IAS; Abraham Flexner to Louis Bamberger, December 19, 1935, IAS.

40. Oswald Veblen to Frank Aydelotte, February 13, 1936, IAS.

41. Herman Goldstine, interview with Nancy Stern, August 11, 1980, CBI, call no. OH 18.

42. Watson Davis, "Super-University for Super-Scholars," *The Science News-Letter* 23, no. 616 (Jan. 28, 1933): 54; Flexner, *I Remember,* p. 375; ibid., pp. 377–78; Frank Aydelotte to Herbert H. Maass, June 15, 1945, IAS.

43. Thorstein Veblen, *The Higher Learning in America* (New York: B.W. Huebsch, 1918), p. 45.

44. Flexner, *I Remember,* pp. 361 and 375.

45. Abraham Flexner to Frank Aydelotte, November 15, 1939, IAS-BS; Klára von Neumann, *Two New Worlds.*

46. Frank Aydelotte, Report of the Director, May 19, 1941, IAS.

47. Woolf, ed., *A Community of Scholars,* p. 130.

48. Oswald Veblen to Abraham Flexner, March 24, 1937, IAS-BS; J. B. S. Haldane, November 12, 1936, IAS-BS.

49. Deane Montgomery, interview with Albert Tucker and Frederik Nebeker, March 13, 1985.

50. Benoît Mandelbrot, interview with author, May 8, 2004, GBD.

51. P. A. M. Dirac to IAS Trustees, n.d., FJD; J. Robert Oppenheimer to Oswald Veblen, May 27, 1959, IAS.

52. Freeman J. Dyson to S. Chandrasekhar, M. J. Lighthill, Sir Geoffrey Taylor, Sydney Goldstein, and Sir Edward Bullard, October 20, 1954, IAS.


## FOUR: NEUMANN JÁNOS

1. Klára von Neumann, *The Grasshopper*, ca. 1963, KVN.

2. Nicholas Vonneumann, interview with author, May 6, 2004, GBD.

3. Nicholas Vonneumann, *John von Neumann as Seen by His Brother* (Meadowbrook, Pa.: Nicholas Vonneumann, 1987), p. 17.

4. Nicholas Vonneumann, interview with author.

5. Stanislaw Ulam, *Adventures of a Mathematician* (New York: Scribner's, 1976), p. 80; Herman Goldstine, *The Computer from Pascal to von Neumann* (Princeton, N.J.: Princeton University Press, 1972), p. 167.

6. Vonneumann, *John von Neumann as Seen by His Brother*, p. 9.

7. Ibid., p. 10.

8. John von Neumann, statement upon nomination to membership in the AEC, March 8, 1955, VNLC.

9. Nicholas Vonneumann, interview with author.

10. Vonneumann, *John von Neumann as Seen by His Brother*, pp. 23, 16.

11. Ibid., p. 24.

12. Nicholas Vonneumann, interview with author.

13. Stanislaw Ulam, "John von Neumann: 1903–1957," *Bulletin of the American Mathematical Society* 64, no. 3, part 2 (May 1958): 1.

14. Klára von Neumann, *Johnny*, ca. 1963, KVN; Ulam, "John von Neumann: 1903–1957," 2:37.

15. John von Neumann to Stan Ulam, December 9, 1939, SFU; Oskar Morgenstern, in *John von Neumann*, documentary produced by the Mathematical Association of America, 1966.

16. Klára von Neumann, *Johnny*.

17. John von Neumann and Oskar Morgenstern, *Theory of Games and Economic Behavior* (Princeton, N.J.: Princeton University Press, 1944), p. 2; Samuelson, "A Revisionist View of Von Neumann's Growth Model," in M. Dore, S. Chakravarty, and Richard Goodwin, eds., *John von Neumann and Modern Economics* (Oxford: Oxford University Press, 1989), p. 121.

18. Klára von Neumann, *Johnny*.

19. Edward Teller, in Jean R. Brink and Roland Haden, "Interviews with Edward Teller and Eugene P. Wigner," *Annals of the History of Computing* 11, no. 3 (1989): 177.

20. Herman H. Goldstine, "Remembrance of Things Past," in Stephen G. Nash, ed., *A History of Scientific Computing* (New York: ACM Press, 1990), p. 9.

21. Klára von Neumann, *Johnny*; Cuthbert C. Hurd, interview with Nancy Stern, January 20, 1981, CBI, call no. OH 76.

22. Klára von Neumann, *Johnny.*

23. Françoise Ulam, "From Paris to Los Alamos," unpublished, July 1994, SFU; Klára von Neumann, *Johnny.*

24. Herman Goldstine, interview with Albert Tucker and Frederik Nebeker; Nicholas Vonneumann, interview with author.

25. Ulam, *Adventures of a Mathematician,* pp. 65, 79; Vincent Ford to Stan Ulam, May 18, 1965, SUAPS.

26. Martin Schwarzschild, interview with William Aspray, November 18, 1986, CBI, call no. OH 124.

27. Paul R. Halmos, "The Legend of John von Neumann," *American Mathematical Monthly* 80, no. 4 (April 1973): 394; ibid.; Eugene Wigner, "Two Kinds of Reality," *The Monist* 49, no. 2 (April 1964), reprinted in *Symmetries and Reflections* (Cambridge, Mass.: MIT Press, 1967), p. 198.

28. Raoul Bott, interview with author, March 10, 2005, GBD.

29. Ulam, "John von Neumann: 1903–1957," 2:2; Eugene P. Wigner, *The Recollections of Eugene P. Wigner, as Told to Andrew Szanton* (New York and London: Plenum Press, 1992), p. 51.

30. Theodore von Kármán (with Lee Edson), *The Wind and Beyond: Theodore von Kármán, Pioneer in Aviation and Pathfinder in Space* (Boston and Toronto: Little, Brown and Co., 1967), p. 106.

31. Abraham A. Fraenkel to Stan Ulam, November 11, 1957, SUAPS.

32. Ulam, "John von Neumann: 1903–1957," 2:11–12.

33. Ibid., 2:12.

34. Paul Halmos, in *John von Neumann,* documentary.

35. Samuelson, "A Revisionist View of Von Neumann's Growth Model," p. 118.

36. Klára von Neumann, *Two New Worlds.*

37. Wigner, *Recollections of Eugene P. Wigner,* p. 134.

38. Klára von Neumann, *Two New Worlds;* John von Neumann to Oswald Veblen, January 11, 1931, OVLC.

39. Klára von Neumann, *Johnny.*

40. John von Neumann to Oswald Veblen, April 3, 1933, OVLC.

41. Klára von Neumann, *Two New Worlds.*

42. Klára von Neumann, *Johnny;* Marina von Neumann Whitman, interview with author, May 3, 2010, GBD.

43. John von Neumann to Klára von Neumann, n.d., evidently summer 1949, KVN.

44. Israel Halperin, interview with Albert Tucker, May 25, 1984, in *Princeton Mathematics Community in the 1930s,* transcript 18.

45. Robert D. Richtmyer, "People Don't Do Arithmetic," unpublished, 1995; Morgenstern, in *John von Neumann,* documentary; Richtmyer, "People Don't Do Arithmetic."

46. Klára von Neumann, *Two New Worlds;* Abraham Flexner to Oswald Veblen, July 26, 1938, in Stern, *A History of the Institute for Advanced Study, 1930–1950,* 1:396.

47. Marina von Neumann Whitman, interview with author, May 3, 2010; John von Neumann to Klára von Neumann, October 25, 1946, KVN.

48. Cuthbert C. Hurd, interview with Nancy Stern.

49. Marina von Neumann Whitman, interview with author, February 9, 2006, GBD; Herman Goldstine, interview with Albert Tucker and Frederik Nebeker.

50. John von Neumann to F. B. Silsbee, July 2, 1945, VNLC; Ulam, *Adventures of a Mathematician*, p. 78; Goldstine, *The Computer from Pascal to von Neumann*, p. 176.

51. Ulam, *Adventures of a Mathematician*, pp. 231–32.

52. John von Neumann to Saunders Mac Lane, May 17, 1948, VNLC; Ulam, "John von Neumann: 1903–1957," 2:5; Lewis Strauss to Stanislaw Ulam, November 12, 1957, SUAPS.

53. John von Neumann to Stan Ulam, November 8, 1940, SFU.

54. John von Neumann to J. Robert Oppenheimer, February 19, 1948, VNLC; John von Neumann to L. Roy Wilcox, December 26, 1941, KVN.

55. John von Neumann, "Theory of Shock Waves," Progress Report to the National Defense Research Committee, August 31, 1942, reprinted in *Collected Works*, vol. 6: *Theory of Games, Astrophysics, Hydrodynamics and Meteorology*, p. 19.

56. Martin Schwarzschild, interview with William Aspray.

57. John von Neumann, "Oblique Reflection of Shocks," Explosives Research Report No. 12, Navy Dept., Bureau of Ordnance, October 12, 1943, reprinted in *Collected Works*, vol. 6: *Theory of Games, Astrophysics, Hydrodynamics and Meteorology*, p. 22.

58. Klára von Neumann, *Johnny*.

59. John von Neumann to John Todd, November 17, 1947, in John Todd, "John von Neumann and the National Accounting Machine," *SIAM Review* 16, no. 4 (October 1974): 526.

60. Nicholas Metropolis and E. C. Nelson, "Early Computing at Los Alamos," *Annals of the History of Computing* 4, no. 4 (October 1982): 352.

61. John von Neumann to Klára von Neumann, September 22, 1943, KVN; John von Neumann to Klára von Neumann, September 24, 1943, KVN.

62. Nicholas Metropolis and Francis H. Harlow, "Computing and Computers: Weapons Simulation Leads to the Computer Era," *Los Alamos Science* 7 (Winter/Spring 1983): 132.

63. Richard P. Feynman, "Los Alamos from Below: Reminiscences of 1943–1945," *Engineering and Science* 39, no. 2 (January–February 1976): 25.

64. Metropolis and Harlow, "Computing and Computers," p. 134.

65. Feynman, "Los Alamos from Below," p. 25.

66. Metropolis and Nelson, "Early Computing at Los Alamos," p. 351.

67. Feynman, "Los Alamos from Below," p. 28.

68. Klára von Neumann, *Johnny*.

69. "Allocution Pronounced by the Reverend Dom Anselm Strittmatter at the Obsequies of Professor John von Neumann, in the chapel of Walter Reed Hospital, February 11, 1957," in Vonneumann, *John von Neumann as Seen by His Brother*, p. 64.

70. Vonneumann, *John von Neumann as Seen by His Brother*, pp. 14–15.

71. Klára von Neumann, *Johnny*.

72. Ibid.

73. Marina von Neumann to Klára von Neumann, August 28, 1945, KVN.

FIVE: MANIAC

1. Minutes of the Institute for Advanced Study Electronic Computer Project, Meeting #1, November 12, 1945, IAS.

2. Vladimir Zworykin, unpublished autobiography, n.d., ca. 1975, p. 24 (in Bogdan Maglich, unpublished Zworykin biography, n.d., courtesy of Bogdan Maglich).

3. Record for Dr. Craig Waff of the conversation with Dr. Zworykin, September 4, 1976, in Maglich, unpublished Zworykin biography.

4. Jan Rajchman, "Vladimir Kosma Zworykin, 1889–1982," *Biographical Memoirs of the National Academy of Sciences*, vol. 88 (Washington, D.C.: National Academies Press, 2006), p. 12.

5. Herbert H. Maass to Frank Aydelotte, October 17, 1945, IAS.

6. FBI SAC (special agent in charge) Newark to Director, FBI, December 6, 1956, after Albert Abramson, *Zworykin, Pioneer of Television* (Urbana and Chicago: University of Illinois Press, 1995), p. 199.

7. Vladimir K. Zworykin, "Some Prospects in the Field of Electronics," *Journal of the Franklin Institute* 251, no. 1 (January 1951): 235–36.

8. Jan Rajchman, "Early Research on Computers at RCA," in Metropolis, Howlett, and Rota, eds., *A History of Computing in the Twentieth Century* (New York: Academic Press, 1980), p. 465.

9. Jan Rajchman, interview with Richard R. Mertz, October 26, 1970, National Museum of American History Computer Oral History Collection, Washington, D.C.

10. Richard L. Snyder Jr. and Jan A. Rajchman, *Calculating Device*: Patent No. 2,424,389, patented July 22, 1947, application July 30, 1943; Jan Rajchman, interview with Richard R. Mertz.

11. Jan Rajchman, "The Selectron," in Martin Campbell-Kelly and Michael R. Williams, eds., *The Moore School Lectures (1946)*, Charles Babbage Institute Reprint Series No. 9 (Cambridge, Mass: MIT Press, 1985), p. 497.

12. Rajchman, "Early Research on Computers at RCA," p. 466; Jan Rajchman, interview with Richard R. Mertz; Jan A. Rajchman, *Electronic Computing Device*, U.S. Patent Office Patent Number 2,428,811, application October 30, 1943, patented October 14, 1947, assigned to Radio Corporation of America.

13. Jan Rajchman, interview with Richard R. Mertz.

14. Goldstine, August 16, 1944, in *The Computer from Pascal to von Neumann*, p. 166.

15. Herman Goldstine, interview with Albert Tucker and Frederik Nebeker; ibid.

16. "Report on History" ("setting forth very briefly the relationship of the ENIAC, EDVAC and the Institute machine"), from Herman H. Goldstine and John von Neumann to Colonel G. F. Powell, February 15, 1947, IAS.

17. Willis Ware, interview with author; J. Presper Eckert, interview with Nancy Stern, October 28, 1977, CBI, OH 13.

18. John W. Mauchly, "The Use of High Speed Vacuum Tube Devices for Calculating," August 1942, reprinted in Brian Randell, ed., *The Origins of Digital Computers: Selected Papers* (New York: Springer-Verlag, 1982), pp. 355–58.

19. Nicholas Metropolis, "The Beginning of the Monte Carlo Method," in *Los Alamos Science*, no. 15, Special Issue: *Stanislaw Ulam, 1909–1984* (1987): 125.

20. John G. Brainerd, "Genesis of the ENIAC," *Technology and Culture* 17, no. 3. (July 1976): 487.

21. Harry L. Reed, November 14, 1996, in Thomas Bergin, ed., *50 Years of Army Computing*, p. 153; Jan Rajchman, interview with Mark Heyer and Al Pinsky, July 11, 1975, IEEE Oral History Project.

22. J. Presper Eckert, "The ENIAC," in Metropolis, Howlett, and Rota, eds., *A History of Computing in the Twentieth Century,* p. 528; Karl Kempf, *Electronic Computers Within the Ordnance Corps* (Aberdeen Proving Ground, Md.: History Office, November 1961).

23. John von Neumann, Memo on Mechanical Computing Devices, to Col. L. E. Simon, Ballistic Research Laboratory, January 30, 1945, VNLC; Nicholas Metropolis to Klára von Neumann, February 15, 1949, KVN; Nicholas Metropolis, "The Los Alamos Experience, 1943–1954," in Nash, ed., *A History of Scientific Computing,* p. 237.

24. Brainerd, "Genesis of the ENIAC," p. 488; Goldstine, *The Computer from Pascal to von Neumann,* p. 149.

25. U.S. Army War Department, Bureau of Public Relations, Press Release, *Ordnance Department Develops All-Electronic Calculating Machine,* February 16, 1946; Samuel H. Caldwell to Warren Weaver, January 16, 1946, RF.

26. Herman Goldstine, interview with Albert Tucker and Frederik Nebeker; Herman Goldstine, November 13, 1996, in Bergin, ed., *50 Years of Army Computing,* p. 33.

27. Eckert, "The ENIAC," p. 525.

28. Goldstine, "Remembrance of Things Past," p. 9.

29. J. Presper Eckert, interview with Nancy Stern.

30. John W. Mauchly, "The ENIAC," in Metropolis, Howlett, and Rota, eds., *A History of Computing in the Twentieth Century,* p. 545; ibid. pp. 547–48.

31. John von Neumann, lecture at the University of Illinois, December 1949, in Arthur Burks, ed., *Theory of Self-Reproducing Automata* (Urbana: University of Illinois Press, 1966), p. 40.

32. Arthur W. and Alice R. Burks, interview with Nancy Stern, June 20, 1980, CBI, call no. OH 75.

33. Summary of *Honeywell Inc. v. Sperry Rand Corp.,* No. 4-67 Civ. 138, Decided Oct. 19, 1973, in *United States Patents Quarterly* 180 (March 25, 1974); 682, 693–94.

34. Mauchly, "The ENIAC," in Metropolis, Howlett, and Rota, eds., *A History of Computing in the Twentieth Century,* p. 547.

35. "Report on History."

36. John von Neumann to Warren Weaver, November 2, 1945, RF; M. H. A. Newman, quoted by I. J. Good in "Turing and the Computer," *Nature* 307 (February 1, 1984): 663.

37. "Report on History."

38. Jan Rajchman, interview with Richard R. Mertz.

39. John von Neumann to J. Robert Oppenheimer, August 1, 1944, LA.

40. John W. Mauchly, letter to the editor, *Datamation* 25, no. 11 (1979).

41. Herman Goldstine, interview with Nancy Stern.

42. John von Neumann, "First Draft of a Report on the EDVAC," Contract No. W-670-ORD-4926 between the U.S. Army Ordnance Department and the University of Pennsylvania, Moore School of Electrical Engineering, University of Pennsylvania, June 30, 1945, p. 1.

43. John von Neumann to M. H. A Newman, March 19, 1946, VNLC.

44. Julian Bigelow, interview with Nancy Stern, August 12, 1980, CBI, call no. OH3.

45. John W. Mauchly, letter to the editor.

46. J. Presper Eckert, interview with Nancy Stern.

47. John von Neumann to Stanley Frankel, October 29, 1946, VNLC.
48. John von Neumann, deposition concerning EDVAC report, n.d., 1947, IAS.
49. Willis H. Ware, "The History and Development of the Electronic Computer Project at the Institute for Advanced Study," RAND Corporation Memorandum P-377, March 10, 1953, p. 6; Arthur W. Burks, interview with William Aspray, June 20, 1987, CBI, call no. OH 136.
50. Willis H. Ware, interview with Nancy Stern.
51. Retainer agreement between von Neumann and IBM, May 1, 1945, VNLC; J. Presper Eckert, interview with Nancy Stern.
52. John von Neumann to Stanley Frankel, October 29, 1946, VNLC.
53. Von Neumann, "First Draft of a Report on the EDVAC," p. 74.
54. Julian H. Bigelow, "Report on Computer Development at the Institute for Advanced Study," for the International Research Conference on the History of Computing, Los Alamos, June 10–15, 1976, draft, n.d. (quoted text was deleted from the published version), JHB; Norbert Wiener to John von Neumann, March 24, 1945, VNLC.
55. James B. Conant to Frank Aydelotte, October 31, 1945, IAS; James Alexander to Frank Aydelotte, August 25, 1945, IAS.
56. Julian Bigelow, interview with Richard R. Mertz, January 20, 1971, Computer Oral History Collection, Archives Center, National Museum of American History, Washington, D.C.; Frank Aydelotte to James W. Alexander, August 22, 1945, IAS; Report of the Anglo-American Committee of Inquiry, April 20, 1946 (Washington, D.C.: Department of State, 1946). Transcript at http://avalon.law.yale.edu/subject_menus/angtoc.asp.
57. Klára von Neumann, *Johnny*; E. A. Lowe to Frank Aydelotte, October 10, 1947, IAS.
58. Frank Aydelotte to John von Neumann, January 22, 1946, IAS; minutes of the School of Mathematics, June 2, 1945, IAS.
59. John von Neumann to Frank Aydelotte, August 5, 1945, IAS.
60. Frank Aydelotte to Samuel S. Fels, September 12, 1945, IAS.
61. Warren Weaver to Frank Aydelotte, October 1, 1945, IAS.
62. John von Neumann to Warren Weaver, November 2, 1945, RF.
63. Marston Morse to Warren Weaver, January 15, 1946, RF.
64. Samuel H. Caldwell to Warren Weaver, January 16, 1946, RF.
65. John von Neumann to Lewis L. Strauss, October 20, 1945, IAS.
66. John von Neumann to Lewis L. Strauss, October 24, 1945, IAS.
67. Herman H. Goldstine, Memo to Mr. Fleming, April 20, 1951, IAS.
68. James Pomerene, interview with Nancy Stern, September 26, 1980, CBI, call no. OH 31.
69. J. Presper Eckert, interview with Nancy Stern, October 28, 1977, CBI, call no. OH 13.
70. Stanley Frankel to Brian Randell, 1972, in Brian Randell, "On Alan Turing and the Origins of Digital Computers," *Machine Intelligence* 7 (1972): 10.
71. Klára von Neumann, *The Computer*, ca. 1963, KVN.

SIX: FULD 219

1. Abraham Flexner to Herbert Maass, December 15, 1937, IAS; Abraham Flexner to Louis Bamberger, December 1, 1932, IAS.
2. Klára von Neumann, *Two New Worlds*.
3. Abraham Flexner to Oswald Veblen, January 6, 1937, IAS; Abraham Flexner to Frank Aydelotte, August 7, 1938, IAS.
4. James Hudson, *Clouds of Glory: American Airmen Who Flew with the British During the Great War* (Fayetteville and London: University of Arkansas Press, 1990), p. 34.
5. Minutes of the Meeting of the Standing Committee of the Faculty, February 18, 1946, IAS.
6. Bernetta Miller, quoted by Joseph Felsenstein, interview with author, March 20, 2007, GBD; Joseph Felsenstein, interview with author.
7. Bernetta A. Miller, "Report on IAS Food Conservation," May 17, 1946, IAS.
8. Bernetta Miller to Frank Aydelotte, September 3, 1946, IAS; Bernetta Miller to Frank Aydelotte and J. Robert Oppenheimer, September 24, 1947; Bernetta Miller to J. Robert Oppenheimer, December 3, 1947, IAS.
9. Bernetta Miller, quoted by Joseph Felsenstein, interview with author.
10. Oswald Veblen, "Remarks on the Foundations of Geometry" (December 31, 1924), in *Bulletin of the American Mathematical Society* 31, nos. 3–4 (1925): 141.
11. Stanislaw Ulam, "Conversations with Gian-Carlo Rota" (unpublished transcripts by Françoise Ulam, compiled 1985), SFU.
12. John von Neumann to Kurt Gödel, November 30, 1930, in Solomon Feferman, ed., *Collected Works*, vol. 5 (Oxford: Oxford University Press, 2003), p. 337.
13. Ulam, *Adventures of a Mathematician*, p. 76.
14. John von Neumann, remarks made at the presentation of the Einstein Award to Kurt Gödel at the Princeton Inn, March 14, 1951, VNLC.
15. Kurt Gödel, "Über formal unentscheidbare Sätze der Principia Mathematica und verwandter Systeme I," *Monatshefte für Mathematik und Physik* 38 (1931), translated as "On Formally Undecidable Propositions of *Principia Mathematica* and Related Systems I," in Kurt Gödel, *Collected Works*, vol. 1 (Oxford: Oxford University Press, 1986), p. 147.
16. Frank Aydelotte to Dr. Max Gruenthal, December 5, 1941, IAS; Marston Morse, Minutes of the Meeting of the IAS School of Mathematics, February 14, 1950, IAS.
17. A. M. Warren to Abraham Flexner, October 10, 1939, IAS.
18. John von Neumann to Abraham Flexner, September 27, 1939, IAS.
19. Kurt Gödel to Frank Aydelotte, January 5, 1940, IAS.
20. Stan Ulam to John von Neumann, June 18, 1940, VNLC.
21. Frank Aydelotte to Herbert Maass, September 29, 1942, IAS; Bernetta Miller to the Department of Motor Vehicles, June 4, 1943, IAS.
22. Frank Aydelotte, Memo for the Standing Committee, December 25, 1941, IAS.
23. Kurt Gödel to Earl Harrison, Department of Justice, March 12, 1942, IAS.
24. Earl G. Harrison to Kurt Gödel, March 19, 1942, IAS; Frank Aydelotte to Benjamin F. Havens, March 21, 1942, IAS.
25. Benjamin F. Havens to Frank Aydelotte, March 27, 1942, IAS.
26. Alan M. Turing to Institute Secretary (Gwen) Blake, December 16, 1941, IAS.

27. Frank Aydelotte to Max Gruenthal, December 5, 1941, IAS.

28. Frank Aydelotte to Max Gruenthal, December 2, 1941, IAS; Max Gruenthal to Frank Aydelotte, December 4, 1941, IAS.

29. Frank Aydelotte to the Selective Service Board, April 14, 1943, IAS.

30. Cevillie O. Jones to Frank Aydelotte, April 20, 1943, IAS.

31. Frank Aydelotte to the Selective Service Board, May 19, 1943, IAS.

32. John von Neumann to Oswald Veblen, November 30, 1945, OVLC.

33. "Notes on Kurt Gödel," March 17, 1948, IAS.

34. Kurt Gödel to J. Robert Oppenheimer, September 6, 1949, IAS.

35. Stanislaw Ulam to Solomon Feferman, July 13, 1983, SUAP; John von Neumann to Oswald Veblen, November 30, 1945, OVLC.

36. Arthur W. and Alice R. Burks, interview with Nancy Stern.

37. Oswald Veblen to Frank Aydelotte, September 12, 1941, IAS.

38. Frank Aydelotte, Appendix to the Report of the Director, February 24, 1941, IAS.

39. Minutes of the Meeting of the Professors of the School of Mathematics, February 13, 1946, IAS; Arthur W. and Alice R. Burks, interview with Nancy Stern.

40. Arthur W. and Alice R. Burks, interview with Nancy Stern.

41. Arthur W. Burks, Herman H. Goldstine, and John von Neumann, *Preliminary Discussion of the Logical Design of an Electronic Computing Instrument* (Princeton, N.J.: Institute for Advanced Study, 1946), p. 53.

42. Herman H. Goldstine to Colonel G. F. Powell, Office of the Chief of Ordnance, May 12, 1947, IAS.

43. Norbert Wiener, "Back to Leibniz!" *Technology Review* 34 (1932): 201; Norbert Wiener, "Quantum Mechanics, Haldane, and Leibniz," *Philosophy of Science* 1, no. 4 (October 1934): 480.

44. G. W. Leibniz to Henry Oldenburg, December 18, 1675, in H. W. Turnbull, ed., *The Correspondence of Isaac Newton*, vol. 1 (Cambridge, UK: Cambridge University Press, 1959), p. 401; G. W. Leibniz, supplement to a letter to Christiaan Huygens, September 8, 1679, in Leroy E. Loemker, trans. and ed., *Philosophical Papers and Letters*, vol. 1 (Chicago: University of Chicago Press, 1956), pp. 384–85.

45. G. W. Leibniz to Nicolas Remond, January 10, 1714, in Loemker, trans. and ed., *Philosophical Papers and Letters*, 2:1,063; G. W. Leibniz, ca. 1679, in Loemker, trans. and ed., *Philosophical Papers and Letters*, 1:342.

46. G. W. Leibniz, ca. 1679, in Loemker, trans. and ed., *Philosophical Papers and Letters*, vol. 1, p. 344.

47. G. W. Leibniz, 1716, *Discourse on the Natural Theology of the Chinese* (translation of *Lettre sur la philosophie chinoise à Nicolas de Remond*, 1716), Henry Rosemont Jr. and Daniel J. Cook, trans. and eds., Monograph of the Society for Asian and Comparative Philosophy, no. 4 (Honolulu: University of Hawaii Press, 1977), p. 158.

48. G. W. Leibniz, "De Progressione Dyadica—Pars I" (MS, March 15, 1679), published in facsimile (with German translation) in Erich Hochstetter and Hermann-Josef Greve, eds., *Herrn von Leibniz' Rechnung mit Null und Einz* (Berlin: Siemens Aktiengesellschaft, 1966), pp. 46–47. English translation by Verena Huber-Dyson, 1995.

49. Burks, Goldstine, and von Neumann, *Preliminary Discussion*, p. 9.

50. John von Neumann and Herman H. Goldstine, "On the Principles of Large Scale Computing Machines," talk given to the Mathematical Computing Advisory Panel, Office

of Research and Inventions, Navy Dept., Washington D.C., May 15, 1946, reprinted in *Collected Works*, Vol. 5: *Design of Computers, Theory of Automata and Numerical Analysis* (Oxford: Pergamon Press, 1963), p. 32.

51. Julian Bigelow, interview with Nancy Stern.

52. Martin Davis, *The Universal Computer: The Road from Leibniz to Turing* (New York: W. W. Norton, 2000), p. 113.

53. Kurt Gödel to Arthur W. Burks, n.d., in Burks, ed., *Theory of Self-Reproducing Automata,* p. 56.

54. G. W. Leibniz to Rudolph August, Duke of Brunswick, January 2, 1697, as translated in Anton Glaser, *History of Binary and Other Non-decimal Numeration* (Los Angeles: Tomash, 1981), p. 31.

55. Kurt Gödel to Marianne Gödel, October 6, 1961, in Solomon Feferman, ed., *Collected Works*, vol. 4 (Oxford: Oxford University Press, 2003), pp. 437–38.

SEVEN: 6J6

1. Alice Bigelow, interview with author, May 24, 2009, GBD.

2. Julian Bigelow, interview with Richard R. Mertz.

3. Ibid.

4. Julian Bigelow, interview with Walter Hellman, June 10, 1979, in Walter Daniel Hellman, "Norbert Wiener and the Growth of Negative Feedback in Scientific Explanation," PhD thesis, Oregon State University, December 16, 1981, p. 148.

5. Norbert Wiener, *Ex-Prodigy*, pp. 268–69; Julian Bigelow to John von Neumann, November 26, 1946, VNLC.

6. Norbert Wiener to Vannevar Bush, September 21, 1940, in Pesi R. Masani, ed., Norbert Wiener, *Collected Works*, vol. 4 (Boston: MIT Press, 1985), p. 124.

7. Norbert Wiener, "Principles Governing the Construction of Prediction and Compensating Apparatus," submitted with S. H. Caldwell, Proposal to Section D2, NDRC, November 22, 1940, in Pesi R. Masani, *Norbert Wiener: 1894–1964* (Basel: Birkhauser, 1990), p. 182.

8. Norbert Wiener and Julian H. Bigelow, "Report on D.I.C. Project #5980: Anti-Aircraft Directors: Analysis of the Flight Path Prediction Problem, including a Fundamental Design Formulation and Theory of the Linear Instrument," Massachusetts Institute of Technology, February 24, 1941, pp. 38–39, JHB.

9. Norbert Wiener, "Extrapolation, Interpolation, and Smoothing of Stationary Time Series, with Engineering Applications," classified report to the National Defense Research Committee, February 1, 1942, declassified edition (Boston: MIT Press, 1949), p. 2.

10. Norbert Wiener, *I Am a Mathematician* (New York: Doubleday, 1956), p. 243; Alice Bigelow, interview with author.

11. Jule Charney, "Conversations with George Platzman," recorded August 1980, in R. Lindzen, E. Lorenz, and G. Platzman, eds., *The Atmosphere—A Challenge: The Science of Jule Gregory Charney* (Boston, Mass.: American Meteorological Society, 1990), p. 47.

12. Julian Bigelow, interview with Flo Conway and Jim Siegelman, October 30, 1999 (courtesy of Flo Conway and Jim Siegelman).

13. Ibid.

14. Julian Bigelow to Warren Weaver, December 2, 1941, JHB.
15. Ibid.
16. Ibid.
17. Julian Bigelow, interview with Flo Conway and Jim Siegelman.
18. Ibid.
19. Norbert Wiener, *I Am a Mathematician*, p. 249.
20. George Stibitz, "Diary of Chairman, July 1, 1942," in Peter Galison, "The Ontology of the Enemy: Norbert Wiener and the Cybernetic Vision," *Critical Inquiry* 21, no. 1 (Autumn 1994): 243.
21. Julian Bigelow, Arturo Rosenblueth, and Norbert Wiener, "Behavior, Purpose and Teleology," *Philosophy of Science* 10, no. 1 (1943): 9 and 23–24.
22. Warren S. McCulloch, "The Imitation of One Form of Life by Another—Biomimesis," in Eugene E. Bernard and Morley R. Kare, eds., *Biological Prototypes and Synthetic Systems, Proceedings of the Second Annual Bionics Symposium Sponsored by Cornell University and the General Electric Company, Advanced Electronics Center, Held at Cornell University, August 30–September 1, 1961*, vol. 1 (New York: Plenum Press, 1962), p. 393.
23. W. A. Wallis and Ingram Olkin, "A Conversation with W. Allen Wallis," *Statistical Science* 6, no. 2 (May 1991): 124.
24. Wiener, *I Am a Mathematician*, p. 243.
25. Frank Aydelotte to Julian Bigelow, September 3, 1946, IAS.
26. Verena Huber-Dyson, note for Julian Bigelow memorial, March 29, 2003, GBD.
27. Willis H. Ware, interview with Nancy Stern.
28. Julian Bigelow, interview with Nancy Stern.
29. Ralph Slutz, interview with Christopher Evans, June 1976, CBI, call no. OH 086.
30. Willis H. Ware, interview with Nancy Stern.
31. Akrevoe Kondopria Emmanouilides, interview with author, January 22, 2004, GBD.
32. Frank Aydelotte to John von Neumann, June 4, 1946, IAS.
33. Julian Bigelow, interview with Richard R. Mertz.
34. Bigelow, draft "Report on Computer Development at the IAS."
35. Willis H. Ware, interview with Nancy Stern; Willis H. Ware, "History and Development of the Electronic Computer Project," p. 8.
36. Willis H. Ware, interview with Nancy Stern; Ralph Slutz, interview with Christopher Evans; Bernetta Miller, "Electronic Computer Project Statement of Expenditures from Beginning November 1945 to May 31, 1946," June 4, 1946, IAS.
37. Ware, "History and Development of the Electronic Computer Project," p. 8.
38. Klára von Neumann, *The Computer;* Julian Bigelow, interview with Richard R. Mertz.
39. Benjamin D. Merritt to Frank Aydelotte, August 29, 1946, IAS.
40. Willis H. Ware, interview with Nancy Stern.
41. Julian Bigelow, interview with Richard R. Mertz.
42. Ibid.
43. Herman H. Goldstine to John von Neumann, July 28, 1947, IAS.
44. Frank Aydelotte to Herbert H. Maass, May 26, 1946, IAS; Klára von Neumann, *The Computer.*
45. Arthur W. Burks, interview with William Aspray; Frank Aydelotte to H. Chandlee Turner, July 2, 1946, IAS.
46. Frank Aydelotte to Colonel G. F. Powell, June 25, 1946, IAS.

47. Julian Bigelow, interview with Richard R. Mertz; Willis Ware, interview with author.

48. Willis Ware, interview with author.

49. Morris Rubinoff, interview with Richard Mertz, May 17, 1971, Archives Center, National Museum of American History.

50. J. Robert Oppenheimer to John von Neumann, February 11, 1949, IAS.

51. Jack Rosenberg, interview with author, February 12, 2005, GBD.

52. Julian Bigelow, "Computer Development at the Institute for Advanced Study," in Metropolis, Howlett, and Rota, eds., *A History of Computing in the Twentieth Century*, p. 293.

53. Herman H. Goldstine to John von Neumann, July 19, 1947, IAS; Julian Bigelow, interview with Nancy Stern.

54. Ralph Slutz, interview with Christopher Evans; Herman Goldstine, interview with Nancy Stern.

55. Herman Goldstine, interview with Nancy Stern.

56. Julian Bigelow, interview with Nancy Stern; Julian Bigelow, interview with Richard R. Mertz.

57. J. H. Bigelow, J. H. Pomerene, R. J. Slutz, and W. Ware, "Interim Progress Report on the Physical Realization of an Electronic Computing Instrument," Institute for Advanced Study, Princeton, N.J., January 1, 1947, p. 12.

58. Ralph Slutz, interview with Christopher Evans.

59. James Pomerene, interview with Nancy Stern; Bigelow, "Computer Development," p. 309.

60. Willis Ware, interview with author; Julian Bigelow, interview with Richard R. Mertz; Bigelow, "Computer Development," p. 308.

61. "Report on Tubes in the Machine," February 8, 1953, IAS; Bigelow, "Computer Development," p. 307.

62. Bigelow, Pomerene, Slutz, and Ware, "Interim Progress Report," pp. 82–83.

63. Jack Rosenberg, interview with author.

64. Jack Rosenberg, "The Computer Project," unpublished draft, February 2, 2002.

65. Ibid.

66. Ralph Slutz, interview with Christopher Evans.

67. Bigelow, Pomerene, Slutz, and Ware, "Interim Progress Report," pp. 15–16.

68. James Pomerene, interview with Nancy Stern.

EIGHT: V-40

1. Andrew and Kathleen Booth, interview with author, March 11, 2004, GBD.

2. Ibid.

3. Ibid.

4. Ibid.

5. Ibid.

6. Ibid.

7. Herman Goldstine to John von Neumann, February 25, 1947, IAS; Andrew Booth to George Dyson, February 26, 2004.

8. Marston Morse, Minutes of the Meeting of the Standing Committee, March 18, 1946, IAS.

9. Frank Aydelotte, Minutes of the Meeting of the Standing Committee, June 27, 1946, IAS.

10. Ibid.

11. Frank Aydelotte, Report of the Director, October 18, 1946, IAS.

12. Stanley C. Smoyer, Memorandum to the Trustees of the IAS, August 7, 1946, IAS-BS.

13. Julian H. Bigelow to Frank Aydelotte, July 3, 1947, IAS.

14. Bernetta A. Miller to Frank Aydelotte, September 19, 1947, IAS; Morris Rubinoff, interview with Richard Mertz.

15. Freeman Dyson, comments at Julian Bigelow memorial, March 29, 2003.

16. Morris Rubinoff, interview with Richard Mertz; Thelma Estrin, interview with Frederik Nebeker, IEEE History Center, Rutgers University, August 24–25, 1992.

17. Gerald and Thelma Estrin, interview with author, April 14, 2005, GBD.

18. James Pomerene, interview with Nancy Stern.

19. Andrew D. Booth and Kathleen H. V. Britten, "General Considerations in the Design of an All-Purpose Electronic Digital Computer," 1947, JHB.

20. Bigelow, "Computer Development," p. 297.

21. James Pomerene, interview with Nancy Stern.

22. John von Neumann to Marston Morse, April 1, 1946, IAS; Institute for Advanced Study Electronic Computer Project, Agreement Concerning Inventions, n.d., 1946, IAS.

23. Julian Bigelow, interview with Nancy Stern; Abraham Flexner, "University Patents," p. 325.

24. Herman Goldstine to Bigelow, Hildebrandt, Melville, Pomerene, Slutz, Snyder, and Ware, June 6, 1947, IAS; Herman Goldstine to Patent Branch, Office of the Chief of Ordnance, May 10, 1947, IAS; Deposition of Arthur W. Burks, Herman H. Goldstine, and John von Neumann, n.d., June 1947, IAS.

25. Julian Bigelow, interview with Nancy Stern.

26. Ibid.

27. I. J. Good, "Some Future Social Repercussions of Computers," *International Journal of Environmental Studies* 1 (1970): 69.

28. William F. Gunning, "Rand's Digital Computer Effort," Rand Corporation Memorandum P-363, February 23, 1953, p. 4.

29. "Institute for Advanced Study Electronic Computer Project Monthly Progress Report: July and August 1947," p. 2, IAS; "Institute for Advanced Study Electronic Computer Project Monthly Progress Report: February 1948," p. 2, IAS.

30. Bigelow, Pomerene, Slutz, and Ware, "Interim Progress Report," p. 8; John von Neumann, Memorandum to Commander R. Revelle, Office of Naval Research, on the Character and Certain Applications of a Digital Electronic Computing Machine, October 21, 1947, VNLC.

31. "Institute for Advanced Study Electronic Computer Project Monthly Progress Report, March 1948," p. 2, IAS.

32. "Institute for Advanced Study Electronic Computer Project Monthly Progress Report: April 1948," p. 2, IAS; Jack Rosenberg, Memo to Julian Bigelow, April 10, 1950, IAS.

33. Willis Ware, interview with author.

34. J. H. Bigelow, H. H Goldstine, R. W. Melville, P. Panagos, J. H. Pomerene, J. Rosenberg, M. Rubinoff, and W. H. Ware, "Fifth Interim Progress Report on the Physical Realization of an Electronic Computing Instrument, January 1, 1949," p. 31, IAS.

35. Herman Goldstine to John von Neumann, July 2, 1947, IAS.

36. F. C. Williams and T. Kilburn, "A Storage System for Use with Binary-Digital Computing Machines," draft, December 1, 1947, p. 1, JHB.

37. Burks, Goldstine, and von Neumann, *Preliminary Discussion,* p. 8; Williams and Kilburn, "A Storage System," p. 1.

38. Williams and Kilburn, "A Storage System."

39. Ibid.

40. Julian Bigelow, interview with Richard R. Mertz.

41. Julian Bigelow to F. C. Williams, September 11, 1952, JHB.

42. Bigelow, Goldstine, Melville, Panagos, Pomerene, Rosenberg, Rubinoff, and Ware, "Fifth Interim Progress Report on the Physical Realization of an Electronic Computing Instrument," p. 2.

43. Ibid., p. 4.

44. Bigelow, "Computer Development," p. 304; "Institute for Advanced Study Electronic Computer Project Monthly Progress Report, August 1949," p. 4, IAS.

45. Jack Rosenberg, interview with author; Julian Bigelow to Warren Weaver, December 2, 1941, JHB.

46. Morris Rubinoff, interview with Richard Mertz.

47. J. H. Bigelow, T. W. Hildebrandt, P. Panagos, J. H. Pomerene, J. Rosenberg, R. J. Slutz, and W. H. Ware, "Fourth Interim Progress Report on the Physical Realization of an Electronic Computing Instrument, July 1, 1948," p. II-16-17, IAS.

48. Leon D. Harmon, "Report of Tests Made on Two Groups of 'Round Robin' Williams Storage Tubes at IAS," July 6, 1953, IAS.

49. James Pomerene, interview with Nancy Stern.

50. Julian Bigelow, interview with Nancy Stern.

51. F. J. Gruenberger, "The History of the Johnniac," RAND Memorandum RM-5654-PR (Santa Monica, Calif.: RAND Corporation, October 1968), p. 22.

52. Jan A. Rajchman, "Memo to V. K. Zworykin re: Status of work on Selectron up to Oct. 5, 1948," October 5, 1948, RCA.

53. Willis H. Ware, interview with Nancy Stern.

54. Herman H. Goldstine to Mina Rees, October 7, 1947, JHB.

55. Jan Rajchman, interview with Richard R. Mertz; Gruenberger, "The History of the Johnniac," p. 25.

56. Julian Bigelow, interview with Nancy Stern.

57. Julian Bigelow, interview with Richard R. Mertz.

58. Willis Ware, interview with author.

59. James Pomerene, interview with Nancy Stern.

60. "Power Supply and Cooling System for Electronic Computer Project," n.d., 1953, IAS.

61. Institute for Advanced Study Electronic Computer Project Machine and General Arithmetic Operating Logs, IAS.

62. John von Neumann, Memorandum to Commander R. Revelle, Office of Naval Research, October 21, 1947, VNLC.

63. Institute for Advanced Study Electronic Computer Project Machine and General Arithmetic Operating Logs, IAS.

64. Ibid.

65. Morris Rubinoff, interview with Richard Mertz.

66. James Pomerene, interview with Nancy Stern.

67. Willis H. Ware, interview with Nancy Stern.

68. Willis H. Ware, interview with Nancy Stern; James Pomerene, interview with Nancy Stern.

69. Atle Selberg, interview with author, May 11, 2004, GBD.

70. Willis H. Ware, interview with Nancy Stern.

71. Morris Rubinoff, interview with Richard Mertz.

72. Julian Bigelow, interview with Nancy Stern.

73. Julian Bigelow, interview with Richard R. Mertz.

74. Ibid.

75. Ibid.

76. Bigelow, "Computer Development," p. 291.

### NINE: CYCLOGENESIS

1. Frank Aydelotte to John von Neumann, June 5, 1947, IAS.

2. Philip Duncan Thompson, "A History of Numerical Weather Prediction in the United States," *Bulletin of the American Meteorological Society* 64, no. 7 (July 1983): 757.

3. Philip Duncan Thompson, in John. M. Lewis, "Philip Thompson: Pages from a Scientist's Life," *Bulletin of the American Meteorological Society* 77, no. 1 (January 1966): 107–8.

4. Lewis Richardson, as quoted by Ernest Gold, "Lewis Fry Richardson, 1881–1953," *Obituary Notices of Fellows of the Royal Society* 9 (November 1954): 230.

5. Ibid., p. 222.

6. Meaburn Tatham and James E. Miles, eds., *The Friends' Ambulance Unit 1914–1919: A Record* (London: Swarthmore Press, 1920), p. 212.

7. Olaf Stapledon to Agnes Miller, December 8, 1916, in Robert Crossley, ed., *Talking Across the World: The Love Letters of Olaf Stapledon and Agnes Miller, 1913–1919* (Hanover and London: University Press of New England, 1987), pp. 192–93.

8. Stapledon to Miller, December 26, 1917, in Crossley, ed., *Talking Across the World*, pp. 264–65.

9. Stapledon to Miller, January 12, 1918, in Crossley, ed., *Talking Across the World*, p. 270.

10. Lewis Fry Richardson, *Weather Prediction by Numerical Process* (Cambridge, UK: Cambridge University Press, 1922), p. 219.

11. Lewis F. Richardson, "The Approximate Arithmetical Solution by Finite Differences of Physical Problems Involving Differential Equations, with an Application to the Stresses in a Masonry Dam," *Philosophical Transactions of the Royal Society of London A* 210 (1911): 307.

12. Richardson, *Weather Prediction by Numerical Process*, p. xi.

13. Ibid., p. xiii.

14. Ibid., pp. 219 and xi.

15. Thompson, "A History of Numerical Weather Prediction," p. 757.

16. Ibid., p. 758.

17. Philip Duncan Thompson, "Charney and the Revival of Numerical Weather Prediction," in R. Lindzen, E. Lorenz, and G. Platzman, eds., *The Atmosphere—A Challenge: The Science of Jule Gregory Charney* (Boston: American Meteorological Society, 1990), p. 98.

18. Akrevoe Kondopria Emmanouilides, interview with author, June 3, 2003, GBD; Philip Thompson, interview with William Aspray, December 5, 1986, CBI, call no. OH 125.

19. Frank Aydelotte, Minutes of the Meeting of the Standing Committee, May 13, 1946, IAS; Minutes of the Meeting of the Standing Committee, May 20, 1946, IAS; Aydelotte, Minutes of the Meeting of the Standing Committee, May 13, 1946, IAS.

20. Jule Charney to Stan Ulam, December 6, 1957, SUAPS.

21. Lewis Strauss, *Men and Decisions* (Garden City, N.Y.: Doubleday, 1962), pp. 232–33.

22. Vladimir K. Zworykin, "Outline of Weather Proposal," RCA Princeton Laboratories, October 1945, pp. 1, 4; John von Neumann to Vladimir Zworykin, October 14, 1945, RCA.

23. Strauss, *Men and Decisions*, pp. 233–34.

24. Sidney Shalett, "Electronics to Aid Weather Figuring," *New York Times*, January 11, 1946, p. 12.

25. Ibid.

26. Proposal submitted by Frank Aydelotte (written by John von Neumann) to Lt. Commander D. F. Rex, U.S. Navy Office of Research and Inventions, May 8, 1946, IAS; John von Neumann to Lewis Strauss, May 4, 1946, VNLC; John von Neumann, "Can We Survive Technology?" *Fortune*, June 1955, p. 151.

27. Institute for Advanced Study, Conference on Meteorology, August 29–30, 1946, undated summary, p. 3, VNLC.

28. Herman H. Goldstine, "Report on the Housing Situation for Meteorology Personnel," July 15, 1946, IAS.

29. John von Neumann, "Meteorology Project Progress Report for the period of November 15, 1946, to April 1, 1947," April 8, 1947, p. 4, IAS.

30. Philip Thompson and John von Neumann, "Meteorology Project Report of Progress During the Period from April 1, 1947, to December 15, 1947," p. 2, IAS.

31. Philip Thompson, interview with William Aspray.

32. Jule Charney to Stanislaw Ulam, December 6, 1957, SUAPS.

33. Jule G. Charney, "Numerical Methods in Dynamical Meteorology," *Proceedings of the National Academy of Sciences* 41, no. 11 (November 1955): 799.

34. Jule Charney to Stanislaw Ulam, December 6, 1957.

35. Joseph Smagorinsky, interview with author, May 4, 2004, GBD; Jule Charney, "Progress Report of the Meteorology Group at the IAS, June 1, 1948, to June 30, 1949," p. 2, IAS.

36. Jule Charney to Philip Thompson, February 12, 1947, in Lindzen, Lorenz, and Platzman, eds., *The Atmosphere—A Challenge*, p. 114.

37. Philip Thompson and John von Neumann, "Meteorology Project Report of Progress During the Period from April 1, 1947, to December 15, 1947," p. 10, IAS.

38. Margaret Smagorinsky, interview, May 4, 2004, GBD.

39. George W. Platzman, "The ENIAC Computation of 1950: Gateway to Numerical Weather Prediction," *Bulletin of the American Meteorological Society* 60, no. 4 (April 1979): 307.

40. Jule Charney to George Platzman, April 10, 1950, in Platzman, "The ENIAC Computation of 1950," pp. 310–11.

41. Platzman, "The ENIAC Computation of 1950," p. 310; John von Neumann, J. G. Charney, and R. Fjørtoft, "Numerical Integration of the Barotropic Vorticity Equation," *Tellus* 2 (1950): 275.

42. Charney, "Numerical Methods in Dynamical Meteorology," p. 800.

43. Thelma Estrin, interview with Frederik Nebeker, August 24–25, 1992, IEEE History Center, Rutgers University, New Brunswick, N.J.; Raoul Bott, interview with author.

44. Jule Charney, "Numerical Prediction of Cyclogenesis," *Proceedings of the National Academy of Sciences* 40 (1954): 102.

45. Clarence D. Smith, "The Destructive Storm of November 25–27," *Monthly Weather Review* 78 (November 1950): 204.

46. Jule Charney, "Conversations with George Platzman," recorded August 1980, in Lindzen, Lorenz, and Platzman, eds., *The Atmosphere—A Challenge*, p. 54.

47. Charney, "Numerical Prediction of Cyclogenesis," p. 102.

48. Electronic Computer Project machine log, May 27, 1953, IAS.

49. Norman A. Phillips, "Progress Report of the Meteorology Group at the IAS, July 1, 1952 to September 30, 1952," p. 4, IAS; Harry Wexler to chief of U.S. Weather Bureau, June 11, 1953, in Joseph Smagorinsky, "The Beginnings of Numerical Weather Prediction and General Circulation Modeling: Early Recollections," *Advances in Geophysics* 25 (1983): 23.

50. Philip Thompson, interview with William Aspray.

51. Smagorinsky, "Beginnings of Numerical Weather Prediction," p. 25; Institute for Advanced Study Electronic Computer Project Monthly Progress Report: September, 1954, p. 3, IAS.

52. Jule Charney to Stanislaw Ulam, December 6, 1957, SUAPS.

53. Richard L. Pfeffer, ed., *Dynamics of Climate: The Proceedings of a Conference on the Application of Numerical Integration Techniques to the Problem of the General Circulation, Held October 26–28, 1955, at the Institute for Advanced Study, Princeton, New Jersey* (New York: Pergamon Press, 1960), p. 3.

54. John von Neumann, "Some Remarks on the Problem of Forecasting Climatic Fluctuations," in ibid., pp. 10–11.

55. Pfeffer, ed., *Dynamics of Climate*, p. 132.

56. Ibid., pp. 133–36.

57. Pfeffer, ed., *Dynamics of Climate*, pp. 133–36.

58. Charney, "Conversations with George Platzman," in Lindzen, Lorenz, and Platzman, eds., *The Atmosphere—A Challenge*, pp. 57–58.

59. Jule Charney and Walter Munk, "The Applied Physical Sciences," talk for Institute for Advanced Study Electronic Computer Project 25th Anniversary, 1972, IAS.

60. Ibid.

61. Von Neumann, "Can We Survive Technology?" p. 151.

62. Smagorinsky, "Beginnings of Numerical Weather Prediction," p. 29.

TEN: MONTE CARLO

1. Klára von Neumann, *Johnny*, ca. 1963.

2. Klára von Neumann, *The Grasshopper*.

3. Ibid.

4. Ibid.

5. Ibid.

6. Ibid.

7. Ibid.

8. John Wheeler to Carl Eckart, November 23, 1963, KVN.

9. Klára von Neumann, *The Grasshopper*.

10. Ibid.

11. Mariette von Neumann to John von Neumann, September 22, 1937, in Frank Tibor, "Double Divorce: The Case of Mariette and John von Neumann," *Nevada Historical Society Quarterly* 34, no. 2 (1991): 361.

12. Mariette von Neumann to John von Neumann, n.d., 1937, in ibid.

13. Klára von Neumann to John von Neumann, November 11, 1937, KVN.

14. John von Neumann to Stanislaw Ulam, April 22, 1938, SFU.

15. Klára von Neumann, *Two New Worlds*.

16. John von Neumann to Klára von Neumann, September 14, 1938, KVN.

17. John von Neumann to Klára von Neumann, September 6, 1938, KVN.

18. John von Neumann to Klára von Neumann, September 5, 1938, KVN; John von Neumann to Klára von Neumann, September 13, 1938, KVN.

19. John von Neumann to Klára von Neumann, September 18, 1938, KVN.

20. Morgenstern, in *John von Neumann*, documentary.

21. Klára von Neumann, *Two New Worlds*.

22. Harry E. King, Credit Manager, Essex House and Casino-on-the-Park, New York, to the Institute for Advanced Study, December 24, 1938, IAS.

23. Klára von Neumann, *Two New Worlds*.

24. Willis Ware, interview with author.

25. John von Neumann to Klára von Neumann, August 10, 1939, KVN.

26. Jack Rosenberg, interview with author, February 12, 2005.

27. Richtmyer, "People Don't Do Arithmetic"; Ulam, *Adventures of a Mathematician*, p. 79; Ulam, "Conversations with Gian-Carlo Rota."

28. Jack Rosenberg, interview with author, February 12, 2005; Marina von Neumann Whitman, interview with author, February 9, 2006.

29. Klára von Neumann to John von Neumann, n.d., ca. 1949, KVN.

30. Klára von Neumann, *Johnny*.

31. Ibid.

32. John von Neumann and Oswald Veblen to Frank Aydelotte, March 23, 1940, IAS.

33. Ibid.

34. Klára von Neumann, *Johnny*.

35. John von Neumann to Stanislaw Ulam, April 2, 1942, VNLC; John von Neumann to Clara [Klára] von Neumann, April 13, 1943, KVN; S. W. Hubbel [Office of Censorship] to Clara [Klára] von Neumann, April 13, 1943, IAS.

36. Klára von Neumann, *Johnny*.

37. John von Neumann to Klára von Neumann, May 8, 1945, KVN; John von Neumann to Klára von Neumann, May 11, 1945, KVN.

38. Nicholas Metropolis, "The MANIAC," in Metropolis, Howlett, and Rota, eds., *A History of Computing in the Twentieth Century*, p. 459; Edward Teller to John von Neumann, August 9, 1945, IAS.

39. John von Neumann to Klára von Neumann, October 4, 1946, KVN.

40. Klára von Neumann, *Johnny.*

41. John von Neumann to Klára von Neumann, December 15, 1945, KVN.

42. Klára von Neumann, *The Computer.*

43. James Pomerene, interview with Nancy Stern.

44. John von Neumann, Memo to Colonel L. E. Simon, Ballistic Research Laboratory, on Mechanical Computing Devices, January 30, 1945, VNLC.

45. John von Neumann to Carson Mark, March 13, 1948, VNLC.

46. Stanislaw Ulam, 1983, in Roger Eckhardt, "Stan Ulam, John von Neumann, and the Monte Carlo Method," in "Stanislaw Ulam, 1909–1984," *Los Alamos Science* 15, Special Issue (1987): 125.

47. Ulam, *Adventures of a Mathematician,* p. 197; Stanislaw Ulam, "Random Processes and Transformations," in *Proceedings of the International Congress of Mathematicians, Cambridge, Mass., August 3–September 6, 1950,* vol. 2 (Providence, R.I.: American Mathematical Society, 1952), p. 266.

48. Ulam, *Adventures of a Mathematician,* p. 197.

49. Ibid.

50. Marshall Rosenbluth, "Genesis of the Monte Carlo Algorithm for Statistical Mechanics," talk at Los Alamos National Laboratory, June 9, 2003 (draft courtesy of Marshall Rosenbluth).

51. Andrew W. Marshall, "An Introductory Note (on Monte Carlo Method)," in Herbert A. Meyer, ed., *Symposium on Monte Carlo Methods, March 16 and 17, 1954, Held at the University of Florida, Sponsored by the Wright Air Development Center of the U.S. Air Force Air Research and Development Command* (New York: Wiley, 1956), p. 14.

52. Ulam, *Adventures of a Mathematician,* p. 199; ibid.; Richtmyer, "People Don't Do Arithmetic"; Stanislaw Ulam, *Testimony, United States District Court, District of Minnesota, Fourth Division, 4-67 Civil 138: Honeywell, Inc., Plaintiff, v. Sperry Rand Corporation and Illinois Scientific Developments, Inc., Defendants,* Transcript of Proceedings, vol. 47, Minneapolis, Minnesota, Tuesday, September 7, 1971, pp. 7427–28.

53. John von Neumann to Robert Richtmyer, March 11, 1947, in Stanislaw Ulam, "Statistical Methods in Neutron Diffusion," LAMS-551 (Los Alamos, N.M: Los Alamos Scientific Laboratory, April 9, 1947), p. 13; John von Neumann to Robert Richtmyer, March 11, 1947, in Ulam, "Statistical Methods in Neutron Diffusion," p. 6.

54. John von Neumann to Norris Bradbury, July 18, 1950, LANL archives, B-9 Files, Folder 635, Drawer 181, quoted in Anne Fitzpatrick, *Igniting the Light Elements: The Los Alamos Thermonuclear Weapon Project, 1942–1952,* LA-13577-T (Los Alamos, N.M.: Los Alamos Scientific Laboratory, 1999), p. 148.

55. Richard F. Clippinger, "A Logical Coding System Applied to the ENIAC," BRL report 673, Ballistic Research Laboratories, Aberdeen Proving Ground, September 29, 1948.

56. Herman H. Goldstine to General Leslie R. Groves, Remington Rand Incorporated, Laboratory of Advanced Research, June 14, 1949, VNLC.

57. John and Klára von Neumann, "Actual Technique: The Use of the ENIAC," MS, n.d., ca. 1947, VNLC; Robert D. Richtmyer, "The Post-War Computer Development," *American Mathematical Monthly* 72, no. 2, part 2: *Computers and Computing* (February 1965): 11.

58. Eckert, "The ENIAC," p. 529.

59. Eckert, "The ENIAC," p. 529; Metropolis, "The MANIAC," p. 459.

60. Metropolis, "The MANIAC," p. 459.

61. Harris Mayer, interview with author, April 14, 2006, GBD; Harris Mayer, interview with author, May 13, 2011, GBD.

62. Klára von Neumann to Françoise and Stan Ulam, n.d., March 1948, SUAPS; Stanislaw Ulam to John von Neumann, May 12, 1948, SUAPS.

63. John von Neumann to Stanislaw Ulam, May 11, 1948, SFU.

64. John von Neumann to Klára von Neumann, December 7, 1948, KVN.

65. John and Klára von Neumann, "Actual Technique."

66. John von Neumann to Edward Teller, April 1, 1950, VNLC.

67. Richtmyer, "People Don't Do Arithmetic."

68. John von Neumann to Klára von Neumann, January 14, 1948, KVN.

69. Richtmyer, "People Don't Do Arithmetic."

70. Klára von Neumann to Harris Mayer, April 8, 1949, KVN.

71. John von Neumann, "Various Techniques Used in Connection with Random Digits," in A. S. Householder, ed., *Monte Carlo Method, Proceedings of a Symposium held June 29, 30 and July 1, 1949, in Los Angeles, California, under the sponsorship of the RAND Corporation, and the National Bureau of Standards, with the cooperation of the Oak Ridge National Laboratory,* National Bureau of Standards Applied Mathematics Series 12, issued June 11, 1951, p. 36

72. *A Million Random Digits with 100,000 Normal Deviates* (Santa Monica, Calif.: RAND Corporation, 1955), p. xii.

73. Klára von Neumann to Stan Ulam, May 15, 1949, SUAPS.

74. Klára von Neumann to Carson Mark, June 28, 1949, KVN.

75. Herman Kahn, "Use of Different Monte Carlo Sampling Techniques," in Meyer, ed., *Symposium on Monte Carlo Methods,* p. 147.

ELEVEN: ULAM'S DEMONS

1. Ulam, January 14, 1974, in "Conversations with Gian-Carlo Rota."

2. Ulam, *Adventures of a Mathematician,* p. 10; Françoise Ulam, "From Paris to Los Alamos"; Mitchell Feigenbaum, "Reflections of the Polish Masters: Interview with Stan Ulam and Mark Kac," n.d., *Los Alamos Science* (Fall 1982): 57.

3. Françoise Ulam, "From Paris to Los Alamos."

4. Bruno Augenstein, interview with author, June 9, 1999, GBD; Françoise Ulam, interview with author, September 17, 1999, GBD; Claire Ulam, in "Stanislaw Ulam, 1909–1984," 1.

5. Françoise Ulam, in "Stanislaw Ulam, 1909–1984," p. 6; Gian-Carlo Rota, "The Barrier of Meaning," *Letters in Mathematical Physics* 10 (1985): 97.

6. Ulam, in "Conversations with Gian-Carlo Rota"; Ulam, *Adventures of a Mathematician,* p. 114.

7. Françise Ulam, "From Paris to Los Alamos."

8. Ibid.

9. Ibid.

10. Ibid.

11. Stanislaw Ulam to John von Neumann, n.d., 1941, VNLC.

12. John von Neumann to Stanislaw Ulam, April 2, 1942, VNLC; Ulam, *Adventures of a Mathematician*, p. 141.

13. John von Neumann to Stanislaw Ulam, November 9, 1943, SFU; Ulam, *Adventures of a Mathematician*, p. 144.

14. Françoise Ulam, "From Paris to Los Alamos."

15. Ibid.

16. Ulam, *Adventures of a Mathematician*, p. 155; ibid., p. 156.

17. Harris Mayer, "People of the Hill: The Early Days," *Los Alamos Science*, no. 28 (2003): 9.

18. Françoise Ulam, "From Paris to Los Alamos."

19. Ibid.

20. Ulam, *Adventures of a Mathematician*, pp. 147–48.

21. Ulam, "Conversations with Gian-Carlo Rota."

22. Françoise Ulam, "From Paris to Los Alamos."

23. Norris Bradbury, at Los Alamos Coordinating Council, October 1, 1945, in David Hawkins, ed., *Manhattan District History: Project Y, the Los Alamos Project*, vol. 1: *Inception Until August 1945* (Los Alamos, N.M.: Los Alamos Scientific Laboratory, 1946–1947; declassified as LAMS-2532, vol. 1, 1961), pp. 120–21.

24. Françoise Ulam, "From Paris to Los Alamos."

25. Edward Teller, February 13, 1945, in ibid.; Françoise Ulam, "From Paris to Los Alamos."

26. Françoise Ulam, "From Paris to Los Alamos"; Ulam, *Testimony, Honeywell, Inc., Plaintiff, v. Sperry Rand Corporation and Illinois Scientific Developments, Inc., Defendants*, p. 7349.

27. H. G. Wells, *The World Set Free* (New York: Dutton, 1914), pp. 114–15.

28. Stanislaw Ulam, "Thermonuclear Devices," in R. E. Marshak, ed., *Perspectives in Modern Physics: Essays in Honor of Hans Bethe* (New York: Wiley Interscience, 1966), p. 593.

29. Edward Teller, "The Work of Many People," *Science* 121 (February 25, 1955): 269.

30. Memorandum to the Secretary of War from Vannevar Bush and James B. Conant, "Supplementary Memorandum Giving Further Details Concerning Military Potentials of Atomic Bombs and the Need for International Exchange of Information," September 30, 1944, in JCAE declassified General Subject Files, Box 60, NARA. After Fitzpatrick, *Igniting the Light Elements*, p. 103.

31. Teller, "The Work of Many People," p. 268.

32. Ibid., p. 269.

33. Edward Teller, *Testimony, United States District Court, District of Minnesota, Fourth Division, 4-67 Civil 138: Honeywell, Inc., Plaintiff, v. Sperry Rand Corporation and Illinois Scientific Developments, Inc., Defendants*, Transcript of Proceedings, vol. 47, Minneapolis, Minn., Monday, August 30, 1971, p. 6702.

34. Hans A. Bethe, "Comments on the History of the H-Bomb," written in 1954, declassified in 1980, with a new introduction by Hans Bethe, in *Los Alamos Science* (Fall 1982): 47.

35. Teller, *Testimony, Honeywell, Inc., Plaintiff, v. Sperry Rand Corporation and Illinois Scientific Developments, Inc., Defendants*, p. 6771.

36. E. Bretscher, S. P. Frankel, D. K. Froman, N. Metropolis, P. Morrison, L. W. Nordheim, E. Teller, A. Turkevich, and J. Von Neumann, "Report on the Conference on the Super," LA-575, February 16, 1950.

37. John von Neumann and Klaus Fuchs, "Improvements in Method and Means for Utilizing Nuclear Energy," U.S. Office of Scientific Research and Development, Office for Emergency Management, Invention Disclosure, May 28, 1946.

38. Françoise Ulam, "From Paris to Los Alamos."

39. J. Robert Oppenheimer, General Advisory Committee, AEC, to David Lilienthal, Chairman, AEC, October 30, 1949, with attached General Advisory Committee's Majority and Minority Reports on Building the H-Bomb, reprinted in Herbert F. York, *The Advisors: Oppenheimer, Teller, and the Superbomb* (San Francisco: W. H. Freeman, 1976), pp. 150–59.

40. James B. Conant, Hartley Rowe, Cyril Stanley Smith, L. A. DuBridge, Oliver E. Buckley, J. R. Oppenheimer, and I. I. Rabi, General Advisory Committee to the U.S. Atomic Energy Commission, Report of October 30, 1949, reprinted in Herbert F. York, *The Advisors: Oppenheimer, Teller, and the Superbomb* (San Francisco: W. H. Freeman, 1976), p. 157; John von Neumann to Joe Mayor, February 3, 1950, VNLC.

41. Françoise Ulam, "From Paris to Los Alamos."

42. Ulam, *Testimony, Honeywell, Inc., Plaintiff, v. Sperry Rand Corporation and Illinois Scientific Developments, Inc., Defendants*, p. 7401.

43. John von Neumann, *Testimony Before AEC Personnel Security Board, 27 April 1954, In the Matter of J. Robert Oppenheimer* (Washington, D.C.: Government Printing Office, 1954), p. 655.

44. Ralph Slutz, interview with Christopher Evans.

45. Françoise Ulam, "From Paris to Los Alamos"; Ulam, *Adventures of a Mathematician*, pp. 216–17; Stan Ulam to John von Neumann, January 27, 1950, VNLC.

46. Ulam, "Thermonuclear Devices," p. 597.

47. John von Neumann to Stan Ulam, February 7, 1950, SUAPS.

48. Françoise Ulam, "From Paris to Los Alamos."

49. Edward Teller, interview with author.

50. Bethe, "Comments on the History of the H-Bomb," pp. 44 and 49.

51. Stan Ulam to Hans Bethe, October 29, 1954, Cornell University/PM.

52. Edward Teller and Stanislaw Ulam, "On Heterocatalytic Detonations I, Hydrodynamic Lenses and Radiation Mirrors," LAMS-1225, March 9, 1951.

53. Mayer, "People of the Hill," p. 25; Harris Mayer, interview with author, May 25, 2011, GBD.

54. Theodore Taylor, interview with Kenneth W. Ford, February 13, 1995, Niels Bohr Library, American Institute of Physics, Washington, D.C.

55. Gordon Dean, *Testimony Before AEC Personnel Security Board, April 19, 1954, In the Matter of J. Robert Oppenheimer* (Washington, D.C.: Government Printing Office, 1954), p. 305.

56. J. Robert Oppenheimer, *Testimony Before the AEC Personnel Security Board, April 16, 1954, In the Matter of J. Robert Oppenheimer* (Washington, D.C.: Government Printing Office, 1954), p. 251.

57. Marshall Rosenbluth, "Genesis of the Monte Carlo Algorithm for Statistical Mechanics" (text of talk at LANL, June 9, 2003), courtesy Marshall Rosenbluth.

58. Bill Borden, Memorandum for the File, August 13, 1951, concerning a conversation with Admiral Strauss, p. 1, NARA-JCAE/PM.

59. Klára von Neumann, *Johnny*.

60. Taylor, interview with Kenneth W. Ford.

61. John S. Walker, Memorandum to the files, Subject: Thermonuclear Matters and the Department of Defense, October 3, 1952, NARA/PM; Klára von Neumann, *Johnny.*

62. J. Robert Oppenheimer, letters on contracts, March 14 and 17, 1950, IAS-BS.

63. Bigelow, "Computer Development," p. 308.

64. Françoise Ulam, "From Paris to Los Alamos."

65. C. J. Everett and S. M. Ulam, "On a Method of Propulsion of Projectiles by Means of External Nuclear Explosions," Los Alamos Scientific Laboratory Report, LAMS–1955, August 1955, pp. 3–5.

66. Stanislaw Ulam, *Testimony, January 22, 1958, Before Senator Albert Gore and Representative James T. Patterson, in Outer Space Propulsion by Nuclear Energy, Hearings Before the Subcommittee on Outer Space Propulsion of the Joint Committee on Atomic Energy,* held Jan. 22, 23 and Feb. 6, 1958, Eighty-fifth Congress, second session, p. 48.

67. James Clerk Maxwell, *Theory of Heat* (London: Longman's, 1871), p. 308.

68. Stanislaw M. Ulam, "On the Possibility of Extracting Energy from Gravitational Systems by Navigating Space Vehicles," LAMS-2219 (written April 1, 1958, distributed June 19, 1958), pp. 3–7.

69. Ibid.

70. Stan Ulam to John von Neumann, February 29, 1952, VNLC.

71. Maxwell, *Theory of Heat,* pp. 288–89.

72. Stan Ulam to John von Neumann, February 7, 1949, VNLC.

73. Stan Ulam to Arthur W. Burks, January 27, 1961, SUAPS; Nicholas Metropolis to Stan Ulam, June 7, 1948, SUAPS.

74. "Notes on Meeting of 25 August 1951 on a Site for a Super Test," edited by William Ogle, Los Alamos Scientific Laboratory, J-Division Experiment Planning, September 8, 1951, p. 20, available at http://www.osti.gov/opennet/servlets/purl/16001505 -Pm1e3p/16001505.pdf.

75. Franke E. Moore Jr. and H. Gordon Bechanan, "History of Operation Ivy, 1951–1952," n.d., p. 274, available at www.hss.doe.gov/HealthSafety/IHS/marshall/collection/ data/ihp1d/59438e.pdf.

76. Lauren R. Donaldson, *Diary of Operation Ivy, First "H" Test,* Oct. 15 to Nov. 13, 1952, available at http://www.osti.gov/opennet/servlets/purl/16205388-DiXyLw/16205388.pdf.

77. Walter Munk and Deborah Day, "Ivy-Mike," *Oceanography* 17, no. 2 (June 2004): 102.

78. Moore and Bechanan, "History of Operation Ivy," p. 277.

79. Stan Ulam to John von Neumann, November 9, 1952, VNLC.

80. George Gamow to Stan Ulam, July 20, 1953, SUAPS.

81. Stan Ulam to George Gamow, July 20, 1953, SUAPS.

TWELVE: BARRICELLI'S UNIVERSE

1. Nils Aall Barricelli, "Experiments in Bionumeric Evolution Executed by the Electronic Computer at Princeton, N.J.," unpublished, August 1953, pp. 2–3, IAS.

2. Gerald Estrin, interview with author; Nils Aall Barricelli, "Sur le Fondement Théorétique pour l'analyse des Courbes Climatiques," PhD thesis, University of Oslo, 1947; Tor Gulliksen, personal communication, November 22, 1995, GBD.

3. Simen Gaure, personal communication, November 23, 1995, GBD.

4. Kirke Wolfe, interview with author, April 29, 2010, GBD; Nils Aall Barricelli, "Prospects and Physical Conditions for Life on Venus and Mars," *Scientia* 11 (1961): 1.

5. Simen Gaure, personal communication, November 23, 1995, GBD; Kirke Wolfe, interview with author; Barricelli, "Symbiogenetic Evolution Processes Realized by Artificial Methods," p. 307.

6. Frank Stahl, interview with author, February 25, 2007, GBD.

7. Simen Gaure, personal communication, November 23, 1995, GBD.

8. Atle Selberg, interview with author, May 11, 2004, GBD.

9. "Institute for Advanced Study Electronic Computer Project, Monthly Progress Report, March 1953," p. 3, IAS.

10. Nils Aall Barricelli, Application for United States Government Travel Grant for Citizens of Norway, Fulbright Act, to be submitted to United States Educational Foundation in Norway, December 8, 1951, IAS.

11. John von Neumann to Fulbright office, U.S. Educational Office in Norway, February 5, 1952, IAS.

12. Barricelli, "Experiments in Bionumeric Evolution," p. 1; Julian Bigelow, interview with author, November 1997, GBD.

13. Nils A. Barricelli, "Numerical Testing of Evolution Theories: Part II," *Acta Biotheoretica* 16 (1962): 122; Claude E. Shannon, "An Algebra for Theoretical Genetics," PhD dissertation, Department of Mathematics, Massachusetts Institute of Technology, April 15, 1940.

14. John von Neumann to Norbert Wiener, November 29, 1946, VNLC; John von Neumann to Mina Rees, Office of Naval Research, January 20, 1947, VNLC.

15. Nils A. Barricelli, "Numerical Testing of Evolution Theories: Part I," *Acta Biotheoretica* 16 (1962): 70.

16. Institute for Advanced Study Electronic Computer Project General Arithmetic Operating Log, June 22, 1956.

17. "Institute for Advanced Study Electronic Computer Project, Final Report on Contract No. DA-36-034-ORD-1023, April 1, 1954," p. II-83-85, IAS.

18. Barricelli, "Symbiogenetic Evolution Processes," p. 152.

19. Barricelli, "Experiments in Bionumeric Evolution," p. 2; Barricelli, "Symbiogenetic Evolution Processes," p. 175.

20. Barricelli, "Numerical Testing of Evolution Theories: Part I," p. 72; ibid., p. 94.

21. Ibid., p. 94.

22. Barricelli, "Symbiogenetic Evolution Processes," p. 159.

23. Barricelli, "Numerical Testing of Evolution Theories: Part I," p. 88.

24. Nils Aall Barricelli, "Evolution Processes Realized by Numerical Elements," Institute for Advanced Study Electronic Computer Project Monthly Progress Report, July 1956, pp. 10–11, IAS.

25. Barricelli, "Symbiogenetic Evolution Processes," p. 169.

26. Gerald Estrin, interview with author; Barricelli, "Symbiogenetic Evolution Processes," p. 164.

27. Barricelli, "Experiments in Bionumeric Evolution," p. 12.

28. Barricelli, "Esempi numerici di processi di evoluzione," p. 48.

29. Barricelli, "Symbiogenetic Evolution Processes," p. 180; Barricelli, "Numerical Testing of Evolution Theories: Part II," p. 117.

30. John von Neumann, "Statistical Theories of Information," Lecture at University of Illinois, December 1949, in Burks, ed., *Theory of Self-Reproducing Automata*, p. 60.

31. John von Neumann, "Reliable Organizations of Unreliable Elements," n.d., late 1951, p. 44, VNLC.

32. John von Neumann, "The Role of High and Extremely High Complication," Lecture at University of Illinois, December 1949, in Burks, ed., *Theory of Self-Reproducing Automata*, p. 71.

33. Barricelli, "Numerical Testing of Evolution Theories: Part II," p. 116; Nils A. Barricelli, "Numerical Testing of Evolution Theories," *Journal of Statistical Computation and Simulation*, vol. 1 (1972), p. 122.

34. Barricelli, "Numerical Testing of Evolution Theories: Part II," p. 100; Barricelli, "Numerical Testing of Evolution Theories" (1972): 126.

35. Barricelli, "Numerical Testing of Evolution Theories: Part II," p. 101.

36. Nils Aall Barricelli, Robert Toombs, and Louis Nelson, "Virus-Genetic Theory Testing by Data Processing Machines, Parts 1–3," *Journal of Theoretical Biology* 32, no. 3 (1971): 621.

37. Kirke Wolfe, interview with author.

38. John von Neumann to Hans A. Bethe, November 13, 1953, VNLC.

39. Frank Stahl, interview with author.

40. Barricelli, "Numerical Testing of Evolution Theories: Part I," pp. 69 and 99; ibid., p. 94.

41. Nils Barricelli, "Genetic Language, Its Origins and Evolution," *Theoretic Papers* 4, no. 6 (Oslo: The Blindern Theoretic Research Team, 1986): 106–7.

42. Ibid., p. 107.

43. Barricelli, "On the Origin and Evolution of the Genetic Code, II," pp. 19 and 21.

44. Nils A. Barricelli, "Suggestions for the Starting of Numeric Evolution Processes Intended to Evolve Symbioorganisms Capable of Developing a Language and Technology of Their Own," *Theoretic Papers* 6 (Oslo: The Blindern Theoretic Research Team, 1987): 121.

45. Nils A. Barricelli, "The Functioning of Intelligence Mechanisms Directing Biologic Evolution," *Theoretic Papers* 3, no. 7 (Oslo: The Blindern Theoretic Research Team, 1985): 126.

46. Barricelli, "Numerical Testing of Evolution Theories," (1972): 123–24.

47. Nils Barricelli, "The Intelligence Mechanisms Behind Biological Evolution," *Scientia*, September 1963, pp. 178–79.

48. Barricelli, "Suggestions for the Starting of Numeric Evolution Processes," p. 144.

49. Carl Woese and Nigel Goldenfeld, "How the Microbial World Saved Evolution from the Scylla of Molecular Biology and the Charybdis of the Modern Synthesis," *Microbiology and Molecular Biology Reviews* 73, no. 1 (March 2009): 20.

50. Nigel Goldenfeld and Carl Woese, "Biology's Next Revolution," *Nature*, January 25, 2007, p. 369.

51. Nils Barricelli, in Paul S. Moorhead and Martin M. Kaplan, eds., *Mathematical Challenges to the Neo-Darwinian Interpretation of Evolution: A Symposium Held at the Wistar Institute, April 25–26, 1966* (Philadelphia: Wistar Institute, 1966), p. 67; Alan Turing, "Computing Machinery and Intelligence," *Mind* 59, no. 236 (October 1950): 456.

52. George Church, West Hollywood, Calif., July 26, 2009, EDGE Foundation, "A

Short Course on Synthetic Genomics" (http://edge.org/event/master-classes/
the-edge-master-class-2008-a-short-course-on-synthetic-genomics).

## THIRTEEN: TURING'S CATHEDRAL

1. Sara Turing, *Alan M. Turing* (Cambridge, UK: W. Heffer and Sons, 1959), p. 11.
2. Alan Turing to Sara Turing, aboard Cunard White Start *Berengaria,* September 28, 1936, AMT.
3. Sara Turing, *Alan M. Turing,* p. 11.
4. Ibid., pp. 11, 23, 27, and 29.
5. Alan Turing to Sara Turing, September 28, 1936, AMT.
6. Alan Turing to Philip Hall, November 22, 1936, AMT.
7. John von Neumann to Oswald Veblen, July 6, 1935, OVLC.
8. Ibid.
9. Lynn Newman to parents, late 1937, in William Newman, "Max Newman—Mathematician, Codebreaker, and Computer Pioneer," in Jack Copeland, ed., *Colossus: The Secrets of Bletchley Park's Codebreaking Computers* (Oxford: Oxford University Press, 2006), p. 179.
10. I. J. Good to Sara Turing, December 9, 1956, AMT, Robin Gandy, "The Confluence of Ideas in 1936," in Rolf Herken, ed., *The Universal Turing Machine: A Half-Century Survey* (Oxford: Oxford University Press, 1988), p. 85.
11. Alan Turing, "On Computable Numbers, with an Application to the Entscheidungs-problem," *Proceedings of the London Mathematical Society,* ser. 2, vol. 42 (1936–1937): 230.
12. Ibid., p. 231.
13. Ibid., p. 250.
14. Ibid., p. 241.
15. Newman, "Max Newman—Mathematician, Codebreaker, and Computer Pioneer," p. 178; Max Newman to Alonzo Church, May 31, 1936, in Andrew Hodges, *Alan Turing: The Enigma* (New York: Simon and Schuster, 1983), pp. 111–12.
16. Alan Turing to Sara Turing, October 6, 1936, AMT; Alan Turing to Sara Turing, February 22, 1937, AMT.
17. Freeman Dyson, interview with author, May 5, 2004, GBD; Martin Davis, "Influences of Mathematical Logic on Computer Science," in Herken, ed., *The Universal Turing Machine,* p. 315.
18. Alonzo Church, "Review of A. M. Turing, 'On Computable Numbers, with an Application to the Entscheidungsproblem,'" *Journal of Symbolic Logic* 2, no. 1 (March 1937): 43.
19. Kurt Gödel, "Remarks Before the Princeton Bicentennial Conference on Problems in Mathematics," December 17–19, 1946, in Solomon Feferman, ed., *Collected Works,* vol. 2 (Oxford: Oxford University Press, 1986), p. 150.
20. M. H. A. Newman, "Alan Mathison Turing, 1912–1954," *Biographical Memoirs of Fellows of the Royal Society,* vol. 1 (1955), p. 256; M. H. A. Newman, "Dr. A. M. Turing," *London Times,* June 16, 1954, p. 10.
21. Herman Goldstine, interview with Nancy Stern; Julian Bigelow, interview with Nancy Stern.
22. Julian Bigelow, interview with Nancy Stern.

23. Malcolm MacPhail to Andrew Hodges, December 17, 1977, in Hodges, *Alan Turing*, p. 138.

24. Turing, "Systems of Logic Based on Ordinals," p. 161.

25. Ibid., pp. 172–73.

26. Ibid., pp. 214–15.

27. Ibid., p. 215.

28. Alan Turing to Sara Turing, October 14, 1936, AMT.

29. Alan Turing to Philip Hall, n.d., ca. 1938, AMT.

30. I. J. Good, "Pioneering Work on Computers at Bletchley," in Metropolis, Howlett, and Rota, eds., *A History of Computing in the Twentieth Century*, p. 35.

31. C. Hugh Alexander, "Cryptographic History of Work on the German Naval Enigma," n.d., unpublished (1945), pp. 19–20, AMT.

32. I. J. Good, "A Report on a Lecture by Tom Flowers on the Design of the Colossus," *Annals of the History of Computing* 4, no. 1 (1982): 57–58.

33. I. J. Good, "Enigma and Fish" (revised, with corrections), in F. H. Hinsley and Alan Stripp, eds., *Codebreakers: The Inside Story of Bletchley Park*, 2nd ed. (Oxford: Clarendon Press, 1994), p. 164.

34. Sara Turing, *Alan M. Turing*, pp. 72–73; I. J. Good to Lee A. Gladwin, June 18, 2002, in "Cryptanalytic Co-operation Between the UK and the USA," in Christof Teuscher, ed., *Alan Turing: Life and Legacy of a Great Thinker* (New York: Springer-Verlag, 2002), p. 472.

35. John R. Womersley, Mathematics Division, National Physical Laboratory, "A.C.E. Project: Origin and Early History," November 26, 1946, AMT.

36. Ibid.

37. Ibid.

38. Max Newman to John von Neumann, February 8, 1946, VNLC.

39. Alan Turing, "Report on visit to U.S.A., January 1st–20th, 1947," AMT.

40. Sara Turing, *Alan M. Turing*, p. 56.

41. Alan Turing, "Proposed Electronic Calculator," n.d., ca. 1946, p. 19, AMT.

42. Sara Turing, *Alan M. Turing*, p. 78.

43. Alan Turing, "Proposed Electronic Calculator," p. 47; Alan Turing, "Lecture to the London Mathematical Society on 20 February 1947," p. 9.

44. J. H. Wilkinson, "Turing's Work at the National Physical Laboratory," in Metropolis, Howlett, and Rota, eds., *A History of Computing in the Twentieth Century*, p. 111.

45. Charles G. Darwin [NPL] to Sir Edward V. Appleton, July 23, 1947, AMT.

46. Alan Turing, "Intelligent Machinery," report submitted to the National Physical Laboratory, 1948, p. 1, AMT.

47. Turing, "Lecture to the London Mathematical Society on 20 February 1947," pp. 23–24.

48. Turing, "Intelligent Machinery," p. 2.

49. Turing, "Lecture to the London Mathematical Society on 20 February 1947," p. 2.

50. Turing, "Intelligent Machinery," p. 6.

51. Ibid.

52. Ibid., p. 18.

53. Turing, "Computing Machinery and Intelligence," p. 456; Turing, "Intelligent Machinery," p. 17.

54. I. J. Good to Sara Turing, December 9, 1956, AMT; Lyn Newman to Antoinette Esher,

June 24, 1949, AMT; I. J. Good, "Ethical Machines," prepared for the Tenth Machine Intelligence Workshop, Case Western Reserve University, April 20–25, 1981, unpublished draft, October 7, 1980, p. ix.

55. Alan Turing to I. J. Good, September 18, 1948, AMT.

56. I. J. Good, "Speculations on Perceptrons and Other Automata," IBM Research Lecture RC-115, June 2, 1959, based on a lecture sponsored by the Machine Organization Department, December 17, 1958, p. 6.

57. Turing, "Intelligent Machinery," p. 17.

## FOURTEEN: ENGINEER'S DREAMS

1. Willis Ware, interview with Nancy Stern.

2. Biographical background on J. H. Bigelow, November 14, 1950, IAS.

3. Ibid.

4. Julian Bigelow, interview with Richard R. Mertz.

5. Bigelow, "Computer Development," p. 291.

6. Julian Bigelow, interview with Richard R. Mertz.

7. Herman Goldstine, interview with Nancy Stern.

8. Thelma Estrin, interview with Frederik Nebeker; Julian Bigelow, interview with Nancy Stern; minutes of Special Meeting of the Members of the Corporation, October 25, 1951, IAS.

9. Minutes of Regular Meeting of the Board of Trustees, October 27, 1955, IAS; Julian Bigelow, interview with Richard R. Mertz.

10. Freeman J. Dyson to S. Chandrasekhar, M. J. Lighthill F.R.S., Sir Geoffrey Taylor, Sydney Goldstein, and Sir Edward Bullard, October 20, 1954, IAS.

11. James Lighthill to Freeman Dyson, November 18, 1954, IAS.

12. David J. Wheeler, interview with William Aspray, May 14, 1987, CBI, call no. OH 132.

13. Freeman J. Dyson, "The Future of Physics," Lecture given at the dedication of Jadwin and Fine halls, Princeton University, March 17, 1970, FJD; John Bahcall, Memo to All Institute Members, September 1976, IAS.

14. Klára von Neumann, *Johnny*.

15. Ibid.

16. Ibid.

17. John von Neumann, *Testimony Before AEC Personnel Security Board, April 27, 1954, in the Matter of J. Robert Oppenheimer* (Washington, D.C.: Government Printing Office, 1954), p. 649.

18. John von Neumann to Klára von Neumann, May 16, 1954, KVN.

19. John von Neumann to Klára von Neumann, May 17, 1954, KVN.

20. Harris Mayer, interview with author, April 14, 2006, GBD.

21. John von Neumann to Klára von Neumann, December 9, 1943, KVN; Klára von Neumann, *Johnny*.

22. John von Neumann, "The Impact of Recent Developments in Science on the Economy and on Economics," Speech to the National Planning Association, Washington, D.C., December 12, 1955, reprinted in *Collected Works*, vol. 1 (Oxford: Pergamon Press, 1963), p. 100.

23. Jule Charney to Stanislaw Ulam, December 6, 1957, SUAPS.

24. Lewis. L. Strauss, in *John von Neumann*, documentary.

25. Julian Bigelow, interview with Nancy Stern.

26. Ulam, *Adventures of a Mathematician*, p. 244.

27. Marina von Neumann Whitman, interview with author, May 7, 2004, GBD.

28. Marina von Neumann Whitman, interview with author, May 3, 2010; Nicholas Vonneumann, *John von Neumann as Seen by His Brother*, p. 17.

29. Marina von Neumann Whitman, interview with author, May 3, 2010; Stanislaw Ulam to Lewis L. Strauss, December 21, 1956, SUAPS.

30. Julian Bigelow to Jule Charney, January 18, 1957, JHB; Klára von Neumann, *Johnny*.

31. Memo on Funeral Arrangements for John von Neumann, February 11, 1957, IAS; Ulam, *Adventures of a Mathematician*, p. 242.

32. Marston Morse to John von Neumann, n.d., quoted in Norman MacRae, *John von Neumann: The Scientific Genius Who Pioneered the Modern Computer, Game Theory, Nuclear Deterrence and Much More* (New York: Pantheon, 1992), p. 379; Morris Rubinoff, interview with Richard Mertz.

33. Martin Davis, interview with author, October 4, 2005, GBD.

34. Julian Bigelow, interview with Flo Conway and Jim Siegelman, October 30, 1999 (courtesy Flo Conway and Jim Siegelman).

35. Julian H. Bigelow, "Theories of Memory," in David L. Arm, ed., *Science in the Sixties: The Tenth Anniversary AFOSR Scientific Seminar, Cloudcroft, New Mexico, June 1965* (Albuquerque: University of New Mexico Press), p. 85.

36. Julian Bigelow, "Physical and Physiological Information Processes and Systems," MS, n.d., JHB.

37. Bigelow, "Theories of Memory," p. 86.

38. Ibid., p. 87.

39. Ibid., p. 86.

40. Ibid., pp. 85–86.

41. Ibid., p. 85.

42. Bigelow, Rosenblueth, and Wiener, "Behavior, Purpose and Teleology," p. 22.

43. Ulam, *Adventures of a Mathematician*, p. 242.

44. John von Neumann, *The Computer and the Brain* (New Haven, Conn.: Yale University Press, 1958), pp. 79–82.

45. Stan Ulam, quoted by Rota, "The Barrier of Meaning," p. 99.

46. John von Neumann, "Problems of Hierarchy and Evolution," Lecture at University of Illinois, December 1949, in Burks, ed., *Theory of Self-Reproducing Automata*, p. 84.

47. Von Neumann, "General and Logical Theory of Automata," p. 31.

48. Good, *The Scientist Speculates*, p. 197; von Neumann, "General and Logical Theory of Automata," p. 21.

49. Von Neumann, "Reliable Organizations of Unreliable Elements," p. 44.

FIFTEEN: THEORY OF SELF-REPRODUCING AUTOMATA

1. Aldous Huxley, *Ape and Essence* (New York: Harper and Brothers, 1948), pp. 38–39.

2. Ibid.

3. Nils A. Barricelli, "On the Origin and Evolution of the Genetic Code, II: Origin of the Genetic Code as a Primordial Collector Language; The Pairing-Release Hypothesis," *BioSystems* 11 (1979): 21–22.

4. John von Neumann to Norbert Wiener, November 29, 1946, VNLC.

5. Ibid.

6. John von Neumann to Irving Langmuir, November 12, 1946, VNLC.

7. Von Neumann, "Problems of Hierarchy and Evolution," p. 84; von Neumann, "General and Logical Theory of Automata," p. 28.

8. Von Neumann, "General and Logical Theory of Automata," p. 31.

9. John von Neumann, "Rigorous Theories of Control and Information," Lecture at University of Illinois, December 1949, in Burks, ed., *Theory of Self-Reproducing Automata*, p. 51.

10. Ibid.

11. Von Neumann, "General and Logical Theory of Automata," p. 31; von Neumann, "Problems of Hierarchy and Evolution," p. 80.

12. John von Neumann, "Problems of Hierarchy and Evolution," lecture at University of Illinois, December 1949, in Arthur W. Burks, ed., *Theory of Self-Reproducing Automata* (Urbana: University of Illinois Press, 1966), p. 78.

13. Ulam, "John von Neumann: 1903–1957," 2:8; John von Neumann, quoted in Claude Shannon, "Von Neumann's Contributions to Automata Theory," in *Bulletin of the American Mathematical Society* 64, no. 3, part 2 (May 1958): 126.

14. John von Neumann, Outline for book (to be co-authored with Stan Ulam) on theory of self-reproducing automata, not dated, ca. 1952, VNLC (partial listing of topics is given here).

15. Ware, "History and Development of the Electronic Computer Project," p. 16.

16. David J. Wheeler, interview with William Aspray; Murray Gell-Mann, interview with author, August 10, 2004, GBD.

17. John von Neumann, "Lectures on Probabilistic Logics and the Synthesis of Reliable Organisms from Unreliable Components," from notes by R. S. Pierce on lectures given at the California Institute of Technology, January 4–15, 1952, p. 1 (later published in C. E. Shannon and J. McCarthy, *Automata Studies* [Princeton, N.J.: Princeton University Press, 1956], pp. 43–99). Murray Gell-Mann, interview with author.

18. John von Neumann, "A Model of General Economic Equilibrium," *Review of Economic Studies* 13 (1945): 1.

19. "Institute for Advanced Study Electronic Computer Project, Monthly Progress Report: June, 1956," pp. 1–2, IAS.

20. Barricelli, "Prospects and Physical Conditions," pp. 1 and 5.

21. Marvin Minsky, in Carl Sagan, ed., *Communication with Extraterrestrial Intelligence: Proceedings of the Conference Held at the Byurakan Astrophysical Observatory, Yerevan, USSR, 5–11 September 1971* (Cambridge, Mass.: MIT Press, 1973), p. 328.

22. Edward Teller, *Memoirs* (Cambridge, Mass.: Perseus Books, 2001), p. 3.

23. Edward Teller, interview with author.

### SIXTEEN: MACH 9

1. Martin Schwarzschild, interview with William Aspray.
2. Ibid.
3. John von Neumann to Klára von Neumann, January 25, 1952, KVN.
4. Ibid.
5. Martin Schwarzschild, interview with William Aspray; Martin Schwarzschild to Subrahmanyan Chandrasekhar, December 3, 1946, Schwarzschild Papers, Princeton University Libraries, Princeton, N.J.
6. "Institute for Advanced Study Electronic Computer Project, Final Report on Contract No. DA-36-034-ORD-1646, Part II—Computer Use, May 1, 1957," p. 21.0, IAS; Martin Schwarzschild, interview with William Aspray.
7. Ingrid Selberg, personal communication, September 9, 2010, GBD.
8. "Institute for Advanced Study Electronic Computer Project, Final Report on Contract No. DA-36-034-ORD-1646, Part II," p. 21.11.
9. Ibid., p. 21.14.0.
10. John Bahcall, interview with author, May 10, 2004, GBD.
11. Martin Schwarzschild, *Structure and Evolution of the Stars* (New York: Dover, 1957), p. 284.
12. John von Neumann, "Discussion on the Existence and Uniqueness or Multiplicity of Solutions of the Aerodynamical Equations," August 17, 1949, in *Problems of Cosmical Aerodynamics, Proceedings of the Symposium on the Motion of Gaseous Masses of Cosmical Dimensions, Held at Paris, France, August 16–19, 1949* (Dayton, Ohio: Central Air Documents Office, 1951), p. 75.
13. Ulam, "John von Neumann: 1903–1957," 2:5.
14. Julian Bigelow to Maurice Wilkes, February 11, 1949, JHB.
15. Julian Bigelow, interview with Flo Conway and Jim Siegelman.
16. Julian Bigelow to Warren Weaver, December 2, 1941, JHB.
17. Stan Ulam to John von Neumann, n.d., ca. 1951, VNLC.
18. Stanislaw Ulam, in Moorhead and Kaplan, eds., *Mathematical Challenges*, 1966, p. 42.

### SEVENTEEN: THE TALE OF THE BIG COMPUTER

1. Von Neumann, "General and Logical Theory of Automata," p. 13.
2. Hannes Alfvén, "Electromagnetic Phenomena in the Motion of Gaseous Masses of Cosmical Dimensions," in *Problems of Cosmical Aerodynamics*, p. 44.
3. Hannes Alfvén, *On the Origin of the Solar System* (Oxford: Oxford University Press, 1954), p. 1.
4. Hannes Alfvén, "Cosmology: Myth or Science?" *Journal of Astrophysics and Astronomy* 5 (1984): 92.
5. John von Neumann to Klára von Neumann, September 11, 1954, KVN.
6. Hannes Alfvén, unpublished new preface for *The Tale of the Big Computer*, February 1981, Alfvén papers, University of California–San Diego Libraries.
7. Ibid.
8. Ibid.

9. Hannes Alfvén [Olof Johannesson], *The Tale of the Big Computer* (New York: Coward McCann, 1968), pp. 19–20.

10. Ibid., p. 76.

11. Ibid., pp. 55–56.

12. Ibid., p. 51.

13. Ibid., p. 96.

14. Ibid., p. 84.

15. Ibid., p. 86.

16. Ibid., p. 105.

17. Ibid., pp. 102–3.

18. Ibid., p. 123.

19. Sean Parker, personal communication, July 17, 2011.

20. Sergey Brin, Google press conference at the San Francisco Museum of Modern Art, 10:41 a.m., September 8, 2010, as reported by Nick Saint (http://www.businessinsider.com/google-search-event-live-2010-9).

21. Alfvén, *The Tale of the Big Computer*, p. 125.

22. William C. Dement, "Ontogenetic Development of the Human Sleep-Dream Cycle," *Science* 152, no. 3722 (April 29, 1966): 604.

23. Nathaniel Hawthorne, *The House of the Seven Gables* (Boston: Ticknor, Reed, and Fields, 1851), p. 283; Turing, "Computing Machinery and Intelligence," p. 433.

24. Alfvén, *The Tale of the Big Computer*, p. 116.

25. Ibid., pp. 117–18.

26. Eva Wisten, personal communication, October 25, 2005, GBD.

27. Alfvén, *The Tale of the Big Computer*, p. 119.

28. Ibid., p. 126.

EIGHTEEN: THE THIRTY-NINTH STEP

1. Klára von Neumann, *The Computer*.

2. Julian Bigelow, interview with Richard R. Mertz.

3. "Institute for Advanced Study Electronic Computer Project, Final Report on Contract No. DA-36-034-ORD-1646 Part II—Computer Use, May 1, 1957," p. 10.0, IAS.

4. Harris Mayer, interview with author, May 25, 2011, GBD.

5. Hans J. Maehly to J. Robert Oppenheimer, August 21, 1957, IAS.

6. Hans Maehly, "Institute for Advanced Study Electronic Computer Project, Final Report on Contract No. DA-36-034-ORD-1646 Part II—Computer Use, for the period 1 July 1954 to 31 December 1956," May 1957, p. 14.0, IAS.

7. Bryant Tuckerman, "Report on Post-Mortem Routine," n.d., IAS.

8. Maehly, "Institute for Advanced Study Electronic Computer Project, Final Report on Contract No. DA-36-034-ORD-1646 Part II," p. 11.1.

9. Julian Bigelow, interview with Nancy Stern.

10. Burks, Goldstine, and von Neumann, *Preliminary Discussion*, p. 23.

11. "Institute for Advanced Study Electronic Computer Project Monthly Progress Report, January 1957," p. 3, IAS.

12. Henry D. Smyth to Dr. Leonard Carmichael, June 11, 1958, IAS.
13. Julian Bigelow to John R. Pasta, June 6, 1958, JHB.
14. Martin Schwarzschild to Hedi Selberg, June 6, 1958, courtesy of Lars Selberg.
15. Herman Goldstine to Garrett Birkhoff, January 28, 1954, IAS.
16. S. Kidd to R. Vogt, November 30, 1959, JHB.
17. James I. Armstrong to Julian H. Bigelow, January 7, 1960, JHB.
18. Colin S. Pittendrigh to Carl Kaysen, October 31, 1966, IAS.
19. J. Robert Oppenheimer, notation on Roald Buhler to J. Robert Oppenheimer, September 30, 1966, IAS.
20. John von Neumann, Biographical background on J. H. Bigelow, November 14, 1950, IAS.
21. Bigelow, Pomerene, Slutz, Ware, "Interim Progress Report."
22. Willis H. Ware, interview with Nancy Stern.
23. John von Neumann to Klára von Neumann, November 9, 1946, KVN.
24. *History of the National Bureau of Standards Program for the Development and Construction of Large-Scale Electronic Computing Machines,* no author, n.d., evidently late 1949, JHB.
25. Herman Goldstine and John von Neumann to General Leslie R. Groves, June 14, 1949, VNLC.
26. Ralph E. Gomory, "Herman Heine Goldstine, September 13, 1913–June 16, 2004," *Proceedings of the American Philosophical Society* 150, no. 2 (June 2006): 368.
27. Jack Rosenberg, unpublished memoir, May 21, 2008 (courtesy of Jack Rosenberg).
28. Jack Rosenberg, interview with author.
29. Gerald and Thelma Estrin, interview with author.
30. Andrew Booth, personal communication, February 26, 2004, GBD.
31. Harris Mayer, interview with author, May 25, 2011, GBD.
32. FBI SAC (special agent in charge), New York, Memo to Director, FBI (Att: Liaison Section), August 25, 1955, PM.
33. Beatrice Stern, notes on conversation with Jean Flexner Lewinson, October 23, 1955, IAS-BS.
34. Lewis F. Richardson, "The Distribution of Wars in Time," *Journal of the Royal Statistical Society* 107, no. 3/4 (1944): 248.
35. Norbert Wiener, in "Revolt of the Machines," *Time* 75, no. 2, January 11, 1960, p. 32.
36. Stanislaw Ulam, "Further Applications of Mathematics in the Natural Sciences," 1981, reprinted in *Science, Computers and People: From the Tree of Mathematics* (Boston: Birkhauser, 1986), p. 153.
37. Edward Teller, "The Road to Nowhere," *Technology Review,* 1981, reprinted in *Better a Shield Than a Sword: Perspectives on Defense and Technology* (New York: Free Press, 1987), pp. 118–20.
38. Edward Teller, interview with author.
39. JmcD to John von Neumann, "A note regarding what we talked about last Wednesday," n.d., ca. 1956, VNLC.
40. Capt. I. R. Maxwell to Klára von Neumann, March 24, 1957, KVN.
41. Verna Hobson, IAS, notes of telephone conversation between J. Robert Oppenheimer and Captain I. Robert Maxwell, October 2, 1957, KVN.
42. The San Diego County coroner reported, after Klári's death, that "she was in her fifth marriage," and her autobiography, opening with a statement that "the cat, they

say, has nine lives; I have had only five," mentions a "Maharajah of notorious fame" who "invited me to stay in his palace, this promising adventure however remained unconsummated [due to] the quick and determined action of my more than alarmed, wrathful father."

43. Klára von Neumann, *The Grasshopper.*

44. Klára Eckart, age fifty-two, found November 10, 1963, Coroner's Investigative Report No. 1772-63, County of San Diego, November 18, 1963.

45. Klára von Neumann, *The Grasshopper.*

46. Paul Baran, interview with Judy O'Neill, March 5, 1990. CBI, OH no. 182.

47. Paul Baran, "On Distributed Communications," RAND Corporation Memorandum RM-3420-PR, August 1964 (11 parts).

48. Paul Baran, interview with Judy O'Neill.

49. J. D. Williams to John von Neumann, October 18, 1951, VNLC.

50. Harris Mayer, interview with author, May 25, 2011, GBD.

51. Robert Richtmyer to Nicholas Metropolis, January 11, 1956, VNLC.

52. Richtmyer, "The Post-War Computer Development," p. 14.

53. Freeman Dyson, "Birds and Frogs," *Notices of the American Mathematical Society* 56, no. 2 (February 2009): 220.

54. Benoît Mandelbrot, interview with author.

55. John von Neumann to Klára von Neumann, September 8, 1954, KVN.

56. Saunders Mac Lane, "Oswald Veblen, June 24, 1880–August 10, 1960," *Biographical Memoirs of the National Academy of Sciences* 37 (1964), p. 334.

57. Klára von Neumann, *Two New Worlds.*

58. Marston Morse to Frank Aydelotte, June 5, 1941, IAS.

59. Marston Morse, "Mathematics and the Arts," read at a conference in honor of Robert Frost, Kenyon College, October 8, 1950, IAS.

60. Kurt Gödel to John von Neumann, March 20, 1956, in Solomon Feferman, ed., *Collected Works*, vol. 5 (Oxford: Oxford University Press, 2003), p. 375 (German original in VNLC).

61. Julian Bigelow memorial service, March 29, 2003, GBD.

62. Ibid.

63. Rush Taggart, interview with author, May 19, 2005, GBD.

# INDEX

Aberdeen Proving Ground (U.S. Army),
    20–3, 36, 56, 58, 64, 69–70, 72–4, 109,
    166, 187, 194, 198, 287, 294
absolute addressing, 189
addressing (of memory), 5–6, 275–7, 309
    Gödel and, 94, 105–6
    as switching problem, 81, 105, 147–8
address matrix, ix, x, xi, xvi, 5–6, 10, 81, 105,
    263, 275, 279, 309, 336, 337–8
Advanced Research Projects Agency
    (ARPA), 217
Aiken, Howard, 132
Air Force, U.S., 8, 70, 155, 169, 197, 203,
    215–16, 217, 221, 270, 271, 287, 295
Air Matériel Command (U.S.), 122, 321
Alamogordo Bombing Range (New
    Mexico, site of Trinity nuclear test), 61
Alexander, Hugh, 254
Alexander, James, 32, 52, 82, 88
Alfvén, Hannes (1908–1995), 303–8, 313–14
Alfvén waves, 304–5
"An Algebra for Theoretical Genetics"
    (Shannon, 1940), 228
algorithms, 22, 75, 104, 191, 247, 263, 264, 280,
    309–10, 318
    see also codes and coding
Amazon.com, 28, 308
ambiguity, 94, 223, 234, 279, 310
American Commission for the Protection
    and Salvage of Artistic and Historic
    Monuments in War Areas, 97
analog computing, 5–6, 69–70, 71, 280–1
    and weather prediction, 160, 163
    and Web 2.0, 280

analog vs. digital, 5–6, 9, 71, 148, 163, 234,
    280–1
Android (operating system), 308
Anschluss (1938), 96, 179
antiaircraft fire control, 5, 68, 73, 109–14,
    258
anti-Semitism, 33, 43, 95, 127, 133, 180,
    185
Ape and Essence (Huxley, 1948), 282–3
apps, see codes and coding
Aquitania, 179, 201
Army, U.S., 6, 41, 55–6, 59, 68, 70, 85–6, 91,
    99–100, 101, 121–2, 127, 138, 182, 204–5,
    216, 271, 294, 333
    see also Aberdeen Proving Ground;
    ENIAC; Los Alamos
Army Air Corps, U.S., 154–5
artificial intelligence, 10, 228, 240, 252,
    259–65, 310–14
Atanasoff, John Vincent (1903–1995), 72–3
Atlas computer (Manchester University),
    235
Atomic Energy Commission, U.S. (AEC),
    9, 205–6, 209–11, 214–16, 234, 269, 271,
    285, 289, 295, 318, 323, 331, 332
    and IAS, 6–7, 214, 267, 297, 333
Atomic Energy for Military Purposes (Smyth,
    1945), 127, 318
Auerbach, Anna, 200
Augenstein, Bruno (1923–2005), 201
automata, see cellular automata; self-
    reproducing automata
Automatic Computing Engine (ACE), 132,
    259–60, 289, 312

## ABOUT THE AUTHOR

George Dyson is a historian of technology whose interests have included the development (and redevelopment) of the Aleut kayak (*Baidarka*, 1986), the evolution of digital computing and telecommunications (*Darwin Among the Machines*, 1997), and the exploration of space (*Project Orion*, 2002).

## A NOTE ON THE TYPE

This book was set in Monotype Dante, a typeface designed by Giovanni Mardersteig (1892–1977). Conceived as a private type for the Officina Bodoni in Verona, Italy, Dante was originally cut only for hand composition by Charles Malin, the famous Parisian punch cutter, between 1946 and 1952. Its first use was in an edition of Boccaccio's *Trattatello in laude di Dante* that appeared in 1954. The Monotype Corporation's version of Dante followed in 1957. Although modeled on the Aldine type used for Pietro Cardinal Bembo's treatise *De Aetna* in 1495, Dante is a thoroughly modern interpretation of the venerable face.

*Composed by Scribe, Philadelphia, Pennsylvania*

*Printed and bound by Berryville Graphics, Berryville, Virginia*

*Book design by Robert C. Olsson*